Risk Analysis and Management of Repetitive Actions

A Guide for Applying the OCRA System (Occupational Repetitive Actions)

THIRD EDITION

Ergonomics Design and Management: Theory and Applications

Series Editor
Waldemar Karwowski
Industrial Engineering and Management Systems
University of Central Florida (UCF) – Orlando, Florida

Published Titles

Application of Systemic-Structural Activity Theory to Design and Training
Gregory Z. Bedny

Applying Systemic-Structural Activity Theory to Design of Human–Computer Interaction Systems
Gregory Z. Bedny, Waldemar Karwowski, and Inna Bedny

Ergonomics: Foundational Principles, Applications, and Technologies
Pamela McCauley Bush

Aircraft Interior Comfort and Design
Peter Vink and Klaus Brauer

Ergonomics and Psychology: Developments in Theory and Practice
Olexiy Ya Chebykin, Gregory Z. Bedny, and Waldemar Karwowski

Ergonomics in Developing Regions: Needs and Applications
Patricia A. Scott

Handbook of Human Factors in Consumer Product Design, 2 vol. set
Waldemar Karwowski, Marcelo M. Soares, and Neville A. Stanton

Volume I: Methods and Techniques

Volume II: Uses and Applications

Human–Computer Interaction and Operators' Performance: Optimizing Work Design with Activity Theory
Gregory Z. Bedny and Waldemar Karwowski

Human Factors of a Global Society: A System of Systems Perspective
Tadeusz Marek, Waldemar Karwowski, Marek Frankowicz, Jussi I. Kantola, and Pavel Zgaga

Knowledge Service Engineering Handbook
Jussi Kantola and Waldemar Karwowski

Trust Management in Virtual Organizations: A Human Factors Perspective
Wiesław M. Grudzewski, Irena K. Hejduk, Anna Sankowska, and Monika Wańtuchowicz

Manual Lifting: A Guide to the Study of Simple and Complex Lifting Tasks
Daniela Colombiani, Enrico Ochipinti, Enrique Alvarez-Casado, and Thomas R. Waters

Neuroadaptive Systems: Theory and Applications
Magdalena Fafrowicz, Tadeusz Marek, Waldemar Karwowski, and Dylan Schmorrow

Organizational Resource Management: Theories, Methodologies, and Applications
Jussi Kantola

Risk Analysis and Management of Repetitive Actions: A Guide for Applying the OCRA System (Occupational Repetitive Actions), Third Edition
Daniela Colombini and Enrico Occhipinti

Safety Management in a Competitive Business Environment
Juraj Sinay

Self-Regulation in Activity Theory: Applied Work Design for Human–Computer Systems
Gregory Bedny, Waldemar Karwowski, and Inna Bedny

Risk Analysis and Management of Repetitive Actions

A Guide for Applying the OCRA System (Occupational Repetitive Actions)

THIRD EDITION

Daniela Colombini
Enrico Occhipinti

CRC Press
Taylor & Francis Group
Boca Raton London New York

CRC Press is an imprint of the
Taylor & Francis Group, an **informa** business

This book was previously published in Italian entitled, L'Analisi E La Gestione Del Rischio Nel Lavoro Manuale Ripetitivo-Manuale Per I'uso del Sistema OCRA, published by FrancoAngeli s.r.l., Milano Italy, 2014.

CRC Press
Taylor & Francis Group
6000 Broken Sound Parkway NW, Suite 300
Boca Raton, FL 33487-2742

Printed on acid-free paper
Version Date: 20160805

International Standard Book Number-13: 978-1-4987-3662-6 (Paperback)

Dedication

It has always been my greatest wish to thank all those, like me, who have dedicated and dedicate their life to safety and prevention and to these artists of ergonomics I dedicate these beautiful phrases of an artist par excellence, Van Gogh.

"The beginning is probably more difficult than any other thing … be confident … and everything will go well."

"There's safety in the middle of the danger … what would life be … without the courage to try?"

"Looking back … with a soft regret … I'll think of what I could have done … but … now I do what I can."

"I do always what I am not able to … in order to learn how to do it."

from Dani the Jefa

Contents

Preface

This volume represents the translation into English of the third book that the Ergonomics of Posture and Movements: International Ergonomics School (EPMIES) has dedicated to the prevention of risks from repetitive movements. The work, which preserves the structure of an application manual of the Occupational Repetitive Actions (OCRA) method, is an update of previous volumes in Italian.

In recent years, the authors, with the contribution of members of the Ergonomics of Posture and Movements: International Ergonomics School, have implemented the applicability capacity of the Occupational Repetitive Actions method, and now they offer a complete and real system of analysis and management of risk due to biomechanical overload of the upper limbs.

Operational models have been implemented to adapt to the different needs and assessment objectives respecting all intrinsic criteria and principles of the more classic Occupational Repetitive Actions method (now the "preferred" method in the specific International Organization for Standardization (ISO) and the European Committee for Standardization (CEN).

The book presents various models of the Occupational Repetitive Actions checklist, ranging from the more simplified one (the Occupational Repetitive Actions minichecklist) to the most complex and precise one.

A chapter is also devoted to the Occupational Repetitive Actions index, the original calculation method, which has now been made even more accurate and suitable for the design and redesign of working cycles.

Application examples are presented making use of spreadsheets in Excel, prepared by the Ergonomics of Posture and Movements International Ergonomics School (EPMIES), which can be downloaded free from the EPMIES website (www.epmresearch.org).

This volume also addresses, in a more systematic and renovated form, the complex problem of multitask analysis, considering the different cases of daily turnover, of weekly/monthly cycle exposures (e.g., office cleaning, meal preparation in large canteens, supermarkets, and the building sector), and of yearly turnover (typical turnover of agricultural work). Again, application examples are offered with calculations made possible through the use of spreadsheets or specific software.

A chapter is dedicated specifically to the risk analysis of notoriously complex jobs such as office cleaning, meal preparation, patient care in hospital, work in industrial laundries, and work in agriculture, showing in particular the need for a preliminary organizational analysis, especially in complex situations. In particular, a chapter is dedicated to organizational analysis and to risk impact.

The volume is completed with notes on health surveillance techniques and on the role that the occupational physician must assume for the complete and effective management of the risk analysed, now recognized as the first professional risk with consequent professional disease impacts worldwide.

Preface

Author

Daniela Colombini has a degree in medicine, specializing in occupational medicine and health statistics, from the University of Milan (Italy). She is a European ergonomist. She is a professor at the School of Specialization in Occupational Medicine, University of Florence. Since 1985, she has been a senior researcher at the Ergonomics of Posture and Movement Research Unit–Milan, where she developed methods for the analysis, evaluation, and management of risk and damage from occupational biomechanical overload. She is co-author of the OCRA method. She recently founded and leads the EPM International Ergonomics School (which operates in different languages, such as English, Spanish, Portuguese, and French). Schools are already working with accredited native teachers in different countries, such as Spain, most of the Spanish-speaking American countries, Brazil, France, and India. She is actual co-chair of the Technical Committee on the Prevention of Musculoskeletal Disorders at the International Ergonomics Association (IEA).

Enrico Occhipinti has a degree in medicine, with postgraduate specialization in occupational medicine and health statistics from the University of Milan (Italy). He is a certified European ergonomist. He is a professor at the School of Specialization in Occupational Medicine, University of Milan, and director of the Research Unit of Ergonomics of Posture and Movement (EPM) at Fondazione Don Gnocchi ONLUS–Milano. He has devoted more than 30 years to ergonomic issues related to physical ergonomics and the prevention of work-related musculoskeletal disorders. He developed and coauthored the OCRA method. He was coordinator (up to 2012) of the Technical Committee on Working with Musculo-Skeletal Disorders at the International Ergonomics Association (IEA) and represents Italy in international commissions of the European Committee for Standardization (CEN) and the International Organization for Standardization (ISO), working on biomechanics.

1 Introduction

1.1 THE REFERENCE FRAMEWORK

Musculoskeletal (or osteoarticular) diseases and disorders, especially those affecting the upper limbs, are acquiring growing importance in the field of occupational medicine.

Over the course of time, much evidence has suggested a link between such pathologies and working conditions featuring mechanical overload in a wide range of working communities, which also generate major impacts in terms of economic losses and social costs.

These multifactorial work-related diseases and disorders are caused by the way that manual jobs or tasks are performed.

In the scientific literature, reference is made to *overuse syndrome*, *repetitive strain injuries*, *repetitive motion injuries*, *occupational cervico-brachial disorders*, *and cumulative trauma disorders*; while these terms reflect slight conceptual differences, they all fall under the general definition of *upper limb work-related musculoskeletal disorders* or UL-WMSD.

In Italy, these disorders translate literally as "musculoskeletal disorders due to biomechanical overload." This is because they are caused by repetitive and/or forceful movements of the upper limbs performed for prolonged periods in the workplace, which put the joints, muscles, tendons, and other soft tissues under significant strain and can also affect the peripheral nerves.

The latter may include disorders or syndromes of the hand and forearm such as tendinitis and tenosynovitis, De Quervain syndrome, carpal tunnel syndrome, Guyon's canal syndrome, medial and lateral epicondylitis, and rotator cuff syndrome.

Table 1.1 lists the main work-related musculoskeletal diseases of the upper limbs due to biomechanical overload in the 2010 International Labour Office (ILO) list of occupational diseases.

Work-related musculoskeletal diseases are on the rise. According to the latest European data (Eurostat, 2010), they account for more than 55% of all occupational health problems recognized by insurance companies in 15 European Union countries (Figure 1.1). It is worth noting that the Eurostat data refers almost exclusively to diseases and disorders caused by biomechanical overload of the upper limbs, the most common being carpal tunnel syndrome and syndromes caused by overloading tendinous and peritendinous structures.

However, throughout Europe, exposure to these working conditions is widespread.

According to the 5th European Working Condition Survey (EWCS) conducted in 2010 by the European Foundation for the Improvement of Living and Working Conditions (Eurofound, 2012), approximately 63% of European workers (in 27 member countries) perform "repetitive movements of the upper limbs" for at least

TABLE 1.1
Principal Work-Related Musculoskeletal Diseases and Disorders of the Upper Limbs Due to Biomechanical Overload

ILO List of Occupational Diseases (Revised 2010)

	2.3 MUSCULOSKELETAL DISORDER
2.3.1	Radial styloid tenosynovitis due to repetitive movements, forceful exertions, and extreme postures of the wrist
2.3.2	Chronic tenosynovitis of hand and wrist due to repetitive movements, forceful exertions, and extreme postures of the wrist
2.3.3	Olecranon bursitis due to prolonged pressure of the elbow region
2.3.5	Epicondylitis due to repetitive forceful work
2.3.7	Carpal tunnel syndrome due to extended periods of repetitive forceful work, work involving vibration, extreme postures of the wrist, or a combination of the three
2.3.8	Other musculoskeletal disorders not mentioned in the preceding items where a direct link is established scientifically, or determined by methods appropriate to national conditions and practice, between the exposure to risk factors arising from work activities and the musculoskeletal disorder(s) contracted by the worker

Source: ILO, *List of Occupational Diseases (revised 2010).* Occupational Safety and Health Series, No. 74. International Labour Office, Geneva, Switzerland, 2010.

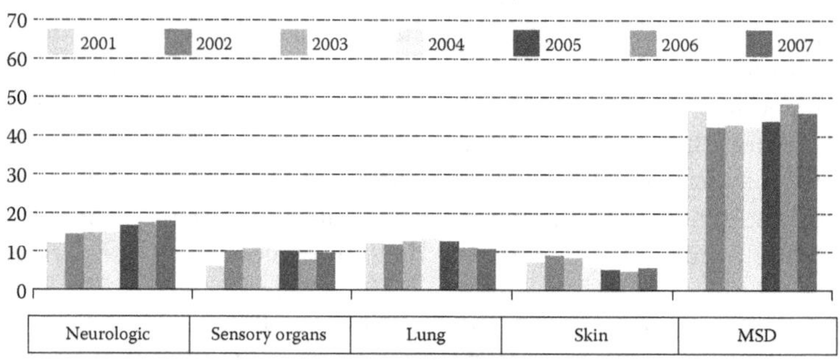

FIGURE 1.1 Main groups of recognized occupational diseases in Europe (in % vs. total) between 2001 and 2007. (From Eurostat, *Health and Safety at Work in Europe (1999–2007). A Statistical Portrait.* Publications Office of the European Union, Luxembourg, 2010.)

one-quarter of their working time. Considering virtually the entire working time, 33% of all workers are exposed (the percentage is the same for both sexes). Figure 1.2 summarizes the results of a more analytical EWCS, the fourth of its kind, conducted in 2005 (Eurofound, 2007).

1.2 A GLANCE AT THE REGULATIONS

1.2.1 European Directives

Current European legislation does not include specific regulations on the prevention of repetitive movements and forceful exertions of the upper limbs, although over the

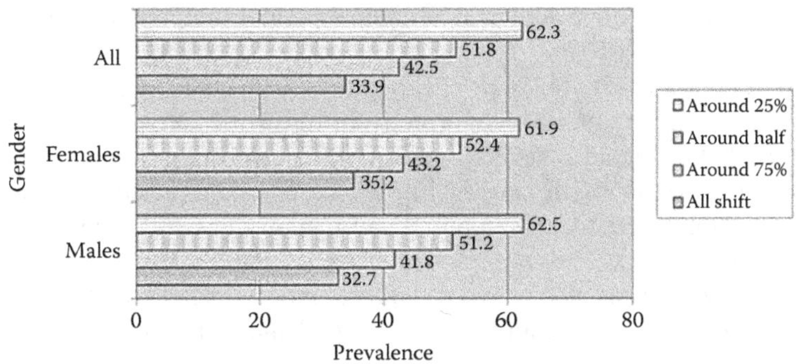

FIGURE 1.2 Prevalence of workers exposed to repetitive movements of the upper limbs, broken down by gender, in Europe (EU-27) in 2005. (From *Eurofound, Fourth European Working Conditions Survey*. European Foundation for the Improvement of Living and Working Conditions, Dublin, 2007.)

past 10 years a directive has been proposed (but not yet finalized) for the prevention of all musculoskeletal pathologies.

Nonetheless, Framework Directive 89/391/CEE does set forth some general "measures to encourage improvements in the safety and health of workers at work."

This directive, while not dealing with specific risks, does call for the employer to adopt a series of general measures to safeguard workers, including "adapting the work to the individual, especially as regards the design of workplaces, the choice of work equipment and the choice of working and production methods, with a view, in particular, to alleviating monotonous work and work at a predetermined work rate and to reducing their effect on health."

Apart from this, the directive stresses that *all the risks for the safety and health of workers* must be evaluated.

Since it has long been stated in the literature, and found in practice, that activities involving repetitive and forceful movements of the upper limbs may represent a health risk (for upper limb musculoskeletal diseases and disorders), it follows that the employer should also assess this type of risk in the workplace, and that if such specific risks are present and are potentially harmful, a plan should be put in place to reduce them to the lowest technically feasible level.

It is equally obvious that once such an evaluation has been made, a preventive plan should be put in place also aiming to prevent occupational risks and deliver information and training, as well as provide the necessary organization and means.

Also, since the matter concerns risks for human health, a specific health surveillance plan for workers must be put in place where necessary.

It should be stressed that while on the one hand there are no specific regulations, allowing the evaluation in question to be carried out using methods chosen "freely" by the employer (provided they are recognized in the literature), on the other, the national legislations of various European countries make frequent reference to their own technical regulations, when available.

In 2007, the International Standards Organization (ISO) adopted a voluntary technical regulation (standard) that could well represent a starting point for evaluating risk and adopting preventive actions relating to repetitive and forceful movements of the upper limbs (ISO, 2007b).

This regulation, ISO 11228-3; Ergonomics—Manual handling—Handling of low loads at high frequency, is perfectly in line with the spirit of Framework Directive 391/89. It includes several steps: risk identification; simple risk estimation; detailed assessment (in certain limited cases) using selected investigation methods, preferably the Occupational Repetitive Actions (OCRA) method; and lastly, risk reduction.

More recently, ISO has also issued a technical report (TR), ISO/TR 12295, which also better specifies the scope and method of implementation of ISO 11228, part 3 (ISO, 2014).

While the TR provides less expert users with a guide for standardizing "risk identification" and "quick assessments" (c.f. Chapter 3 of this volume), it also gives more skilled professionals guidance on how to make more effective use of the methods and tools identified in ISO 12228-3. More specifically, it stresses the usefulness of the OCRA checklist (latest version) as a means of assessing risk and better defines the methods for analyzing rotating repetitive tasks, preferably using the OCRA system (checklist and index). Brief comments are also provided on the not always successful evolution of the other methods mentioned in ISO 11228-3.

1.2.2 The Machinery Directive

The Machinery Directive, issued in accordance with the principle of the free circulation of goods among the member states of the European Union, primarily targets the designers and manufacturers of machinery and equipment and aims to ensure acceptable standards of safety and protection for the health of users.

According to the latest version of the directive, both new machines and existing ones whose design has been revised or that have been assigned to different uses must comply with largely the same safety and ergonomic requirements laid down in principle by the regulation itself and in practice by the so-called harmonized rules issued by the European Committee for Standardization (CEN), which are binding for all member states.

CEN defines a set of standards that relate closely to the prevention of WMSDs. These include

1. Ergonomic design principles that take into account the interaction between the design of machinery and work tasks (EN 614-2)
2. Anthropomorphic principles for the design and arrangement of workstations (EN ISO 14738)
3. Criteria regarding force limits for machinery operation (EN 1005-3)
4. Criteria regarding working postures at workstations (EN 1005-4)
5. Criteria regarding repetitive handling at high frequency (EN 1005-5)

The first four standards have already been issued and are to all effects and purposes harmonized standards. The last one, approved in 2007 and of greater interest here, takes the form of a recommendation and cannot be regarded as a "harmonized" (binding) standard owing to controversies that arose at the time of its ratification. The standard uses the OCRA method of "preventive" risk analysis (manual handling at high frequency), adjusted to more effectively design machinery, workstations, and relevant work processes (CEN, 2007).

Since manufacturing firms not only purchase and install machinery but often also design their own machines or adapt them to their own purposes, they must understand and comply with these standards.

It is equally essential for the standards to be employed in order to implement the necessary structural changes to machines and production lines that are at significant risk of causing injury or disease due to biomechanical overload of the upper limbs.

1.3 APPROACH TO RISK EVALUATION AND MANAGEMENT

In studying UL-WMSDs, many different factors (mechanical, organizational, psychosocial, and individual) have been regarded as important for the purposes of devising general risk analysis, evaluation, and management models and for epidemiological purposes.

Consequently, it is widely agreed that a *holistic* approach toward preventing them is necessary, especially with regard to generating guidelines and action plans at the international level. It should be noted that the term *holistic* refers to something integrated, organic, complex, global, and multifactorial that can and must be approached in an inter- or meta-disciplinary manner.

Nonetheless, almost in contrast with the need for a holistic approach, and considering the significant increase in WMSDs and the many factors that cause them (as stated before, mechanical, organizational, psychosocial, etc.), more and more national and international prevention specialists and agencies are asking for simple tools for assessing and managing this specific risk, and that can also be used by non-experts in both advanced and less advanced countries.

The World Health Organization (WHO) defines a toolkit for WMSDs as "a set of practical risk assessment procedures and related management guidance documents, including advice on simple risk control options" (WHO, 2010). Such a toolkit should provide a comprehensive model for identifying, evaluating, and controlling occupational hazards. It should be simple and practical and also lend itself to utilization by non-experts in small and medium-sized enterprises and developing countries.

As mentioned, ISO is another international body that, as well as defining a set of technical standards on the physical ergonomics of working postures, manual handling, and repetitive manual tasks (ISO 11226 and ISO 11228 series), has now completed a special application document (ISO/TR 12295) to better clarify the application modalities and procedures of the methods reported in such standards, but above all to provide users with the appropriate key enters and quick assessments for the various risks included in the standards.

The proposals presented here (and in later chapters) endeavor to simplify a complex subject based on two main principles:

1. A step-by-step approach, starting with the basic tools and gradually moving to more complex ones only when required for the purposes of prevention

2. An awareness of complexity and of the presence, at every step, of numerous risk factors, albeit explored at different levels of detail

With reference to the first principle, it is a well-known fact that the aforesaid ISO technical reports adopt a general approach toward assessing and managing risk according to four fundamental steps (Figure 1.3):

- Hazard identification
- Risk estimation

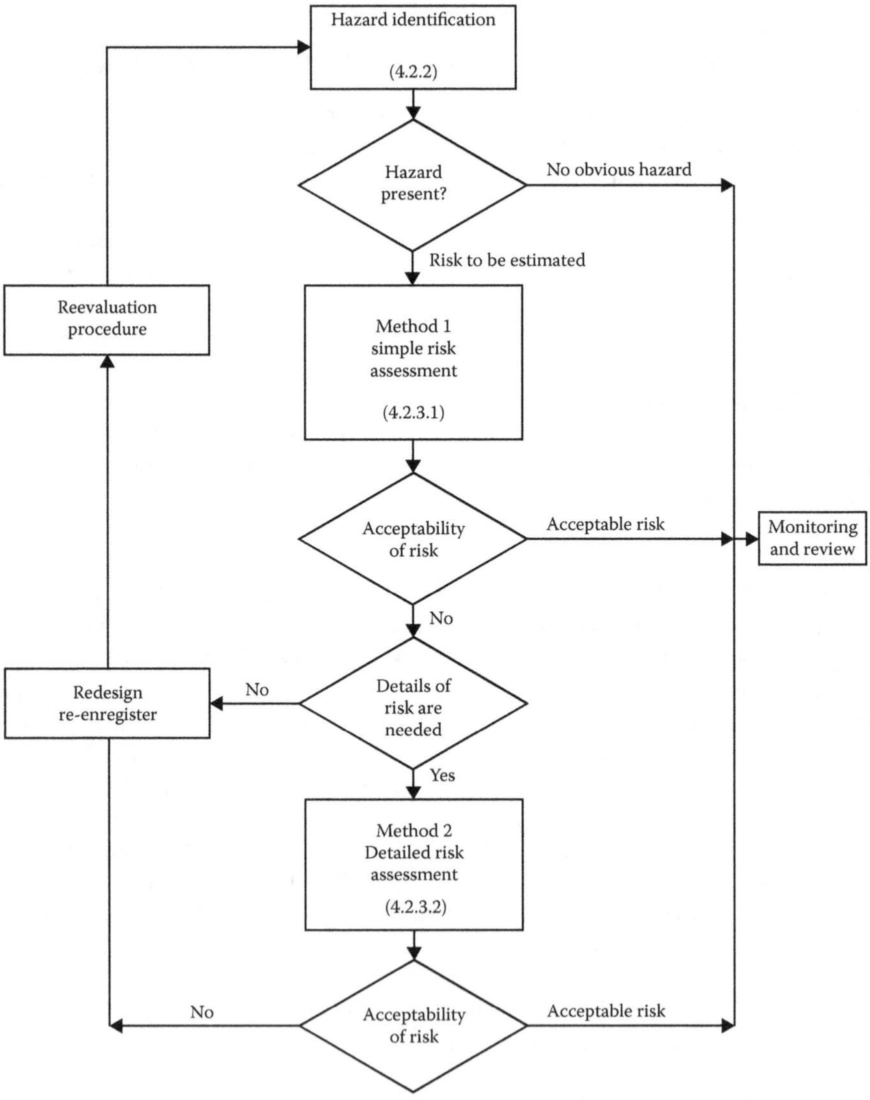

FIGURE 1.3 General scheme for the assessment and management of risk in ISO 11228-3.

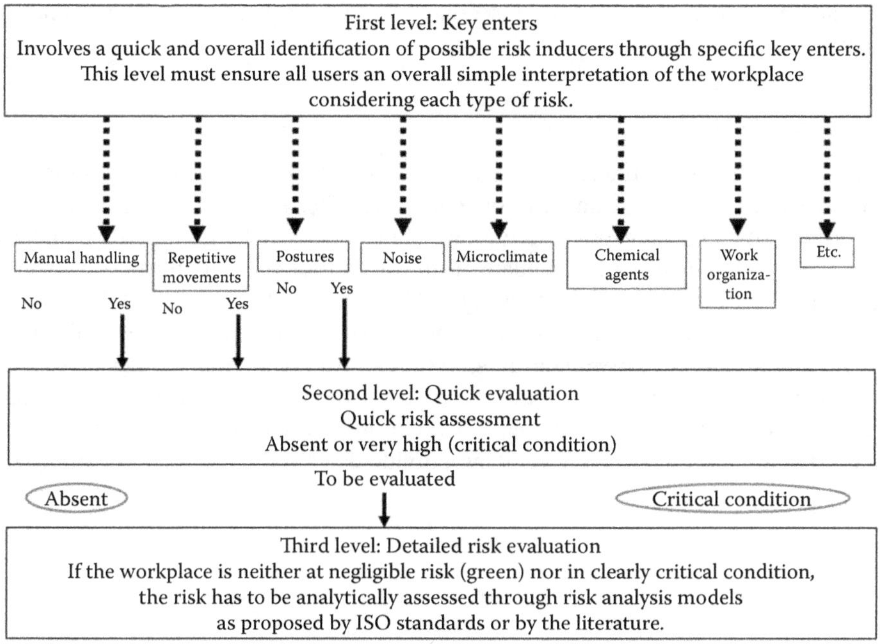

FIGURE 1.4 Levels in the identification, quick assessment, and analysis of workplace risk.

- Detailed risk evaluation
- Risk reduction

This same approach appears in numerous European directives dealing with health and safety in the workplace and is largely endorsed by prevention experts; it is fully in line with all the aims and objectives set forth here and in fact underpins our proposal.

The proposal presented in this volume therefore features the following levels (Figure 1.4):

1.3.1 First Level

Preliminary identification of main hazards (or issues) related to working conditions and priority-setting via specific questions or *key enters*. Ideally, this level singles out possible hazards (or issues) pertaining to ergonomics, industrial health and hygiene, and occupational medicine. The focus here is on the main aspects relating to hazards (or issues) relating to the musculoskeletal system. This level can be handled by non-specialists with limited training.

1.3.2 Second Level

Focus on risk factors for WMSDs and their identification via a quick assessment. This level can be handled by non-specialists with minimal specific training. Detailed explanations concerning first and second-level investigations can be found in Chapter 3.

1.3.3 Third Level

Based on the results of the second-level enquiry, risk is estimated using tools recognized by international standards or guidelines. These tools must be capable of measuring the main risk factors, in our case those associated with repetitive manual work. Only specialists with adequate training should conduct third-level assessments. The leading method for studying WMSDs at this level is the OCRA checklist, which is the subject of this book. If the investigation, whose ultimate aim is to develop and adopt preventive measures, requires additional details, other more analytical methods recommended in the standards or the literature, such as the OCRA index, can be used. However, the more detailed methods should only be used when circumstances call for them and only by individuals with the necessary training.

2 Simplified Analysis of Workplace Organization

From a Macro- to a Micro-Analysis of the Workplace and the Basis for Risk Exposure Assessment among Workers (and Homogeneous Worker Groups)

2.1 INTRODUCTION: SECTORS AND CLASSIFICATION OF WORK ACTIVITIES

Finding an effective procedure for assessing the risk of biomechanical overload of the upper limbs first requires a painstaking organizational study to gather the most objective data concerning the duration and type of worker exposure.

The aim of this chapter is therefore to briefly and simply provide the reader with the rudiments of organizational analysis, including simple definitions and/or flow diagrams, and classic terminology that is often taken for granted but at times needs to be clarified.

The best way to start is by identifying and defining industrial sectors (Figure 2.1).

Three sectors have been recognized:

- *Primary*: This is the most basic form of economic activity performed by man. It includes farming, livestock raising, fishing, mining, and forestry.
- *Secondary*: This sector meets people's secondary needs. It includes building construction, apparel and textiles, metal working, metallurgy, food, and electronics.

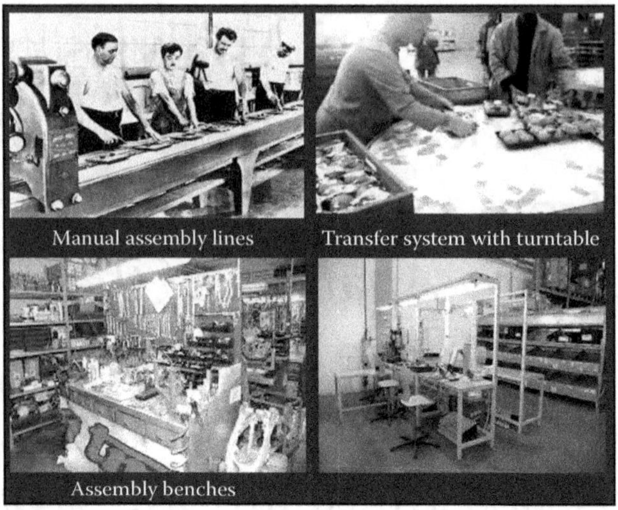

FIGURE 2.1 Example of bench or line assembly stations.

- *Tertiary (or services)*: This sector includes activities that are complementary and auxiliary with respect to the primary and secondary sectors. It includes banking, telecommunications, tourism, marketing and advertising, transportation, and so on.

The *secondary sector*, which is the main focus of this chapter, is the manufacturing industry. The secondary sector develops as an economy grows and is characterized by the extensive use of technology for producing goods.

The principal industries belonging to the secondary sector include

- *Metallurgy*: This sector is specialized in the processing of metals. One of the main products of the metallurgy industry is steel.
- *Chemicals*: This industry produces substances employed by other manufacturing sectors. One of the main products of the chemicals industry is plastics.
- *Metalworking*: The metalworking sector deals with the production of vehicles, automobiles, and machinery.
- *Textiles*: The textiles sector specializes in the production of fabrics and apparel.
- *Agrifood:* This sector converts agricultural products into food, which is then distributed to supermarkets and stores.
- *Building construction*: The construction industry also encompasses the installation of plumbing and electricals and the construction of infrastructure (roads, railways, etc.).

2.2 HOW PRODUCTS ARE MADE

The manufacturing process consists of two activities:

- *Processing:* Processing technology is what guides design and management decisions.
- *Production or manufacturing*: This consists of two stages:
  - *Making parts*: Materials are modified in some way or another in order to obtain the desired morphological characteristics.
  - *Assembling parts*: Two or more parts are joined together to make an assembly that, when put together with other assemblies, makes up the end product.

The output of an industry, be it a product or service, is the result of a set of integrated processes.

2.3 HOW A MANUFACTURING UNIT IS STRUCTURED: DEFINITIONS OF MACHINERY TYPES AND WORKSTATIONS

The following is a list of definitions that apply to the operating areas and assembly stations (Figure 2.1) that may be found in a production plant.

- *Manufacturing cells (or areas)*: A cell (or area) contains all the machinery required to manufacture a group of products that are similar from the standpoint of the production cycle.
- *Manual assembly lines*: Operators are positioned at workstations along the assembly line. A conveyor belt transports the part to be assembled from one workstation to the next. The process follows a fixed sequence.
- *Manual assembly benches*: Each workstation assembles a certain number of parts (by hand or machine). The workstations are linked to one another by a rigid conveyor system. The process does not follow a fixed sequence. Different products may be produced, requiring different parts and components.
- *Automatic assembly lines*: The operators inspect and/or carry out maintenance on the machinery.

There are various types of *product transfer methods* on a production line. The machines consist of a succession of *processing stations*, that is, workstations. Parts are transferred from one station to the next by a system that may be

- *Synchronous*, with parts entering the station at fixed intervals (paced or takt time). Parts are processed and transferred from one station to the next at the same time.
- *Asynchronous*, with the machines not entirely coordinated and parts permitted to transfer between the stations independent of one another, separated

by buffers that hold intermediate inventory. The station transfers the finished part to a downstream buffer and takes a new part from an upstream buffer.

- *Continuous transfer*, which is when operators move along the line at the same time as the product and assemble different components as they go (e.g., the final assembly of an automobile).

2.4 HOW TO ANALYZE A DEPARTMENT: IDENTIFICATION OF PRODUCTION AREAS, MACHINERY, ASSEMBLY STATIONS, AND OPERATOR ALLOCATION

Having provided these short basic definitions, it is necessary, before going on to the risk assessment process, to describe the *department*, starting with the *manufacturing cells* and *types of machinery* and then the *workstations*. Any other *facilities*, such as warehouses, offices, and so on, must also be identified and described.

The analysis of the department starts with a description of the *workforce*, first in terms of the total number and gender of the workers and then in terms of hourly exposure per workstation. Table 2.1 shows a simple reconstruction of a company structure (a kind of index) indicating the approach used for gathering the information required for the subsequent risk mapping exercise (Chapter 7).

Before describing shifts, it might be worth running through the main kinds of shifts used in the manufacturing industry (Figure 2.2):

1. *Single or daily shift*: One 8-h shift only, 4 h in the morning and 4 h in the afternoon (total: 480 min)*; contract does not include meal breaks, which are therefore unpaid.*
2. *Multiple daily 480-min shifts (8 h)*: There are several 8-h shifts; *contract does include meal breaks, which are therefore paid.*
3. *Multiple daily 360-min shifts (6 h)*: There are several 6-h shifts; *no meal breaks.*

2.5 JOB ANALYSIS: A FEW BASIC DEFINITIONS

The next step involves analyzing the job; the focus moves from the machine and type of machinery to the tasks performed to actually making the end product.

The *job* is the set of working activities (tasks) performed by the operator during a shift. The job may consist of one or several tasks.

Figure 2.3 lists various types of tasks that may appear in a work shift or make up a job: repetitive tasks, nonrepetitive tasks, and tasks that may qualify as recovery. The description of the shift also includes breaks and/or other significant official or nonofficial interruptions to the job in order to identify the *net repetitive working time* (Chapters 3 and 4).

The following is the definition of repetitive work, which will be discussed extensively in these two chapters, and the definition of a work cycle.

TABLE 2.1
Identification of Workstations by Department, Area, Line/Bench, and Number/Gender of Workers per Workstation and Shift

Department	Area or Cell	Assembly Line or Bench	Workstation (Workplace) Name	No. of Males per Shift and Workstation			No. of Females per Shift and Workstation		
				No.	Average Age	Average Years in Current Position	No.	Average Age	Average Years in Current Position
Pre-assembly	Line A	LINE A1	L-A1-1	30	35	8	10	40	15
			L-A1-2						
			L-A1-3						
			L-A1-4						
			L-A1-5						
		LINE A2	L-A2-1				3	45	18
			L-A2-2						
			L-A2-3						
	Bench A	No. 3 benches	B-A1	3	43	9			
			B-A2						
			B-A3						
	Department A warehouse			10	45	7			

(*Continued*)

TABLE 2.1 (CONTINUED)
Identification of Workstations by Department, Area, Line/Bench, and Number/Gender of Workers per Workstation and Shift

Department	Area or Cell	Assembly Line or Bench	Workstation (Workplace) Name	No. of Males per Shift and Workstation			No. of Females per Shift and Workstation		
				No.	Average Age	Average Years in Current Position	No.	Average Age	Average Years in Current Position
Assembly C	Line C	LINE C1	L-C1-1	10	20	18			
			L-C1-2						
			L-C1-3						
			L-C1-4						
			L-C1-5						
		LINE C2	L-C2-1				8	40	15
			L-C2-2						
			L-C2-3						
			L-C2-4						
		LINE C3	L-C3-1	6	34	6			
			L-C3-2						
			L-C3-3						
	Department C warehouse			3	45	18			
General warehouse	Incoming goods warehouse			15	43	12			
	Outgoing goods warehouse			18	46	15			

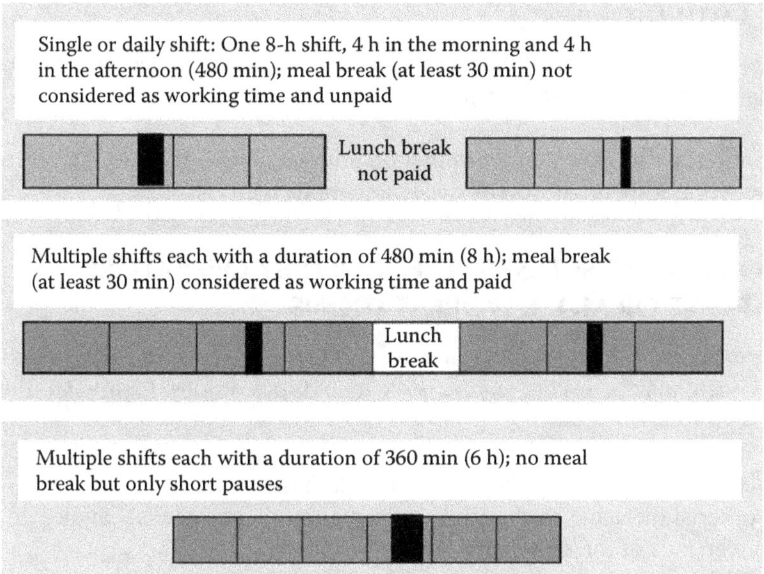

FIGURE 2.2 Example of three common types of shifts.

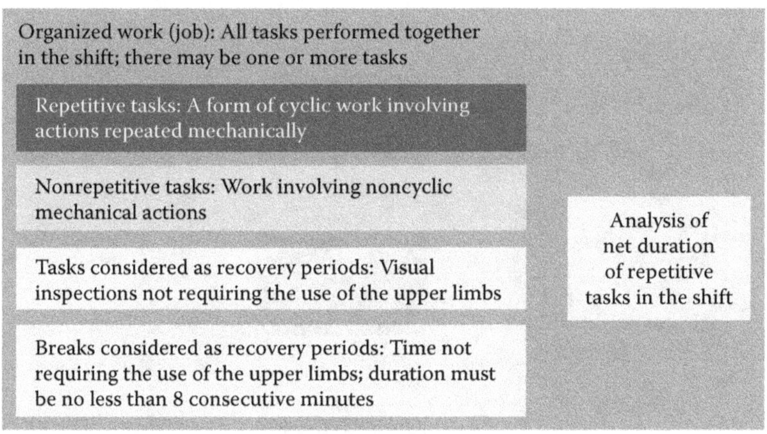

FIGURE 2.3 Classification of various types of tasks that may be performed in a shift and make up a job.

2.5.1 Repetitive Task: Definition

Work is repetitive when

- Tasks consist of *cycles* (regardless of their length).
- The job consists of performing virtually identical technical actions (or sequences of technical actions), which are always repeated in the same way for more than half of the working time under examination.

2.5.2 Cycle: Definition

- A sequence of identical mechanical technical actions, of variable duration, that is repeated several times.

It is important to note that these two definitions merely refer to the presence of repetitive work. Repetitive work is not synonymous with risk.

2.6 JOB ANALYSIS: TASKS PERFORMED BY OPERATORS AT ONE OR MORE WORKSTATIONS

For the purposes of analyzing risk due to exposure to repetitive tasks, a workstation (line or bench) may be said to feature three main types of jobs. Figure 2.4 illustrates these tasks:

- Worker performs only one repetitive task in the shift
- Worker at the same workstation performs several tasks (e.g., handles different versions of the same part)
- Worker rotates between several workstations, performing one or more tasks

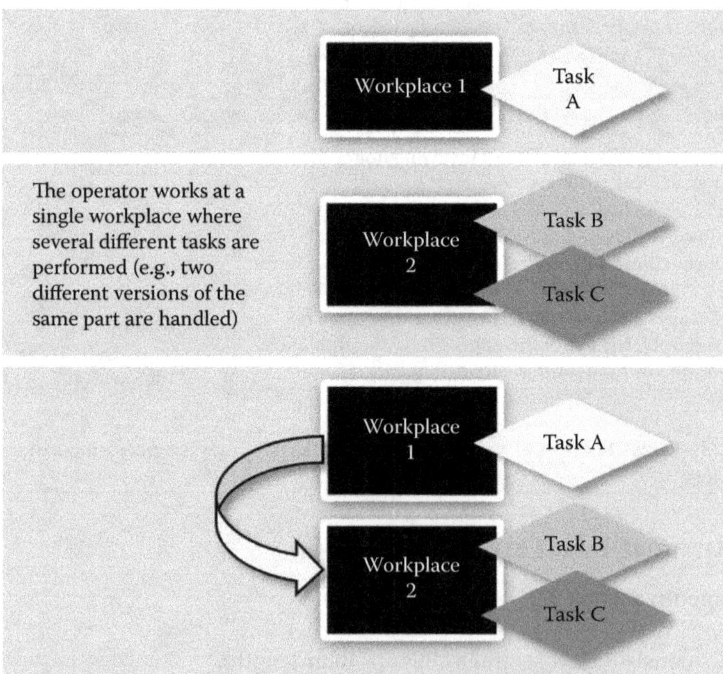

FIGURE 2.4 The three main types of tasks for the purposes of carrying out repetitive task exposure risk studies.

2.7 REPETITIVE TASK ANALYSIS: TYPE AND CONTENT OF THE WORK CYCLE MAKING UP THE TASK

Several more definitions, presented in steps, are now presented that will help to embark on a risk assessment using the Occupational Repetitive Actions (OCRA) method.

The steps entail

a. Identifying the cycle
b. Characterizing the various types of cycles
c. Calculating the cycle duration times
d. Performing a micro-analysis of the structure of the tasks making up the cycle

a. Cycle identification (Figure 2.5a and b)

i. A cycle may correspond to a unit of work (such as processing a part).
ii. A cycle may correspond to a set or series of units of work (or processing several parts).
iii. If the same activities are repeated the same way virtually all the time for the whole task, consider the cycle as a representative period with a duration of between 30 and 120 s.

b. Characterization of different cycles based on the workstation where they are performed

Figure 2.6 schematically illustrates the characteristics of the different cycles, based on the type of workstation and the way objects are transferred.

c. Calculation of cycle duration

In order to correctly assess risk using the OCRA method, it is essential to properly measure the cycle time, since various risk factor scores will depend on the ratio of the individual cycle to the total cycle time.

Considering different types of cycles for different work stations, as shown in Figure 2.6, there are five ways of calculating cycle duration:

c1. Cycle and cycle duration analysis: assembly line workstations with transfers at fixed intervals

When analysts prepare the task to be performed at each workstation along a line, they define it via a set of *micro-phases* (also called *phases* or *elements*). The maximum execution time for a task at a workstation along the line determines the pace of the line. On any given line, at least one workstation will have a cycle duration equal to the pace; this workstation is called the *line bottleneck.*

Figures 2.7 and 2.8 show diagrams for a workstation along a line featuring fixed transfers, including a few basic definitions to help understand how to calculate cycle times.

The *pace* (or total cycle time or takt time) therefore represents the time that elapses between the arrival of one part to be worked at a workstation and the arrival of the next one; it is in fact the machine time that is set for the line.

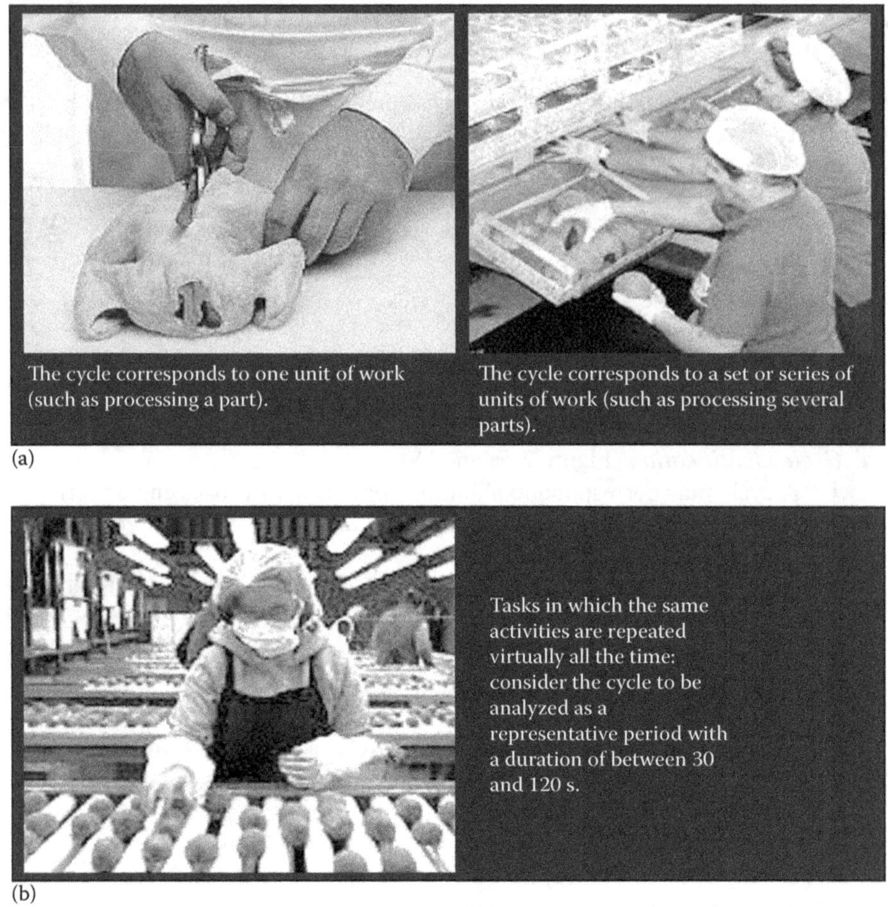

FIGURE 2.5 (a) Examples of work cycles corresponding to one unit of work (part) or set of parts. (b) Types of work cycles: choice of a representative cycle time for a task featuring the same repeated actions.

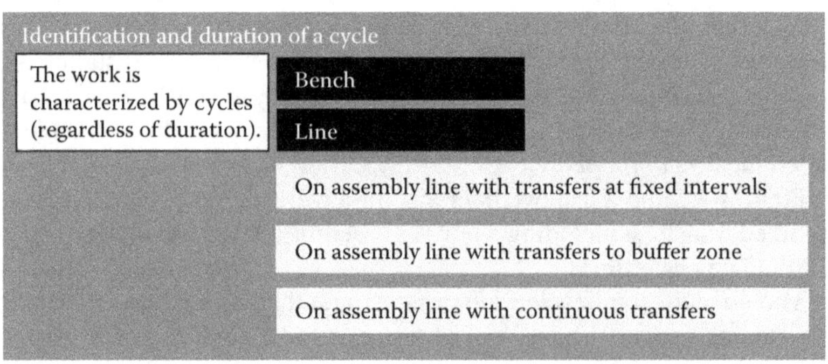

FIGURE 2.6 Types of cycles based on the workstation where they are performed.

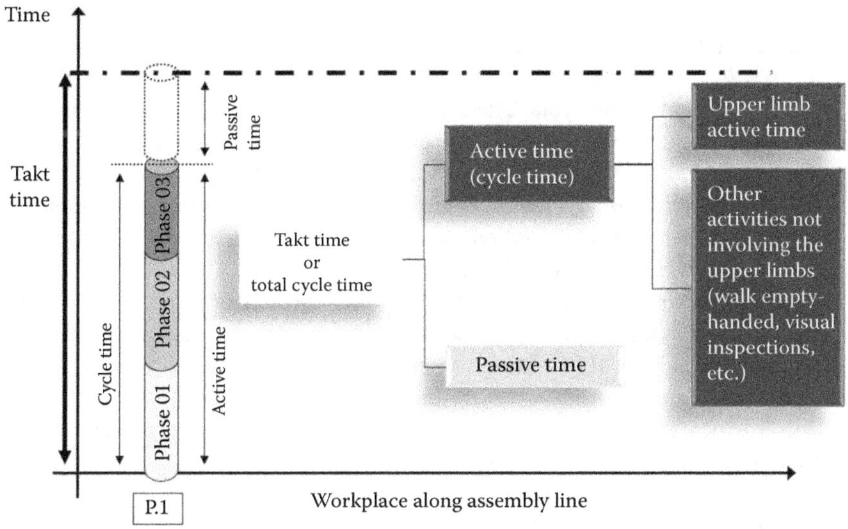

FIGURE 2.7 Workstation along an assembly line, task broken down into micro-phases, line pace, total cycle time (takt time), active time, and passive time.

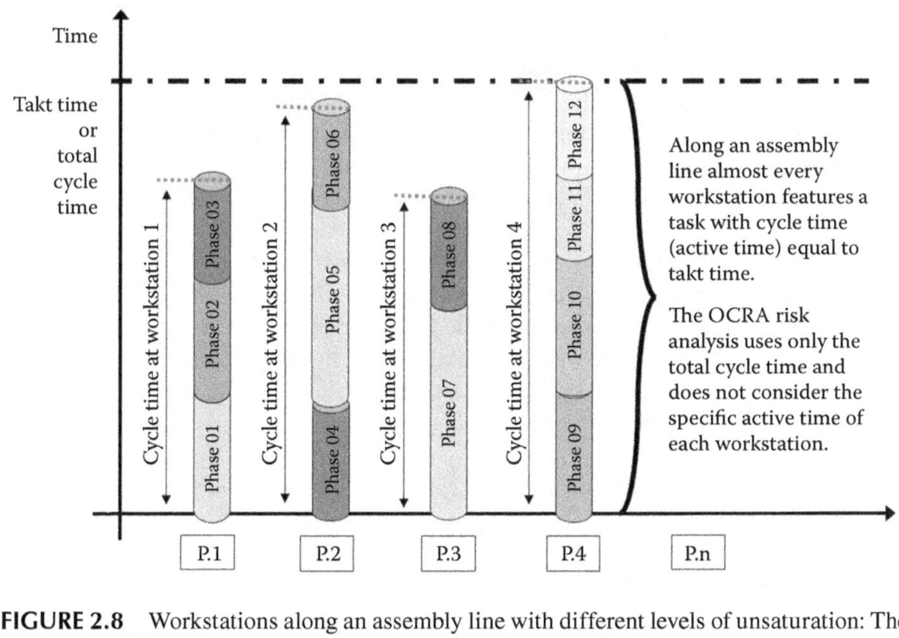

FIGURE 2.8 Workstations along an assembly line with different levels of unsaturation: The workstation with maximum saturation determines the pace of the line.

It is also necessary to define other important times that may make up the total cycle time or pace. These include

i. *Cycle time*: A term commonly used by analysts to define the active component of the cycle. The cycle time is therefore the same as the pace when the allocated operating phases occupy the entire pace.
ii. *Upper limb active time*: Upper limbs are working.
iii. *Other active time*: The upper limbs are not working but micro-phases are performed that do not require them to be used (e.g., walking to fetch an object or performing a visual inspection).
iv. *Passive time*: Difference between pace and active time. This determines the degree of *unsaturation* of the task expressed as a percentage and its reciprocal (% saturation).

To calculate risk indexes with the OCRA method, the pace of the line will always be used, regardless of the different cycle times present at the various workstations along the line.

c2. Cycle and cycle duration analysis: assembly line with buffer

When lines have a buffer, the operator can vary the cycle time—albeit within certain limits—depending on the number of "extra" parts that can be kept in the buffer.

To calculate risk indexes with the OCRA method, the pace of the line will be used here too, regardless of the different times taken by the various workers; this is determined by the number of parts that can be kept in the buffer. Of course, if dozens of parts can be "parked" in the buffer, this unusual workstation would more closely resemble a work bench than a standard assembly line in terms of work organization.

c3. Cycle and cycle duration analysis: assembly line with continuous transfers

In this case, for each workstation there is a *maximum distance* that one operator can cover before entering into the area of the next operator (Figure 2.8). Since the speed of the line is constant, a *maximum total cycle time* is created, determined by the maximum time in which the task must be performed, within the maximum available space. To calculate risk indexes with the OCRA method, the *maximum total cycle time* will always be used.

c4. Cycle and cycle duration analysis: repetitive tasks on bench

The work is not paced, as on an assembly line.

To calculate risk indexes with the OCRA method, the *total cycle time* will be used, as before, which is obtained by dividing the *net duration of repetitive work* (i.e., the shift minus actual breaks and time spent performing nonrepetitive tasks) by the *number of parts or cycles* to be carried out during the shift.

c5. Cycle and cycle duration analysis: repetitive tasks in environments without workstations

If there are no identifiable cycles, repetitive tasks often consist of the repetition of the same actions. To calculate risk indexes with the OCRA method, a representative portion of the task (such as 30 or 60 s) will be taken as the total cycle time.

When there are cycles, the *total cycle time* is obtained by dividing the *net duration of repetitive work* (i.e., the shift minus actual breaks and time spent performing nonrepetitive tasks) by the *number of parts or cycles* to be carried out during the shift, as for bench work.

2.8 REPETITIVE TASK ANALYSIS: MICRO-ANALYSIS OF THE STRUCTURE OF THE TASK WITHIN THE CYCLE

Figures 2.7 and 2.8 show that for expert analysts, the task is actually comprised of micro-phases (also termed phases or elements), like the "bricks" that form a building.

Table 2.2 provides an example of a task broken down into *micro-phases*, *phases*, or *elements* using different analytical methods (methods time measurement [MTM] or chronometric). This type of analysis supplies the intrinsic duration of each phase and is the starting point for risk analysis using the OCRA index (while the OCRA checklist does not require this level of detail). When tasks have very long total cycle times (several minutes) or are complex (featuring elements with different characteristics), such a breakdown can also be useful for applying the OCRA checklist.

Of course, micro-phase duration, active/nonactive upper limb times, other active times, and passive times are all supplied for every phase as well as for the overall task.

TABLE 2.2
Example of Breakdown of a Work Cycle into Micro-Phases (Also Called Phases or Elements)

Task: Fan Assembly: Total Cycle Time 20 s

Name of Phase (Micro-Phase or Element)	Phases Duration (sec.)	Technical Actions	Duration of Upper Limb Technical Actions	Duration of Other Activities	Passive Time	% Unsaturation
1: Position 2 brackets	5	Take, position, turn, fit 2 brackets (8 actions)	4	1		
2: Position 2 washers	3	Take and position 2 washers (4 actions)	3			
3: Position fan	4	Reach, take, turn twice, position (5 actions)	3	1		
4: Position 2 screws	2	Take and position 2 screws (4 actions)	2			
5: Screw	6	Take screwdriver, position, press button, position over second screw, switch on, replace screwdriver (4 actions) (6 actions)	5		1	20%
Total	20	27	17	2	1	5%

3 Initial Identification and Preliminary Assessment (Pre-Mapping) of Potential Risks

3.1 ISO/TR 12295: KEY ENTERS AND QUICK ASSESSMENT

As briefly reported in Chapter 1, the International Standards Organization (ISO) recently published a technical report, ISO/TR 12295 *Ergonomics—Application document for ISO standards on manual handling (ISO 11228-1, ISO 11228-2 and ISO 11228-3) and working postures (ISO 11226)*, the main section of which concerns the identification of risk via *key enters* (i.e., the field of application of different standards) and *quick assessments* (ISO, 2014).

For ISO, a technical report is a sort of guideline representing the "state of the art" in a specific area and provides users with detailed information. That being said, it is also a reliable and useful tool for initially identifying risk, in this case that related to repetitive manual tasks.

Before going into a broader approach toward identifying and pre-assessing potential health risks, it is first necessary to consider what ISO/TR 12295 says about the risk of biomechanical overload.

3.1.1 Key Enters

Key enters are used to detect the existence of an occupational hazard (problem), in this case, the risk of biomechanical overload and work-related musculoskeletal disorders (WMSDs), and decide whether further analyses and assessments are required (*hazard identification*).

Key enters define the field of application of the four sections of the relevant ISO standards.

Table 3.1 lists the key enters for ISO/TR 12295.

The following is an in-depth analysis of one particular aspect of the standard, relating to repetitive manual work.

According to the key enters for the application of ISO 11228-3, a task is classified as repetitive when

- The task features repeated work cycles.
- The same work actions are repeated for more than 50% of the time.

TABLE 3.1
ISO/TR 12295: Key Enters for the Application of ISO 11226 and ISO 11228 (Parts 1-2-3)

1 Application of ISO 11228-1		
Is there any manual lifting/lowering or carrying of an object weighing 3 kg or more?	No	Yes
If No, this standard is not relevant; go to the next "Key Question" regarding other standards. If Yes, go to step 2: "Quick Assessment"		
2 Application of ISO 11228-2		
Is there any two-handed whole-body pushing or pulling of loads?	No	Yes
If No, this standard is not relevant; go to the next "Key Question" regarding other standards. If Yes, go to step 2: "Quick Assessment."		
3 Application of ISO 11228-3		
Are there one or more repetitive tasks involving the upper limbs with a total duration of 1 h or more per shift?	No	Yes
Where the definition of "repetitive task" is a task characterized by repeated work cycles Or tasks during which the same working actions are repeated for more than 50% of the cycle time. If No, this standard is not relevant; go to the next "Key Question" regarding other standards. If Yes, go to step 2: "Quick Assessment."		
4 Application of ISO 11226		
Are there any static or awkward working postures of the Head/Neck, Trunk and/or Upper and Lower Limbs maintained for more than 4 consecutive seconds and repeated for a significant part of the working time?	No	Yes
For example:		
• **Head/Neck** (neck bent back/forward/sideways, twisted)	No	Yes
• **Trunk** (trunk bent forward/sideways/, bent back with no support, twisted)	No	Yes
• **Upper Limbs** (hand(s) at or above the head, elbow(s) at or above shoulder height, elbow/hand(s) behind the body, hand(s) turned with palms completely up or down, extreme elbow flexion-extension, wrist bent forward/back/sideways)	No	Yes
• **Lower Limbs** (squatting or kneeling) for more than 4 consecutive seconds and repeated for a significant part of the working time	No	Yes
If "No," this standard is not relevant. If "Yes," go to step 2: "Quick Assessment."		

Based on these definitions, when the upper limbs perform one or more repetitive tasks for 1 h or more in the shift, the assessment must be pursued further by means of the *quick assessment*.

It should be noted that the only reason for indicating the presence of a repetitive task is to establish that it must be assessed; it does not suggest that there is necessarily any risk. The existence of risk will be established only after the next stage of the evaluation. Conversely, if there are no repetitive tasks, then no further assessments are required.

The same rationale applies to the other aspects dealt with in ISO/TR 12295 (lifting and carrying loads; pushing and pulling; static and/or awkward working postures).

3.1.2 The Quick Assessment

The *quick assessment* consists of rapidly checking for potential risk conditions (in this case, for WMSDs), by asking simple qualitative and quantitative questions. The quick assessment is essentially designed to identify three possible conditions or outcomes:

- Acceptable (green): No further actions are required.
- Critical (purple): The work or process must be immediately redesigned.
- A more detailed analysis is required and a proper risk assessment needs to be carried out using specific analytical tools (as suggested by the relevant standard).

It is worth remembering that when conditions are found to be *acceptable* or *critical*, it is not always necessary to conduct a more in-depth analysis of the exposure level (level 3), especially in the case of critical conditions. Efforts should be directed toward reducing the detected risk rather than being spent on often complex and sometimes useless investigations into the situation.

On the other hand, as happens in the majority of cases, if neither of these two "extreme" conditions are clearly detected, it is essential to assess risk using a simplified or more detailed approach and traditional assessment methods (as will be described in later chapters). This assessment may result in risk being classified as green, yellow, red, or purple, with all the relevant consequences.

Here, only the recommendations featured in ISO/TR 12295 will be reported with regard to the topic of this book, that is, repetitive tasks and standard ISO 11228-3. Other aspects will be mentioned and included in an overview of the pre-mapping tool described in this chapter.

Table 3.2 lists the conditions that must all be present *simultaneously* to classify a repetitive manual task as acceptable (green).

The rapid assessment of acceptability referred to a repetitive task has been drawn from standard ISO 11228-3 (ISO, 2007b) and especially from EN 1005-5 (step 1) (CEN, 2007).

If a repetitive task is assessed as acceptable using the quick assessment procedure, this would be equivalent to classifying it as acceptable using the detailed methods indicated by the relevant standards.

Conversely, Table 3.3 lists situations which, *even if they occur alone*, determine a *critical* condition.

To make a quick assessment of *definitely critical* conditions, it is possible to use the definitions and criteria included in the methods recommended by the standards (starting with the Occupational Repetitive Actions [OCRA] method), which

TABLE 3.2
ISO/TR 12295: Quick Assessment of Repetitive Manual Tasks: Acceptability Criteria (Green)

Are either upper limbs working for less than 50% of the total duration of the repetitive task(s)?	No	Yes
Are both elbows held below shoulder level for almost 90% of the total duration of the repetitive task(s)?	No	Yes
Is moderate force (perceived effort max 3 or 4 on the CR-10 Borg scale) exerted by the operator for no more than 1 h for the duration of the repetitive task(s)?	No	Yes
Absence of peak force (perceived effort 5 or more on the CR-10 Borg scale)?	No	Yes
Presence of breaks (including meal break) lasting at least 8 min every 2 h?	No	Yes
Are repetitive task(s) performed for less than 8 h a day?	No	Yes

- If all the answers to these questions are "Yes," the task is in the Green area (Acceptable) and it is unnecessary to continue the risk evaluation.
- If the answer to at least one of the questions is "No," evaluate the task(s) using ISO 11228-3.

TABLE 3.3
ISO/TR 12295: Quick Assessment of Repetitive Manual Tasks: Criteria for Identifying Critical Conditions

If even only one of the situations listed is present, risk should be considered as Critical and the task must be redesigned Urgently		
Are technical actions performed by a single limb so fast that they cannot be counted by simple direct observation?	No	Yes
Are one or both arms operating with the elbow at shoulder height for half or more of the total repetitive working time?	No	Yes
Is a "pinch" grip (or any kind of grasp using the fingertips) used for more than 80% of the repetitive working time?	No	Yes
Is peak force applied (perceived effort 5 or more on the CR-10 Borg scale) for 10% or more of the total repetitive working time?	No	Yes
Is there no more than one break (including the meal break) in a shift of 6–8 h?	No	Yes
Does the total repetitive working time exceed 8 h within the shift?	No	Yes

- If the answer to at least one of the questions is "Yes," a critical condition is present.
- If a critical condition is present, apply ISO 11228-3 to identify urgent corrective actions.

indicate one or more highly risky elements. These include actions performed with extremely high frequency using the upper limbs or the need for almost high peak force to be used. When a repetitive manual task is found to be critical, even for just one of the situations listed in Table 3.3, the recommendation is to opt for a rapid and substantial corrective action (i.e., risk reduction) without necessarily carrying out further analyses. However, more in-depth investigations might be undertaken at a later date, to monitor the potential effectiveness of the corrective measures.

3.2 PRE-MAPPING OF HAZARDS, DANGERS, AND DISCOMFORT

3.2.1 Foreword

As stated previously, one of the latest developments being pursued by the WHO and other international organizations (International Labour Office [ILO], International Standards Organization [ISO]) in relation to preventing work-related diseases and disorders concerns the creation of toolkits.

The main aim is to rapidly but accurately identify the presence of possible sources of risk using instruments that can easily be used by occupational safety and health (OSH) operators, occupational physicians, employers, workers, trade union representatives, and occupational safety services.

However, this objective also reflects the criteria set forth in ISO/TR 12295 with respect to the risk of biomechanical overload, as previously mentioned.

Against this backdrop, the "problem" of WMSDs must be considered together with other occupational "hazards" (be they physical, chemical, or other) for the more general purposes of prevention.

The aim here is to suggest a methodology and some simple tools for bringing together various parties to undertake a preliminary *mapping of discomfort/danger* (i.e., to identify *risk sources in the work cycle*) in the workplace, especially in small and very small enterprises.

The tool does not pretend to replace the standard risk evaluation process but to support such a process in order to identify hazardous situations in the workplace, on the basis of which to single out emerging problems that need to be submitted to a full risk assessment (in the appropriate order of priority).

The tool can also be used by employers and/or trade union representatives to more readily identify situations that may also call for an occupational health expert to be present during the assessment.

The tool is primarily designed to be used by employers and OSH operators, but it may also be useful for

- Medical staff conducting periodical inspections and drafting health surveillance protocols.
- Work safety officers periodically monitoring hazardous situations in the workplace.
- Supervisory bodies (labor inspectors) conducting inspections in the workplace, needing to rapidly detect potentially dangerous situations requiring specific preventive interventions.

The procedure presented here demands a cooperative approach toward assessing and managing risk, as it also entails interviews with workers.

In accordance with the recommendations of the World Health Organization (WHO, 2010), three main criteria underpin the methodology:

Globality: A global approach is taken to assessing the worker's discomfort, due to either the task or the workplace.

Simplicity: The methodology consists of an easy-to-use model for collecting data.

Priority-setting: The results obtained automatically via dedicated software and depicted clearly in bar graphs will not only help to identify problems but will also offer a scale of priorities for conducting subsequent assessments.

3.2.2 The Pre-Mapping Model

The operation involves two levels of intervention.

Level 1 entails a rapid and general identification of possible risk inducers via the use of specific key enters. This preliminary level ensures that all users (regardless of their skills and education) can simply and generally observe the workplace. The first level is broken down into several "boxes" relating to the main types of risks: *handling loads*, *repetitive movements of the upper limbs*, *postures*, *noise*, *micro-climate*, *chemicals*, *organization of work*, and so on.

Level 2 entails quickly identifying acceptable risk (indicated, using the traditional traffic light method, as *green*) or very high risk (*critical*, or *purple*) using the quick assessment procedure.

If the situation is *code green*, the risk assessment process could even stop here because it means that there is no meaningful occupational risk.

If a *critical code* is detected, then there is definitely a significant occupational risk and immediate corrective actions will be required.

If the quick assessment finds that the risk level at the work station is neither acceptable (green) nor critical (purple) and therefore the situation is intermediate (potentially code yellow or red), then the risk assessment will have to be carried out using the analytical methods suggested by ISO or the accredited literature.

The methodology consists in an easy-to-use computer-based model for collecting data (Excel spreadsheet): The ERGOepm-PREMAPPING-ERGOCHECK-EN software can be downloaded free of charge in Italian or English from www.epmresearch.org.

The methodology calls for the active participation of workers as well as all those involved in workplace health and safety.

The model provides a general preliminary overview of all the main risk factors that may be present, regardless of the size of the manufacturing facility, and it is underpinned by the basic tenets of ergonomics entailing a global interpretation of the worker's discomfort deriving from the task or the workplace.

The main highlights of the model will be described later.

3.2.3 First-Level Pre-Mapping (Key Enters)

3.2.3.1 Details Concerning the Place of Employment (Figure 3.1)

Once the facility has been described, the pre-mapping form is completed for individual workers or groups of workers (i.e., homogeneous groups) who perform the same job or tasks in the shift. The term *job* is used to describe a set of tasks performed in

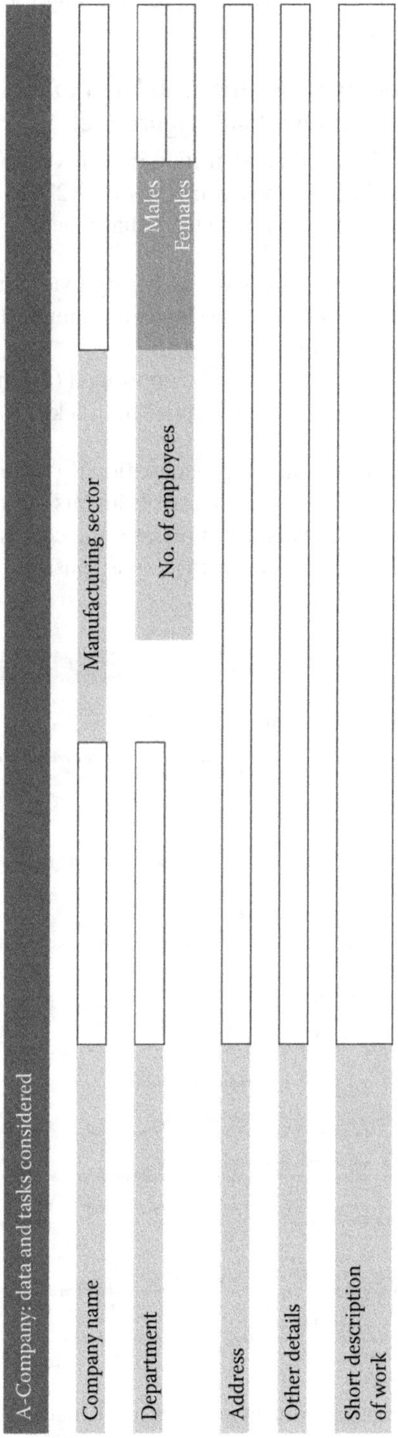

FIGURE 3.1 Description of the workplace and definition of homogeneous group.

a typical shift or longer time frame. A separate pre-mapping form must be used for each homogeneous group of workers.

3.2.3.2 Key Enters for Identifying Priorities Related to the Risk of Biomechanical Overload (Figure 3.2)

The various biomechanical risks are then identified in terms of *presence* or *absence*, using the same key enters as those proposed in ISO/TR 12295. The key enters are actually very simple questions with yes/no answers that reveal

- Whether a potential source of risk (such as lifting loads) in a specific task does not require further analysis (i.e., no loads are manually lifted because the object weighs less than 3 kg).
- Or, alternatively, if the assessment must be continued (for instance, because the load lifted manually is equal to or heavier than 3 kg).

These key enters allow anyone analyzing occupational risk exposure to quickly and unequivocally know whether or not further assessments are required.

Figure 3.2 (showing the Excel spreadsheet) indicates the corresponding key enters for repetitive movements, lifting loads, carrying loads, pushing and pulling loads, and working postures.

B-Biomechanical overload

B1	Biomechanical overload of upper limbs during repetitive tasks		
	Presence of repetitive task: The task is organized in cycles, regardless of duration, or the task is characterized by similar working gestures for over 50% of the time. The term does not indicate the presence of risk.	Yes	
		No	
B2	Biomechanical overload due to manual handling - lifting		
	Presence of objects weighing more than (or equal to) 3 kg to be manually lifted (if the weight is less, no need to continue the investigation).	Yes	
		No	
B3	Biomechanical overload due to manual handling- carrying		
	Presence of objects heavier than 3 kg to be manually carried (if the loads are lighter, no need to continue the investigation).	Yes	
		No	
B4	Biomechanical overload due to manual pushing and/or pulling		
	Is there whole-body pushing or pulling of loads?	Yes	
		No	
B5	Biomechanical overload due to awkward postures-trunk and lower limbs		
	Are there static or awkward working postures of the head/neck, trunk, and/or upper and lower limbs held for more than 4 consecutive seconds and repeated for a significant part of the workng time?	Yes	
	In practice, generally speaking there are no awkward postures (sign no) if the worker: -is sitting with his/her back well supported, with adequate space for the legs and can stand up (change position) at least every hour; -is standing with the trunk erect (no deep bending or twisting) but can walk around or sit down at least every hour (with the back well supported and adequate space for the legs).	No	
	For example		
	Head/neck (neck bent back/forward/sideways, twisted)	Yes	
	Trunk (trunk bent forward/sideways, bent back with no support, twisted)	Yes	
	Upper limbs (hand(s) at or above the head, elbow(s) at or above the shoulders, elbowlhand(s) behind the body, hand(s) turned with palms completely up or down, extreme elbow flexion-extension, wrist bent forward/back/sideways)	Yes	
	Lower limbs (squatting or kneeling) for more than 4 consecutive seconds and repeated for a significant part of the working time	Yes	

FIGURE 3.2 Key enters for setting assessment priorities in relation to biomechanical overload due to repetitive movements of the upper limbs, manual load handling, and working postures.

Trunk posture	
Standing or squatting posture (not seated)	
Nearly always upright	
Frequent moderate flexions	
Frequent torsions	
Frequent major flexions	
Seated posture	
Works leaning on the back	
Works in upright position but there is no backrest	
Works mostly bent forward	
Frequent trunk torsions	
Note: Described time of trunk posture:	

Lower limbs posture	
Standing or squatting posture (not seated)	
Standing posture with possibility of walking around	
Standing fixed posture	
Kneeled or squatted legs	
Seated posture	
Leg space is sufficient	
Leg space is insufficient or very limited	
Leg space is nonexistent	
Note: Described time of lower limbs posture:	

Use of lower limbs	
No acting pedals	
Use of lower limbs for acting pedals	
Note: Described time of use of lower limbs:	100%

FIGURE 3.3 Simplified "indicators" for analyzing awkward postures of the spine and lower limbs.

With regard to biomechanical overload due to awkward postures of the spine and lower limbs, in addition to using the key enters provided by ISO 11226 (for static postures only and in reality with no key-enter criteria) (ISO, 2000), other simple discomfort detection criteria have been used (and can be found in a specific folder of the Excel file), taken and reassembled from numerous ergonomics checklists and/or questionnaires and/or manuals that are readily available in the literature. These "items" have been converted into simple closed-ended questions (Figure 3.3). The form relating to postures is completed, with the help of the attached pictures, by indicating with a cross the main postures adopted by the worker when performing the task (among those listed) and then entering their duration in percentage terms, adding up to a sum total of 100%.

For repetitive movements, manual load handling, and postures, once the presence of risk has been determined, the process calls for completing the specific quick-assessment forms (i.e., the second stage in the assessment process), but for all subsequent physical and organizational risk indicators, the questions are more in-depth, albeit simple, in order to allow assessors to begin prioritizing evaluations and preventive actions.

Returning to Figure 3.3; after completing the form, the software will generate the appropriate color below the light: *green* (no postural problems for the back or lower limbs), *yellow* (slight discomfort), *red* (discomfort present calling for further investigations), or *purple* (critical discomfort present requiring urgent investigation). It was a deliberate decision to avoid assigning a score to these initial indicators and instead to use colors to depict either the absence of the problem or the presence of the problem and the need to deal with its more or less urgently (color-coded priorities).

Figure 3.4 shows an example of how the Excel software changes the color of the traffic light. In this instance, the homogeneous group spends 40% of the time standing,

Trunk posture			
Standing or squatting posture (not seated)			
Nearly always upright	40%	1	0.4
Frequent moderate flexions		2	
Frequent torsions		3	
Frequent major flexions	10%	4	0.8
Seated posture			
Works leaning on the back		0	
Works in upright position but there is no backrest		2	
Works mostly bent forward	50%	3	1.5
Frequent trunk torsions		4	
Note: Described time of trunk posture:		2.7	

FIGURE 3.4 "Concealed" scores listing discomfort from low to high so as to obtain color-coded priorities.

10% with the spine in extreme flexion, and 50% seated with the back slightly bent. The column to the right of the traffic light (under the heading PUNT.), indicates the scores assigned to each posture, ranging from 0 to 4 (although the software masks them): 0 = almost ideal postures, 4 = more tiring postures. These scores should not be regarded as predictors of the likelihood of disease: All they do is "arrange" the level of biomechanical overload from lower to higher. The next column (under the heading PUNT. POND.) weights the scores based on the duration of the postures in percentages. The weighted cumulative score is equal to 2.7. Having established that the *maximum score* is 4, it can be assumed that the posture shown here corresponds to 68% of the maximum score (2.7/4*100). The significance of this percentage will be referred to later, when it is used as a final brief descriptive term for each of the various risk indicators. For scores above 0% but below 30%, the color will be *green*; for scores above 30% but below 50%, the color will be *yellow*; for scores between 50% and 80% the color will be *red*; for scores of more than 80% the color will be *purple*, which is always used for the highest possible score and therefore indicates a critical condition.

Similar masked scores are also used for all the other risk inducers present in the pre-mapping form and therefore will not be specified.

3.2.3.3 Key Enters for Identifying Indoor Lighting Problems (Figure 3.5)

As for all the other sections of the pre-mapping form, these questions are also closed ended and refer to the visual effort required to perform the work in relation to the lighting both at the work station and in the workplace in general.

It is advisable to obtain answers to these questions by interviewing the members of a homogeneous group; each one should be asked to express their judgment regarding the type of lighting, that is, whether it is *sufficient*, *poor*, or *excessive*; this should also be considered in relation to the time frame, in other words *for a few hours a day* or *all day*. The workers are also asked about the surfaces of their work benches, that is, whether they are opaque or shiny.

3.2.3.4 Key Enters for Identifying Problems Relating to Outdoor Work/UV Radiation (Figure 3.6)

With regard to detecting the possible existence of a problem, all that is considered here is the time spent working outdoors.

3.2.3.5 Key Enters for Identifying Problems Relating to Noise (Figure 3.7)

Here, workers are asked whether or not the task *involves* or *does not involve* verbal communications with co-workers or other people (for work-related reasons). In either case, additional information is acquired by asking the following simple questions, which provide an indirect indication of the noise level:

- Noise is slightly bothersome, but it is possible to talk with co-workers (or with the interviewer if there are no co-workers).
- Noise is bothersome, and it is difficult to talk with co-workers (or with the interviewer if there are no co-workers).
- Noise is very loud, and it is impossible to talk with co-workers (or with the interviewer if there are no co-workers).

C: key questions for identifying lighting problems		
General lighting: judgment of visual effort required at work		
Sufficient		
Poor :	For several hours a day	
	All day long	
Excessive:	For several hours a day	
	All day long	
Artificial lighting: Needed but unavailable		
Localized lighting: judgment of visual effort required at work		
Sufficient		
Poor :	For several hours a day	
	All day long	
Excessive:	For several hours a day	
	All day long	
Needed but unavailable		
Kind of surfaces: judgment of visual effort required at work		
Surfaces of work tops	Opaque	
	Bright and shiny	
Surfaces of objects being processed	Opaque	
	Bright and shiny	

FIGURE 3.5 Key enters for identifying indoor lighting problems.

D: key questions for identifying problems relating to outdoor work/UV radiation	
Exposure to UV radiation and/or weather-related factors	
Occasional outside work	
Outside work for a significant part of the year (1/3)	
Outside work for more than half the year (2/3)	
Outside work nearly all year long (3/3)	

FIGURE 3.6 Key enters for identifying problems relating to outdoor work/UV radiation.

E: key questions for identifying problems relating to noise	
The task requires verbal communications with coworkers	
The noise is not bothersome	
The noise is slightly bothersome, but it is possible to talk with coworkers	
The noise is maddening; it is difficult to talk with coworkers	
The noise is very loud; it is impossible to talk with coworkers	
The task does not require verbal communications with coworkers	
The noise is not bothersome	
The noise is slightly bothersome, but it is possible to talk with coworkers	
The noise is maddening, it is difficult to talk with coworkers	
The noise is very loud; it is impossible to talk with coworkers	

FIGURE 3.7 Key enters for identifying problems relating to noise.

3.2.3.6 Key Enters for Identifying Problems Relating to Micro-climate (Figure 3.8)

The key questions are extremely simple: *Does it feel hot? Does it feel cold?* A description of the duration is also asked for: *only in summer, only in winter, all year long.*

The workers are also asked if the work is prevalently outdoors (e.g., farming, building construction), involving exposure to the elements.

3.2.3.7 Key Enters for Identifying Problems Relating to Tools/Equipment (Figure 3.9)

In this section, workers report on the most frequent "intrinsic" problems with a tool or piece of equipment, or with the way they are used. Problems are indicated by placing an X in the corresponding box.

3.2.3.8 Key Enters for Identifying Problems Relating to Exposure to Vibrations (Figure 3.10)

A distinction is made between hand–arm vibrations caused by the use of tools (pneumatic screwdrivers, grinders, cutters, pneumatic drills, and so on. for at least one-third of the time) or whole-body vibrations related to driving vehicles (whole-body vibrations for a large percentage of the time).

3.2.3.9 Key Enters for Identifying Problems Relating to the Use of Machinery or Machine Parts (Figure 3.11)

As in the section on tools, in this section workers report on the most frequent "intrinsic" problems with machinery or parts of machines, or with the way that they are used. Problems are indicated by placing an X in the corresponding box.

3.2.3.10 Key Enters for Identifying Problems Relating to Pollutants (Chemical Risk, Biological Risk) and Other Specific Risk Factors (Figure 3.12)

The key questions in this important section may seem very simple. The questions, however, are designed to detect the presence and quantity of any pollutants. It should be stressed that if the presence of a pollutant is detected, another questionnaire needs to be completed in addition to this one to collect preliminary data concerning the possibility that the pollutant might become a risk inducer (see the Section 3.2.4.5 on quick assessments).

This tool neither represents nor replaces official risk assessment documents. Any report of workplace hazards associated with exposure to chemicals is an alarm signal that must be followed by an in-depth evaluation based on the standards originating from the technical and scientific knowledge existing at the time.

3.2.3.11 Key Enters for Identifying Organizational Problems (Figure 3.13)

Shift type and duration and forced working rate are the main factors that can potentially cause organizational problems.

F: key questions for identifying problems relating to the microclimate		
Quite good all year long		
It is hot:	Only in summer	
	All year long	
It is cold:	Only in winter	
	All year long	

Working out of doors with exposure to the elements

Only in summer	
Only in winter	
All year long	

FIGURE 3.8 Key enters for identifying problems relating to weather and micro-climate.

G: key questions for identifying problems relating to the use of tools/equipment	
Adequate and in good condition (satisfactory maintenance)	
Heavy	
Noisy	
Requires use of force	
Not working efficiently	
Cumbersome and/or hard to grasp	
Not fit for specific use and/or technologically backward	
The equipment rapidly overheats	
Requires excessive attention	
May produce lesions (cuts, abrasions, blisters, burns, etc.)	
Requires use of body parts as equipment, with consequent lesions (calluses, rashes, cuts, etc.)	

FIGURE 3.9 Key enters for identifying problems relating to the use of tools/equipment.

H: key questions for identifying problems relating to vibration	
No exposure to vibration	☐
The task involves the use of vibrating tools	
Occasional use	☐
Use of pneumatic screwdrivers for at least 1/3 of time	☐
Use of grinders/cutters/polishers for at least 1/3 of time	☐
Use of jack hammers for at least 1/3 of time	☐
The task involves driving a vehicle	
Occasional driving	☐
Almost always driving a car, motorcycle, van	☐
Almost always driving a truck, bus	☐
Almost always driving a tractor, agricultural vehicle, scraper, digger	☐

FIGURE 3.10 Key enters for identifying problems relating to exposure to vibration (hand–arm or whole body).

I: key questions for identifying problems relating to machinery	
Adequate and in good condition (satisfactory maintenance)	
Requires use of force	
Requires lifting heavy components	
Noisy	
Not working efficiently	
Not fit for specific use and/or technologically backward	
Requires excessive attention	
Limited space around machinery	
May produce lesions (cuts, abrasions, blisters, burns, etc.)	

FIGURE 3.11 Key enters for identifying problems relating to the use of machinery or parts of machines.

L: key questions for identification of problems associated with pollutants, and biological or other special agents			
No pollutants and biological or other special agents are present			
Dust: type		Present	
		Highly present	
Fumes: type		Present	
		Highly present	
Unpleasant odors: type		Present	
		Highly present	
Chemicals: type		Present	
		Highly present	
Others: type		Present	
		Highly present	

FIGURE 3.12 Key enters for identifying problems relating to pollutants (chemical risk, biological risk) and other special risk factors.

M: key questions for identifying organizational problems		
Shift work		
Shift work	Only one daily shift	
	Several daily shifts	
	Only night shift	
	Several shifts also at night	
Work rate		
Work rate	Free	
	Imposed by machinery	
Duration		
Shift duration	No more than 8 h per shift	
	More than 8 h per shift	

FIGURE 3.13 Key enters for identifying several organizational problems (shift work, work rate imposed by machinery, shift duration).

3.2.3.12 Key Enters for Identifying Potential Work-Related Stress Inducers (Figure 3.14)

With regard to this important and very challenging aspect, the decision was taken to deal with the issue by simply suggesting a limited list of possible stressors. The items on the list should therefore not be used as actual key enters but only as descriptions of situations to be monitored. Unlike in the previous situations, no numbers are used in order to classify risk and set priorities using the traffic light system. There are in fact no simple recommendations in the literature that would serve this purpose. The authors will wait until such time as enough experience has been acquired, following a broader enforcement of work-related stress assessment requirements, to implement the system.

3.2.4 Second-Level Pre-Mapping (Quick Assessment)

Once the potential risk inducers (absent/present) have been identified using the key enters, it is possible to quickly assess the presence of *acceptable* or *critical* risk using simple evaluation methods such as the quick assessment.

This procedure consists of checking whether or not certain essential assumptions and requirements are met, without calculations or equations, and generating three potential scenarios:

- *Critical (purple light)*
- *Acceptable (green light)*
- *Neither critical nor acceptable*: This scenario calls for the use of one of the traditional risk assessment methods listed in the reference standards (level 3).

After applying the procedure, if all the acceptability criteria have been met and no critical codes emerge, the condition is defined as *acceptable*, and no further assessments are required.

If conditions are *critical*, since the information has already been collected evidencing the existence of serious potential risk, it is advisable to begin immediate remedial and corrective actions, based on the reference standards for *risk reduction*. However, at times it is still worth conducting a more detailed evaluation according to the calculations or equations provided by the standards dealing with biomechanical load (ISO and European Committee for Standardization [CEN]).

If conditions are *neither critical* nor *acceptable* (as in a large percentage of cases), it will be necessary to use the classic *third-level* risk assessment procedures and methods, which in turn may determine the presence of acceptable risk (green), borderline risk (yellow), or risk that is present but slight/medium (red) or intense (purple).

In the proposed pre-mapping model, the second-level analysis (*quick assessment*) refers only to biomechanical overload due to repetitive movements and manual load handling, in which the Ergonomics of Posture and Movement (EPM) Research Unit has decades of experience. The criteria suggested by ISO/TR 12295 were found to be of great help, and as such have been used.

N: key enters for identifying generic potential stressors	
Night shift	
Work rate imposed by machinery	
Work shift longer than 8 h	
Uncomfortable work environment due to nonergonomic work place, lighting, microclimate, noise, vibration, etc. (...)	
Prolonged contact with the public	
Contact with people in pain	
Work at high risk of accident or injury	
Work at high risk of physical violence or psychological abuse	
Piece work or highly performance-contingent pay	
Job with high responsibility for others	
Job with high responsibility for productivity	
Use of socially unintegrated workforce	
Other:	
Other:	
Other:	
Notes:	

FIGURE 3.14 Key enters for identifying potential stressors.

With regard to physical risk (noise, vibration, micro-climate, etc.), reference should be made to the excellent evaluation techniques that can be found in the literature. Going forward, other expert groups may eventually come up with genuine quick-assessment techniques.

A detailed, descriptive questionnaire is provided only as regards the indicators for chemical risk, which, after the key enters level, allows for a more detailed analysis of hazardous situations in order to define how urgently the assessment process needs to be pursued.

The quick-assessment questionnaires relative to repetitive movements, manual load handling, and chemical risk are described next.

3.2.4.1 Quick Assessment of Repetitive Tasks

Once key enters have established that the work entails repetitive tasks, the relevant form is completed.

The first part of the questionnaire refers to certain organizational aspects (Figure 3.15): shift duration, number and actual duration of breaks, and duration of nonrepetitive tasks. The purpose of this section is to determine how long the worker spends performing repetitive tasks during the shift, in other words, the *net duration of repetitive work*.

The second part (Figure 3.16a and Table 3.2) proposes a number of pre-defined scenarios which, when present simultaneously, allow the risk associated with repetitive work performed by a homogeneous group of workers to be defined as acceptable (*green* code).

The third part (Figure 3.16b and Table 3.3) proposes more pre-defined scenarios in which, if even only one of the replies is positive, it can be stated that the repetitive work performed by a homogeneous group of workers entails critical high-risk conditions (*purple* code).

In short, the quick-assessment form for investigating repetitive work will lead to one of the following three conclusions:

- Repetitive work is *acceptable* in terms of risk because the scenarios reported in the *code green* section have all been checked

Summary of net duration of repetitive work on a typical day			
Total average shift duration (in minutes)		Total repetitive working time (in minutes)	0

Description of nonrepetitive work, duration, and timing of breaks—total duration	
Fetching supplies	
Cleaning	
Other	
Total duration of nonrepetitive work per shift (in minutes)	0
Total duration of breaks (average) per shift (in minutes) including meal break only if included in the shift duration	
Number of breaks (including meal break) lasting at least 8 min	

FIGURE 3.15 Quick assessment of risk due to repetitive movements: Organizational data.

Acceptable conditions

If all underreported conditions are fulfilled and all replies are "Yes," the risk level is acceptable for repetitive work and it is not necessary to continue the risk evaluation

N.B. reply by placing an X in the appropriate white box

Do either upper limbs work for less than 50% of the total duration of the repetitive task(s)?	No		Yes	
Are both elbows held below shoulder height for at least 90% of the total duration of the repetitive task(s)?	No		Yes	
Is no or moderate force (perceived effort max. 3 or 4 on the CR-10 Borg scale) exerted by the operator for no more than 1 h throughout the duration of the repetitive task(s) and are there no force peaks (perceived effort 5 or more on CR-10 Borg scale)?	No		Yes	
Are there breaks (including meal break) lasting at least 8 min every 2 h and are repetitive tasks performed for less than 8 h a day?	No		Yes	

(a)

Critical conditions

If even only one of the situations listed below is present, risk should be considered as critical and the task must be redesigned urgently

N.B. Reply placing an X in the appropriate white box

Are technical actions of a single limb so fast that they cannot be counted by simple direct observation?	No		Yes	
Are one or both arms operating with the elbow at shoulder height for half or more of the total repetitive working time?	No		Yes	
Is a "pinch" grip (or any kind of grasp using the fingers tips) used for more than 80% of the repetitive working time?	No		Yes	
Is peak force applied (perceived effort 5 or more on the CR-10 Borg scale) for 10% or more of the total repetitive working time?	No		Yes	
Is there is no more than one break (including meal break) in a shift of 6–8 h or does the total repetitive working time exceed 8 h within a shift?	No		Yes	

(b)

FIGURE 3.16 (a) Quick assessment of risk due to repetitive movements: Pre-defined scenarios for determining acceptable risk (code green). (b) Quick assessment of risk due to repetitive movements: pre-defined scenarios for determining the presence of high risk conditions (*critical code, purple*).

- Repetitive work is definitely *at risk* if even only one of the replies in the *critical code* section is positive. In this case, remedial action must be taken urgently, especially with regard to the critical result. A risk assessment should also be conducted using the classic evaluation tools (the OCRA checklist, or the even more precise OCRA index)
- Repetitive work *could be at risk* because there are one or more positive replies in the scenarios listed in the *code green* section. A risk assessment must be performed using the classic evaluation tools (OCRA checklist or, for level 4, the more precise OCRA index); the situation is potentially a code *yellow* or *red*.

In actual fact, when conditions are neither acceptable nor critical, a simple scheme is recommended (Figure 3.17), which leads to a rough preliminary estimate of the OCRA checklist score range that might be expected with a more detailed analysis. As previously seen for the key enters, the results of this quick assessment do not generate a visible score but rather color scales indicating the level of risk and priorities for the corrective action plan.

In order to define the intervals of these color scales, potential OCRA checklist values have been calculated for each scenario, so that when the questionnaires are compiled, the system generates a kind of *pre-index* (not visible to the compiler), where *green* (less than 7.5) means acceptable risk; *yellow* (7.5–11) means potential risk, to be assessed but not high priority; *red* (11.1–22.5) means risk probably present and to be assessed; and *purple* (more than 22.5) means risk definitely present and assessment urgently required. The maximum value for the concealed pre-index is 25, and the percentage of biomechanical load is calculated with respect to this value; this ratio will later be used to briefly define the extent to which the upper limbs are engaged with respect to other risk factors.

3.2.4.2 Quick Assessment of Manual Load Lifting

Once it has been established, using key enters, that the work includes manual load lifting, the relevant quick-assessment form is compiled.

The first part of the questionnaire gathers information concerning certain aspects of the environment and objects that the worker lifts or carries, which might represent an added risk in load-handling work and must be taken into due account in the assessment, as also indicated in an Annex to EU Directive 269/90 (Figure 3.18).

The second part of the questionnaire (Figure 3.19) suggests another set of predefined scenarios. If even only one of the replies is positive, it means that the lifting task entails critical conditions and high risk (*critical code, purple*).

These are the scenarios for which ISO 11228-1, which adopts the Revised NIOSH (National Institute of Occupational Safety and Health) Lifting Equation or RNLE (Waters et al., 1993), suggests using the multiplier "Ø" for calculating the recommended weight. In other words, in the presence of these situations, the recommended weight is equal to Ø kg (i.e., under these conditions, load lifting is unadvisable). Scenarios include situations where the load is held more than 63 cm away from the body, the height of the hands at the beginning/end of lift is over 175 cm, and so forth.

Frequency					
What is the frequency of technical actions with the dominant hand?					
Slow (no more than one action per 2 s)		No		Yes	
Medium (no more than one action per second) or holding an object in the hands for most of the time		No		Yes	x
High (more than one action per second): Difficult to count the actions		No		Yes	
Pace					
Is the pace mainly determined by the machine?		No		Yes	x
Awkward postures					
Shoulder	Are arms operating with elbow at shoulder height from one-third to half of the total repetitive working time?	No		Yes	
Hand	A "pinch" (or any kind of grasp using the finger tips) is used from half to 80% of the repetitive working time?	No		Yes	
Wrist	Are extreme wrist deviations (flexion, extension, or lateral deviations) present for quite all the time?	No		Yes	
Elbow	Are extreme forearm movements (elbow flexion-extension or rotations) present for quite all the time?	No		Yes	
Lack of variation					
Are the same actions or gestures repeated for most of the time? Or is the cycle time very short (less than 8 s)?		No		Yes	
Use of force					
Are peaks of force (perceived effort = 5 or more in CR-10 Borg scale) applied for 1% – 9% of the time?		No		Yes	
Is a moderate force (perceived effort = max. 3 or 4 on CR-10 Borg scale) exerted by the operator?		No		Yes	

FIGURE 3.17 Quick assessment of repetitive movements under conditions that are neither acceptable nor critical.

Additional organizational and environmental risk factors to be considered				
Is the working environment unfavorable for manual lifting and carrying?				
Are there extreme (low or high) temperatures?	No		Yes	
Is there a slippery, uneven, unstable floor?	No		Yes	
Is there insufficient space for lifting and carrying?	No		Yes	
Are the objcets unsuitable for manual lifting and carrying?				
Does the size of the object reduce the operator's view and hinder movement?	No		Yes	
Is the center of gravity of the load unstable (e.g., liquids, loose items inside a container)?	No		Yes	
Does the object have sharp edges, surfaces, or protrusions?	No		Yes	
Are the contact surfaces too cold or too hot?	No		Yes	
Is the manual lifting or carrying task performed for more than 8 h a day?	No		Yes	

FIGURE 3.18 Quick assessment of manual load-lifting and carrying risk: Preliminary evaluation of certain unfavorable characteristics of the work organization, environment, and load.

Critical conditions

If even only one of the conditions listed below is present, risk should be considered as high and the task must be redesigned immediately.

Task layout and frequency					
Vertical location	The hand location at the beginning/end of the lift is higher than 175 cm or lower than 0 cm	No		Yes	
Vertical displacement	The vertical distance between the origin and the destination of the lifted object is more than 175 cm	No		Yes	
Horizontal distance	The horizontal distance between the body and the load is greater than a full arm's length	No		Yes	
Asymmetry	Extreme body twisting without moving the feet	No		Yes	
Frequency	Equal to or above 15 times/min for short duration (max 60 min)	No		Yes	
	Equal to or above 12 times/min for medium duration (max 120 min)	No		Yes	
	Equal to or above 8 times/min for long duration (over 120 min)	No		Yes	

Presence of loads exceeding the following limits					
Men (18–45 years)	25 kg	No		Yes	
Women (18–45 years)	20 kg	No		Yes	
Men (<18 or >45 years)	20 kg	No		Yes	
Women (<18 or >45 years)	15 kg	No		Yes	

FIGURE 3.19 Quick assessment of manual load-lifting risk: pre-defined scenarios for determining the presence of high risk conditions: *Critical code* (purple light).

Moreover, conditions are defined as *critical* when a single worker has to manually lift weights in excess of the maximum lifting limits (Figure 3.19) recommended, for specific gender and age subgroups, by the technical standards ISO 11228-1 and EN1005-2 (ISO, 2003; CEN, 2003).

The third part (Figure 3.20) suggests several more pre-defined scenarios. If even only one of the replies is positive, it means that the lifting work performed by the homogeneous group entails acceptable risk (*code green*).

These scenarios refer to the manual lifting of loads weighing up to 10 kg, under virtually "ideal" lifting frequency and workplace setup conditions.

In short, the quick-assessment form for investigating tasks involving the manual load lifting will, as in the case of repetitive movements, lead to one of the following three conclusions:

- The manual load-lifting work is *acceptable* in terms of risk because the situations reported in the *code green* section have all been checked.
- Repetitive work could be at risk because there are one or more positive replies in the scenarios listed in the code green section. A risk assessment should be conducted using the classic evaluation tools (revised NIOSH lifting equation).
- The manual lifting work is *definitely* at risk if even only one of the replies in the scenarios for the *critical code* section is positive. Upgrades or remedial action must be undertaken urgently with regard to the conditions defined as critical, to be followed by a risk assessment using one of the classic evaluation tools (revised NIOSH lifting equation).

In order to complete the analysis of lifting tasks through the use of *code greens* and *critical codes*, other scenarios have been added to more comprehensively describe the handling of the various weights (Figure 3.21). As in the case of repetitive movements, with this additional information it will be possible to more accurately determine the color scale score and the biomechanical effort of the spine with respect to the pre-defined maximum scores.

If even only one of the scenarios on the *critical code* list is positive, the color scale moves toward *purple*. Having pre-defined the *maximum value* (taking into account problems relating to the environment, load characteristics, workplace layout, and organizational setup), the percentage of biomechanical effort is calculated with respect to this threshold. As for all risk inducers, this ratio is used to briefly define the degree of involvement of the spine in manual load-lifting tasks. For scores between 0% and 30%, the color green will appear; for scores up to 50%, the color yellow will appear; for scores between 50% and 80%, the color will be red; and for scores of 81%–100%, the color will be purple, which is always used to indicate the highest possible score and therefore indicates a critical condition.

3.2.4.3 Quick Assessment of Manual Load-Carrying Tasks (Figure 3.22)

For the initial quick assessment, the calculation of the *cumulative mass* (as per ISO 11228-1) has been used, that is, the total weight of all the loads carried in a shift. The *cumulative mass carried* is compared with the *cumulative mass tolerable* for

Acceptable conditions

If all underreported conditions are fulfilled and replies are all "YES" (using both hands for lifting) for all weight categories present, the risk level is acceptable for manual lifting loads. However, it is advisable to check for additional risk factors (see above).
N.B. reply by placing an X in the appropriate white box

Do loads weigh between 3 and 5 kg?		No		Yes	
Between 3 and 5 kg	Asymmetry (e.g., body rotation, trunk twisting) is absent	No		Yes	
	Load is kept close to the body	No		Yes	
	Load vertical displacement is between hips and shoulders	No		Yes	
	Maximum frequency: Less than 5 lifts per minute	No		Yes	

Do loads weigh between 5 kg and 10 kg?		No		Yes	
Between 5 and 10 kg	Asymmetry (e.g., body rotation, trunk twisting) is absent	No		Yes	
	Load is kept close to the body	No		Yes	
	Load vertical displacement is between hips and shoulders	No		Yes	
	Maximum frequency: Less than five lifts per minute	No		Yes	

No loads exceeding 10 kg are present?	No		Yes	

FIGURE 3.20 Quick assessment of manual load-lifting risk: pre-defined scenarios for determining the presence of acceptable risk: *Code green*.

Characteristics and frequency of load ranges (over 10 kg)					
Do loads weigh between 10 and 15 kg?		No		Yes	
Between 10.5 and 15 kg	Asymmetry (e.g., body rotation, trunk twisting) is absent	No		Yes	
	Load is kept close to the body	No		Yes	
	Load vertical displacement is between hips and shoulders	No		Yes	
	Maximum frequency: Less than one lift every 5 min	No		Yes	
Do loads weigh between 15 and 25 kg?		No		Yes	
Between 10.5 and 25 kg	Asymmetry (e.g., body rotation, trunk twisting) is absent	No		Yes	
	Load is kept close to the body	No		Yes	
	Load vertical displacement is between hips and shoulders	No		Yes	
	Maximum frequency: Less than one lift every 5 min	No		Yes	

FIGURE 3.21 Quick assessment of manual load-lifting risk: pre-defined scenarios for descriptive completion of lifting conditions.

No. of objects exceeding 3 kg carried in a shift	Weight of objects carried	Cumulative mass (kg)
		0
		0
		0
		0
Cumulative mass (total load carried) =		0

Max. distance	

Recommended cumulative mass (total load in kg carried during specific period)					
Is the cumulative mass carried per hour spent carrying loads Heavier than this limit (kg)?	750	No		Yes	
Is the cumulative mass carried per minute spent carrying loads heavier than this limit (kg)?	15	No		Yes	

Are awkward postures adopted while carrying loads?	No		Yes	

FIGURE 3.22 Quick assessment (critical condition) of manual load-carrying risk: calculation of cumulative mass.

8 h, 1 h, and 1 min: If the weight of the mass carried is higher than that of the mass tolerated, then the situation would be deemed as critical (*critical code*, purple light). The *cumulative tolerated mass* varies based on the distance over which the load is carried, from 10,000 kg for 8 h to 6,000 kg for distances of over 10 m. To estimate the cumulative mass carried, the number of objects, weight (at least by category), and distance carried must be entered into the proposed questionnaire; the calculation will be made automatically. The ratio of the cumulative mass carried to the cumulative tolerated mass is calculated but is "concealed" by the system.

3.2.4.4 Quick Assessment of Load Pushing and/or Pulling

Figure 3.23 lists the conditions to be examined first to define the manual pushing or pulling as acceptable. These conditions could in fact represent an additional risk in load-handling work and must be taken into due consideration in the evaluation.

Figure 3.24 describes the conditions for assessing manual pushing and pulling tasks as acceptable. The conditions in Figure 3.24 must all be met (together with the previous ones in Figure 3.23) to conclude that the situation is *code green*.

Figure 3.25 shows a series of conditions where, if even only one of the replies is positive, the manual pulling or pushing task can be defined as critical, regardless of which complex assessment method is used as recommended in ISO 11228-2 (ISO, 2007a) and ISO/TR 12295—Annex B).

It should be noted that the quick assessment of manual pushing and pulling involves the use of the Borg scale (version CR-10) (Borg, 1998) to estimate the important parameter of intensity of effort (or force).

When the level on the Borg scale is equal to or less than 2 (slight force), the conditions may be acceptable, while if the level is equal to or more than 8, the conditions are definitely critical. For intermediate levels (between 3 and 7), the percentage of effort is classified based on a set of percentage ranges.

The perceived effort (ascertained by interviewing workers using the CR-10 Borg scale)	

Additional organizational and environmental risk factors to be considered				
Is the working environment unfavorable for pushing or pulling?				
Are floor surfaces slippery, unstable, uneven, sloping, fissured, cracked, or broken?	No		Yes	
Are paths restricted or constrained?	No		Yes	
Are temperatures in the working environment high?	No		Yes	

Is the object unsuitable for pushing or pulling?				
Does the object (or trolley, transpallet, etc.) limit the vision of the operetor or hinder movement?	No		Yes	
Is the object unstable?	No		Yes	
Does the object (or trolley, transpallet, etc.) have hazardous features, sharp surfaces, projections, etc. that may injure the operator?	No		Yes	
Are the wheels or casters worn, broken or not properly maintained?	No		Yes	
Are the wheels or casters unsuitable for the working conditions?	No		Yes	

FIGURE 3.23 Quick assessment of manual pushing or pulling risk: preliminary evaluation of certain unfavorable characteristics of the work organization, environment, and objects carried.

Acceptable conditions

If all underreported conditions are fulfilled and all replies are "YES," the risk level is acceptable for pushing/pulling tasks. However, it is advisable to check for additional risk factors (see above).

N.B. reply by placing an X in the appropriate white box

	No		Yes	
The perceived effort (ascertained by interviewing workers using the CR-10 Borg scale) shows the presence, during pushing/pulling tasks, of up to slight force exertion (score 2 or less on the Borg CR-10 scale)?	No		Yes	
Does the task involving manual pushing or pulling for up to 8 h a day?	No		Yes	
Is pushing/pulling force applied to the object between hip and mid-chest height?	No		Yes	
Is the pushing/pulling action performed with an upright trunk (not twisted or bent)?	No		Yes	
Are the hands held between shoulder width and in front of the body?	No		Yes	

FIGURE 3.24 Quick assessment of manual pushing or pulling risk: Pre-defined scenarios for determining the presence of acceptable risk conditions: *Code green.*

Critical conditions If even only one of the situations listed below is present, risk should be considered as high and the task must be redesigned immediately				
Does the perceived effort on the CR-10 Borg scale (obtained by interviewing the workers) indicate the presence of high peaks of force (i.e., A score of 8 or more on the Borg CR-10 scale)?	No		Yes	
Is the pushing/pulling action performed with the trunk significantly bent or twisted?	No		Yes	
Is the pushing/pulling action performed in a jerky manner or in an uncontrolled way?	No		Yes	
Are the hands held either beyond shoulder width or not in front of the body?	No		Yes	
Are the hands held higher than 150 cm or lower than 60 cm?	No		Yes	
Does the pushing/pulling action also feature vertical force components ("partial lifting")?	No		Yes	
Does the manual pushing/pulling task last for more than 8 h a day?	No		Yes	

FIGURE 3.25 Quick assessment of manual pushing or pulling risk: Pre-defined scenarios for determining the presence of high risk conditions: *Critical code* (purple light).

This new shared but simpler approach was introduced to avoid having to perform measurements using a dynamometer during manual pushing or pulling actions.

3.2.4.5 Description and Initial Inspection of Chemical Pollutants

Figure 3.26 presents a descriptive diagram for describing potential chemical pollutants; it suggests beginning by looking at the technical specifications on the label, indicated with letters or pictograms.

Once the presence of the chemical has been recognized and classified (qualitative data), the description is completed with the relevant quantitative data (Figure 3.27). The type of exposure may be closed cycle or controlled cycle or may involving direct handling. The frequency of exposure will be reported as *sporadic*, *short but daily*, or *high and daily*.

With regard to the "concealed" scores, descriptive scores are attributed from the highest risk products to the lowest risk products. The scores are also adjusted to the quantitative degree of exposure.

It should be noted that as per the original approach, the questionnaire aims to measure exposure to chemicals or particulates by essential elements.

3.2.5 Summary Results of Pre-Mapping

The results of the pre-mapping exercises carried out via key enters and quick assessments can also be summarized graphically to more comprehensively define the "PRE-MAP" and corrective action priorities. The software described previously and duly programmed generates this summary automatically.

Figure 3.28 shows an example of a summary of the results obtained from "pre-mapping" a group of stonemasons who work serpentine, a stone typical of northern Italy's Valtellina region, which is a costly cladding and roofing material.

The homogeneous group is composed of five workers who all perform the same tasks in the shift. They do not work in a mine but in a small, old-fashioned workshop; the blocks of stone (weighing from 25 to 50 kg) are left outside the workshop after being extracted from the mine.

Manual transpallets are used to transport the blocks into the workshop, where the workers use manual tools, especially sledgehammers and chisels, to produce luxurious tiles (serpentine is a very valuable stone). The larger blocks are broken down into smaller ones, weighing between 10 and 20–25 kg, which are generally handled manually. To reduce the blocks into tiles, the cuts have to follow the natural crystallization of the stone. This is a job that requires great skill and expertise. The stone is a silicate, and in the quarry, serpentine may be found close to asbestos veins. In the example given, a homogeneous group of workers performs several tasks, from transporting stone with transpallets to carrying them manually and cutting the stone (Figure 3.29). However, just one pre-mapping form must be compiled; the questionnaire is not for the entire company or for a single task but should be used to describe a complete task performed by a specific homogeneous group of workers. The work is repetitive and features high-risk conditions (critical code), since it entails the use of peak force. Manually lifted loads may weigh more than 25 kg (critical code). There are also problems associated with pushing and pulling manual transpallets, given the

	Quali-quantitative identification of present chemical agents or agents generated in processing																			
	Health risks from acute exposure						Health risks from chronical exposure					Risks for safety								
	Extremely high	High		Medium	Low	Sensitization risk	Extremely high	High	Medium	Low	Senzitization risk	Extremely high			High		Medium	Low		
	Very toxic	Toxic	Corrosive	Harmful	Irritating	Sensitizing	Cancerogenic; mutagenic, reproductive cycle risk: teratogenic	Cancerogenic; mutagenic, reproductive cycle risk: teratogenic: toxic	Toxic	Irritating	Sensitizing	Explosion	Extremely inflamable	Combustive	Easily inflamable	Explosion	Inflamable	High inflamability point (I.E >70°)		
	H300 H310 H330	H301 H311 H331	H314 H318	H302 H312 H332	H315 H319 H335	H317 H334	H340 H350 H360	H341 H351 H361 H370 H372	H371 H373	H315 H319 H335 or without "H" label	H317 H334	H200 H201 H202 H203 H240 H241	H220 H222 H224 H241 H242 H251 H252	H250	H225 H228	H204	H223 H226 (In certain conditions) H225: se punto di inflammabilità<23 °C e di ebollizione >35 H223 H224 H225 H226 (secondo condizioni)	No label		
Acids																				
Bases																				
Fuels																				
Organic compounds																				
Dusts																				
Gas fumes																				
Plastic materials																				
Metalloids and metals																				
Oxidants																				
Pesticides																				
Solvents																				

FIGURE 3.26 Descriptive diagram of potential pollutants.

	Working process							
	Brief description of the working process for characterizing workers' exposure							
	Way of exposure					Frequency of exposure		
	Closed cycle	Controlled cycle		Open cycle		Occasional (not every day)	Low but every day	High every day
	Complete	Aspiration	Separazione	Direct contact and manipulation	Manipolazion E diretta senza DPI			
Acids								
Bases								
Fuels								
Organic compounds								
Dusts								
Gas fumes								
Plastic materials								
Metalloids and metals								
Odixants								
Pesticides								
Solvents								

FIGURE 3.27 Descriptive diagram of potential pollutants: descriptive quantification of exposure.

uneven ground outside the workshop. Noise and the micro-climate are generally also a problem; the tools are obsolete and could cause accidents or injuries. Silica dust (and possibly also asbestos) is present, although manual sledgehammers are used to split the stone rather than grinders. The workshop has no dust aspirators.

Figure 3.28 shows the final summary results generated automatically by the software, depicting histograms for all possible risks.

The height of each histogram derives from the percentages resulting from the equation:

$$\left(\mathbf{PI}_i / \mathbf{PM}_i\right) * \mathbf{100}$$

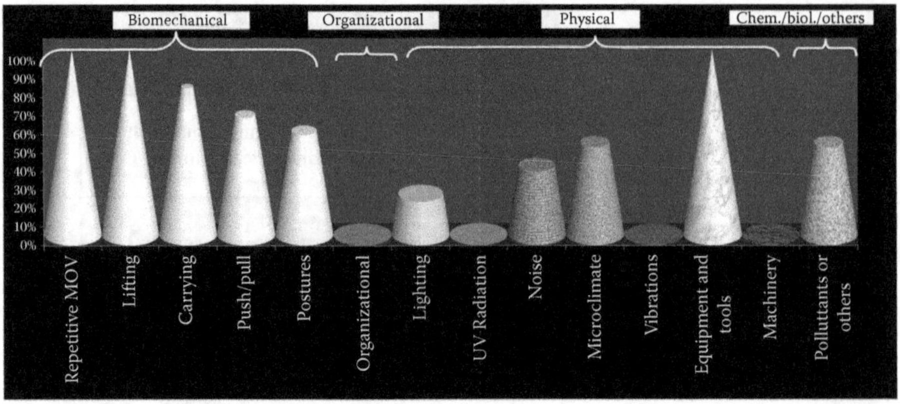

FIGURE 3.28 Final summary results depicted as histograms for all potential risks: Example of a homogeneous group of stonemasons splitting serpentine rock (silicate) in a workshop.

FIGURE 3.29 Splitting serpentine rock.

where:

PI$_i$ = intrinsic score for the risk inducer deriving from the sum of the scores attributed to the individual parameters describing it and emerging from the analysis.

PM$_i$ = pre-defined maximum inducer score.

As already stated, these are merely descriptive scores, to be used to "rank" events from best to worst. The scores do not reflect an analysis or assessment of risk; they are simply descriptive scales designed to help not only identify problems but also set priorities for the analyses and evaluations that will have to be undertaken to adopt immediate measures to reduce risk, especially for conditions defined as *critical.*

3.2.6 Examples of Pre-Mapping Results

This section shows examples of preliminary results obtained by applying the proposed model in just a few of the innumerable small businesses covered by the investigation.

Additional analyses and details can be found in Volume 102, nr. 1, 2011, of *La Medicina del Lavoro.*

3.2.6.1 Pasta Makers: Manual Preparation of Fresh Tagliatelle

The work is performed for 3–4 h a day in a chain of small fresh pasta–making stores. Figure 3.30 shows a part of the activity, and Figure 3.31 shows the summary results of the relative pre-mapping exercise.

3.2.6.2 Manufacturing of Parmigiano Reggiano Cheese

This work is carried out by craftsmen even in the largest cheese-making factories.

The most strenuous tasks involve lifting the curd mass, or "twins," from the vat, changing the cloth wraps, and positioning the forms in brine (Figure 3.32). Before maturing, the forms weigh around 40 kg.

Figure 3.33 shows the results of pre-mapping applied to the manual processing of Parmigiano Reggiano cheese.

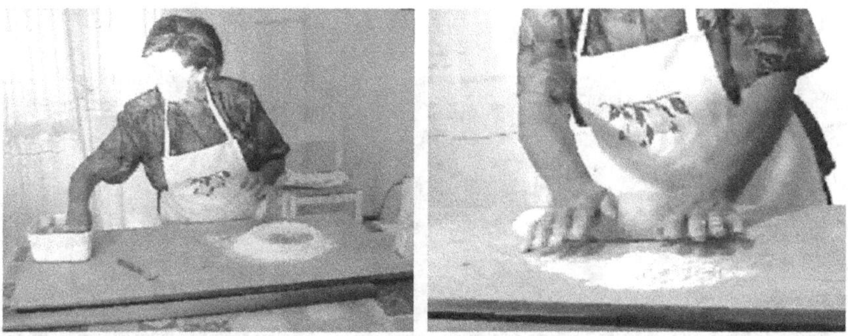

FIGURE 3.30 Steps in the manual preparation of fresh tagliatelle.

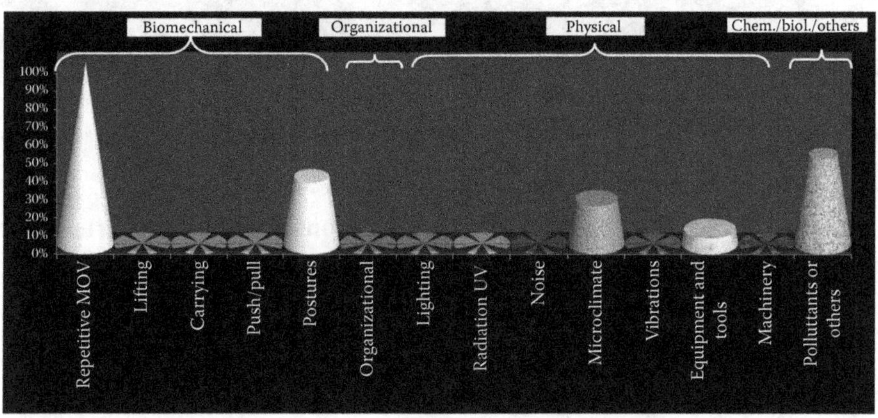

FIGURE 3.31 Results of pre-mapping applied to the manual preparation of fresh tagliatelle.

FIGURE 3.32 Steps in the processing of Parmigiano Reggiano cheese.

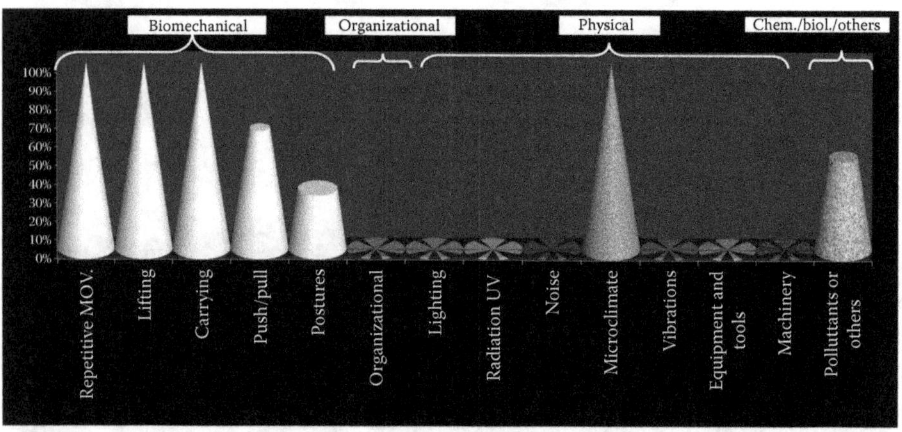

FIGURE 3.33 Results of pre-mapping applied to the manual processing of Parmigiano Reggiano cheese.

3.2.6.3 Manual Manufacture of Piano Accordions (Mechanical Parts)

The manufacturing process involves fitting mechanical parts to the lacquered wood frame of the piano accordion. The manual mechanical assembly work is meticulous and involves the use of small tools. Each worker performs a specific finishing task. The instruments (weighing between 5 and 10 kg) are carried manually from one work bench to the next (Figure 3.34).

Figure 3.35 shows the results of pre-mapping in the production of hand-crafted piano accordions.

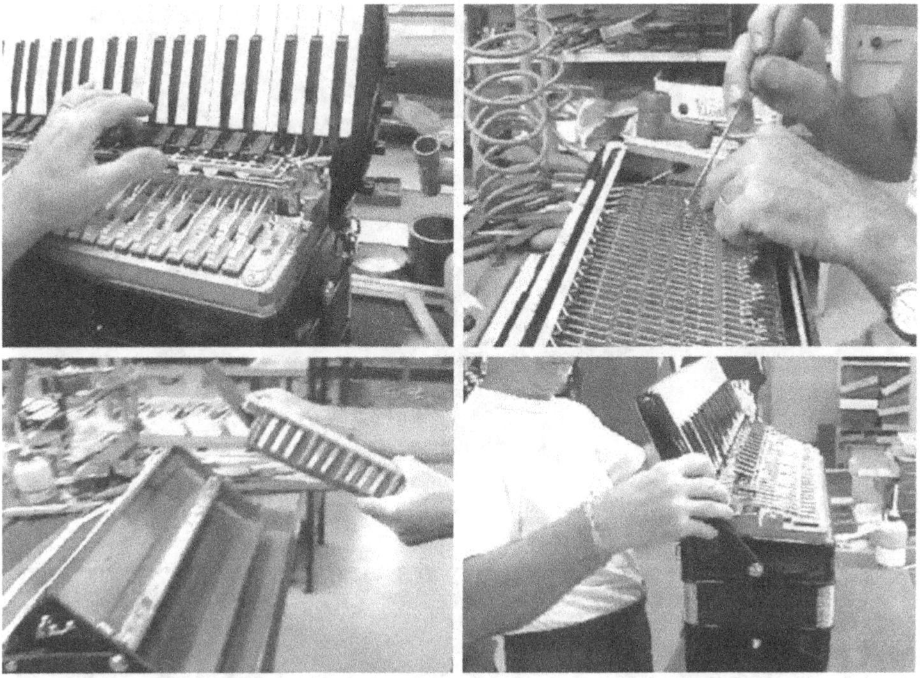

FIGURE 3.34 Steps in the mechanical assembly of handmade piano accordions (Pigini, Castelfidardo).

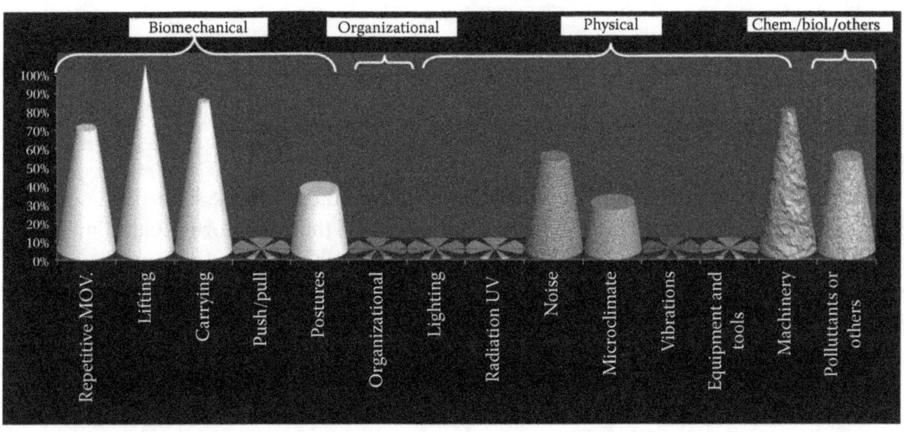

FIGURE 3.35 Results of pre-mapping in the production of hand-crafted piano accordions.

FIGURE 3.36 Enamelling and decoration of artistic stoneware.

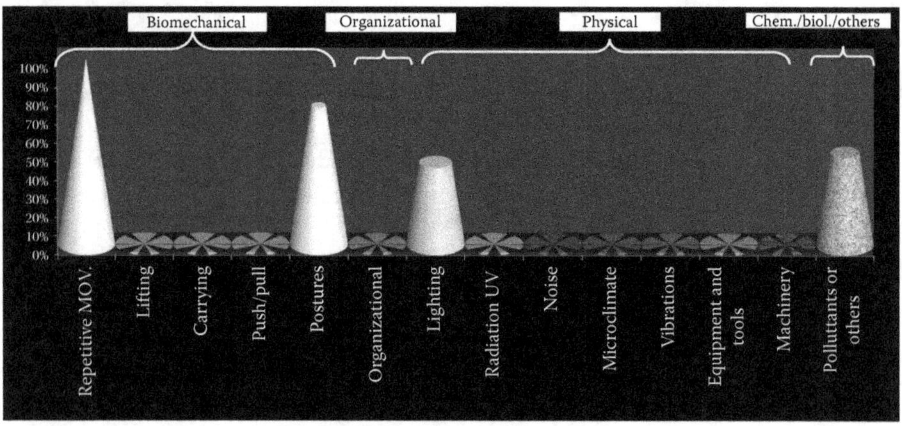

FIGURE 3.37 Results of pre-mapping in the enameling and decoration of artistic stoneware.

3.2.6.4 Manually Decorated Stoneware

This work includes enameling and painting.

Enameling involves using a set of tweezers to manually dip the kiln-fired ceramic parts into an enamel bath; for the decoration process, the operator is seated and uses a brush to paint each individual piece. The task must be performed confidently and precisely, since errors are not allowed (Figure 3.36).

Figure 3.37 shows the results of pre-mapping in the production of hand-decorated stoneware.

4 Procedures and Criteria for Applying the OCRA Checklist

4.1 CONTENTS AND APPLICATION METHODS

4.1.1 Introduction: The OCRA Checklist System for Analyzing the Risk of Biomechanical Overload of the Upper Limbs: Requirements and General Contents

Before illustrating the criteria and techniques for applying the classic Occupational Repetitive Actions (OCRA) checklist, it is necessary to briefly revise the approach toward assessing risk associated with repetitive movements of the upper limbs.

It might be useful to refer to the OCRA system, because while there is just one OCRA method and approach, *different tools can be used to assess risk at different levels in order to achieve different specific objectives.*

Three specific tools are now available for this purpose (Figure 4.1):

- The OCRA index (Occhipinti and Colombini, 1996; Occhipinti, 1998; Colombini et al., 2002), which meets the need to provide an analytical risk assessment tool and is advisable for designing or redesigning workplaces and examining various aspects pertaining to the organization of work.
- The classic OCRA checklist (Occhipinti and Colombini, 2004a; Colombini et al., 2005), which is the ideal tool for preliminary risk mapping; it "weighs" the risk associated with repetitive tasks. Mapping is useful for defining what proportion of workstations can be classified as green (no risk), yellow (very low to borderline risk), red, or purple (low, medium, or high risk). While mapping may be completed quickly, it is not overly accurate since the approach employs "stepwise" scoring and is not as analytical as the OCRA index.
- The OCRA minichecklist: The latest version (Colombini and Occhipinti, 2011), presented in this volume, provides an even faster (and therefore less precise) assessment than the classic OCRA checklist. It is better suited, and probably accurate enough, for assessments in special settings such as in small and very small enterprises (craftwork, agriculture, etc.), where the work does not follow the same pace, timing, and cycles as in larger factories.

The Ergonomics of Posture and Movement (EPM) Research Unit has developed simple Excel programs for each of the aforesaid tools, to collect data and estimate exposure or final risk indicators (see Figure 4.20).

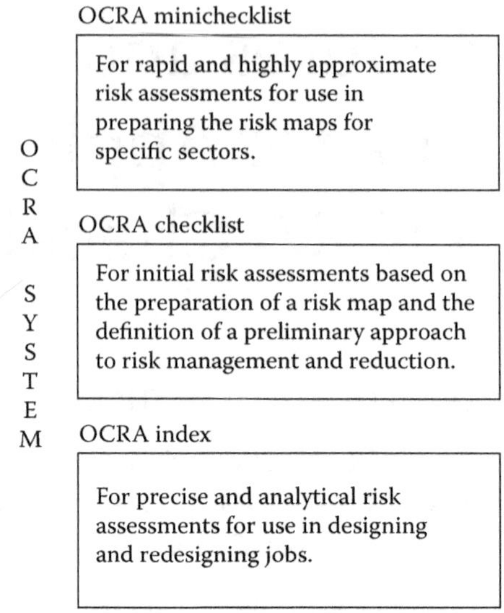

FIGURE 4.1 The three principal tools used by the OCRA system.

4.1.2 The OCRA Checklist: General Criteria

The OCRA checklist is a simplified tool (based on the OCRA index) for measuring the risk of biomechanical overload of the upper limbs, which can be used both in the initial stage of estimating risk levels in a certain industrial setting (i.e., mapping), or later for managing the aforesaid risk.

The OCRA checklist consists of five parts that focus on the four main risk factors (lack of recovery time, frequency, force, awkward posture/stereotyped movement) and a number of additional risk factors (vibration, low temperatures, precision work, repeated impacts, etc.), and also factor in the net duration of repetitive jobs on the final estimate of risk. The entire "hard-copy" model of the OCRA checklist, which is to be completed manually, is included in Annex 4.1.

The OCRA checklist can also be completed by watching the worker carry out his or her job, but as for the OCRA index, it is easier to perform the analysis by looking at films of the specific task as performed by workers.

The classic analysis proposed by the OCRA checklist entails using preassigned scores (the higher the score, the higher the risk) to define the risk associated with each of the aforementioned factors.

The sum and product of the partial values generate a final score that estimates the exposure level based on the OCRA index, featuring four different levels (green, yellow, red, and purple) (Occhipinti and Colombini, 2004a). The calculation procedure for reaching the final result (Figure 4.2) shows how all the risk factors are included: The *lack of recovery period* factor is a multiplier that is to be applied, along with the *duration factor*, to the sum of the scores for the other risk factors.

It is worth mentioning that this method is not only useful for fairly accurately measuring the risk of biomechanical overload of the upper limbs but also for gathering

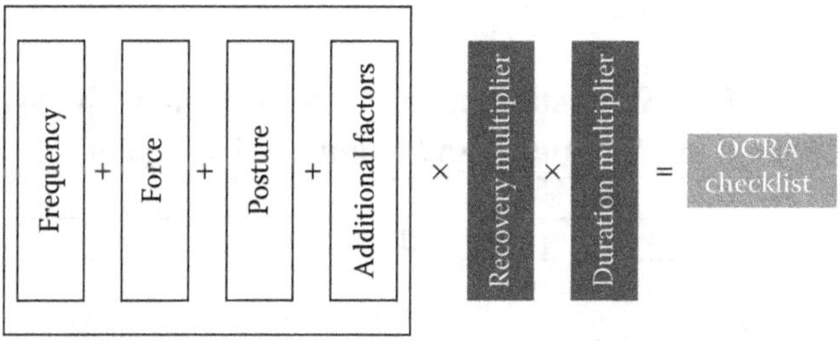

FIGURE 4.2 The new final OCRA checklist calculation procedure.

vital information for the purposes of risk management (such as corrective actions, job rotation, etc.) and damage limitation (e.g., returning workers to the workforce).

All the individual sections of the OCRA checklist will now be described analytically.

4.1.3 Description of the Task and of the Work Organization (Annex 1: Page 1, Part 1)

To start with, the OCRA checklist is used to describe a workstation and estimate the exposure level embedded in the task as if this task were the only one performed by a single worker for the entire duration of the shift. The procedure identifies the workstations in the factory that, for structural and organizational reasons, feature *no*, *low*, *medium*, or *high* risk exposure levels, regardless of workers rotating over different workstations/tasks. This analytical method provides the basis for building a specific workstation risk map in respect of biomechanical overload of the upper limbs.

However, the method cannot be used to determine exposure indexes for workers performing multiple tasks. These indexes can be calculated later on, after mapping workstations where workers carry out repetitive tasks, using methods described elsewhere by the authors (Occhipinti and Colombini, 2009; Occhipinti et al., 2009) and discussed later in this book.

The first part of the OCRA checklist (Table 4.1) calls for a short description of the workstation and the work performed there.

The OCRA checklist applies to repetitive tasks in which

- Work is characterized by cycles (regardless of their duration).
- Work is characterized by a series of practically identical technical actions that are repeated for more than half of the working time.

It is important to note that these two definitions only identify repetitive work in which the term "repetitive" is not synonymous with risk; analyzing the work using the OCRA checklist will define the relevant risk level or attest to its absence.

Additional information that should be provided:

- Number of workstations identical to the one described.
- Number of shifts that use the workstation(s).

TABLE 4.1
Brief Description of the Workstation, Task/s Performed At the Workstation, Number of Identical Workstations, and Number of Workers

Company		Department				
Line/workstation/job/task		No. Workers		M		F
Brief task description						

-How many workstations are identical or very similar..
-How many shifts are worked in a day ....
-How many people work at these workstations during one day and considering all identical workstations..

- Total number of workers assigned to identical workstation(s) and shifts and their gender (no. of males versus no. of females)
- Percentage of time that each workstation is actually utilized during a shift (It may happen that a workstation is operational (or utilized) for only part of the shift.)

Before going on to analyze the various risk factors, in order to assess risk accurately, it is necessary to estimate the *net duration of the repetitive work*, as was also proposed to calculate the OCRA index. The form illustrated in the first part of the OCRA checklist (Table 4.2) helps the compiler to calculate this number, which is

TABLE 4.2
OCRA Checklist: Calculation of the Net Duration of Repetitive Work (Form 1, First Part)

	Description	Actual Duration (Min)
Shift duration (for the duration consider the Real duration)	Official	
	Actual	
Official breaks	Contractual	
Actual breaks (for the duration and numbers consider the Actual duration and the actual numbers)	Actual	
Meal break (i.e., the actual duration of the meal break. To calculate the net duration of repetitive task, consider only if included in the shift duration)	Official Actual	
Nonrepetitive tasks (e.g., cleaning and fetching supplies) (i.e., the actual duration of nonrepetitive tasks)	Official Actual	
Net duration of repetitive task/net duration of repetitive task/s		

obtained by subtracting the following times from the *total shift time* or *time for which the worker is paid*:

- Total duration of breaks, whether official or otherwise
- Actual duration of the meal break (if included in the shift and therefore paid)
- Estimated duration of nonrepetitive task/work and of tasks not involving the upper limbs (i.e., visual control)

In some workplaces, breaks are not scheduled at specific times, in which case it is important to analyze the "modal behavior" of workers (by involving both management and workers) in respect of "bathroom" breaks or other additional breaks so as to enter the duration into the checklist.

To finish estimating the net duration of the repetitive work, the following information should also be collected:

- Actual beginning of the work time at the workstation under examination; any minutes lost to reach the workstation, get dressed, and so on should be taken into account
- Mean duration of scheduled breaks or other additional interruptions (modal behavior of workers)
- Actual duration of the worker's absence from the workstation to go to the canteen or changing room at the end of the shift

It is worth remembering that if the start of the shift is delayed or the shift ends earlier, this will decrease the *net duration of the repetitive work* but cannot be regarded as an additional break for the purposes of calculating the *lack of recovery time* risk factor.

Once the net duration of the repetitive work has been calculated, it is possible to estimate the *net total cycle time or rate* (in seconds) (Table 4.3). This value is calculated by considering the actual number of units (or number of cycles) that the worker must complete during the shift, using the following formula:

$$\text{Net total time of cycle} = \frac{(\text{Net repetitive work time in minutes} \times 60)}{\text{No. of units (or cycles)}}$$

The net total cycle time thus calculated must then be compared with the *observed total cycle time* (measured through direct observation at the workstation or via a video, using a stopwatch); if the values are the same, then the next assessments listed on the checklist can be performed. If there is a significant difference (over 5%) between the two cycle times, the actual content of the shift should be reconsidered in terms of duration of breaks, time spent on nonrepetitive tasks, number of units or cycles actually worked, and so on, until the behavior of the worker during the shift is accurately reconstructed.

It is crucial for the calculated and the observed net total cycle time to correspond in order to accurately assess the worker's risk exposure level; failure to comprehend

TABLE 4.3
OCRA Checklist: Calculation of Net Total Cycle Time for Repetitive Work (Form 1, First Part)

Evaluation of Net Duration of Repetitive Task/s (min)	
No. of units (or cycles)	Planned
(consider actual numbers)	Actual
Net calculated cycle time (s)	
Observed cycle time (s)	
% Difference between observed and computed cycle time (accepted limit 5%)	

the actual operating content of the shift will lead to under- or overestimating the risk level. For example, a 5% discrepancy for a *net duration of repetitive work* of 400 min means not knowing what the worker is doing for 20 min; a period of up to 20 min is acceptable.

But in order to avoid mistakes, it is worth remembering that the *total observed cycle time* begins when one piece reaches a specific point and ends when the next one reaches the same point. Therefore, this time includes

- Periods of activity involving the upper limbs.
- Periods of activity not involving the upper arms (e.g., walking without carrying loads).
- Periods of nonactivity (e.g., short waiting times before the arrival of the next piece).

4.1.4 Duration Multiplier for Repetitive Work

If the *net duration of the repetitive work* in the shift lasts less than 420 min or more than 481 min, the final OCRA checklist score must be corrected accordingly and must factor in the actual duration of the tasks; the purpose is to weight the final risk index for the actual duration of the repetitive tasks performed by the worker (Figure 4.2).

The duration multipliers shown in Table 4.4 can be seen to increase with each additional hour of exposure.

If the net time of repetitive work is greater than 480 min, the multipliers grow exponentially. When the work performed is characterized by the presence of many tasks, for the calculation of the final risk index with special mathematical models (see Chapter 6), multipliers are necessary even for very short partial durations of the single tasks.

In order to properly apply the *duration multiplier*, the following methods can be used:

a. *Manual:* Once the *net duration of the repetitive work* has been calculated (Table 4.2), it must be matched with the corresponding multiplier in

TABLE 4.4
Duration Multiplier for Calculating the Final OCRA Checklist Score Based on the Net Duration of the Repetitive Work

Multiplier for the Net Duration of the Repetitive Work Performed During the Shift			
Net Duration of Repetitive Work (min)	**Duration Multiplier**	**Net Duration of Repetitive Work (min)**	**Duration Multiplier**
60–120	0.5	301–360	0.925
121–180	0.65	361–420	0.95
181–240	0.75	421–479	1
241–300	0.85		
	Less than 60 min (only for multitask analysis)		
Up to 1.87	0.01	7.6–15	0.1
1.88–3.75	0.02	15.1–30	0.2
3.76–7.5	0.05	31–59	0.35
	More than 480 min		
480–539	1.2	660–719	2. 8
540–599	1.5	720 or more	4
600–659	2		

Overall shift duration (minutes)	480	Effective shift duration (minutes)	480
Duration of nonrepetitive tasks (e.g., cleaning, fetching supplies, etc.,) in minutes			10
No. of actual breaks (recovery periods) during the shift, with a duration of at least 8 min (excluding meal break) that can be considered as recovery periods			2
Overall duration of all actual breaks (excluding meal break) in minutes			20
Actual duration of meal break if included in shift duration (minutes)			
No. of other breaks (i.e., meal break not included in working time; travel time to/from different company locations). Mark one number only when these break last at least 30 min.			1
Description of repetitive task			
There are identified cycles: report the number of units per worker per shift		Net duration of repetitive tasks in the shift (in minutes)	450

FIGURE 4.3 Example 1: automatic calculation of net duration of repetitive work and corresponding duration multiplier.

Table 4.4 and used as the multiplier for the final checklist score, according to the calculation model shown in Figure 4.2.

b. *Automatic:* Once all the organizational data required to calculate the *net duration of the repetitive work* has been entered, the software tool automatically calculates this value and proposes the corresponding duration multiplier (Figure 4.3).

There are identified cycles: report the number of units per worker per shift	450	Net duration of repetitive task in the shift (in minutes)	450	
There are identified cycles: report the observed cycle time (in seconds)	58			
There are no identified cycles but the same actions are repeated all the time: report the time (in seconds) of your representative observation.		Net cycle time duration (computed) (seconds)	60.0	Nonjustified minutes in the shift
There are recovery times within the cycle (cross if yes)		% difference between observed and calculated cycle time (accepted limit 5%)	3%	15

FIGURE 4.4 Example 1: automatic calculation of net total cycle time (or rate) versus observed total cycle time.

The software tool also automatically calculates the following (Figure 4.4):

- *Net total cycle time*, entering the actual number of cycles performed in the shift. It is worth remembering that the cycle number may correspond to the number of units worked or alternatively to the number of packages containing multiple units.
- Difference, in percentage terms, between the net total calculated and observed cycle time and the total number of *nonjustified* minutes, that is, minutes devoted to unknown operations.

4.1.5 Lack of Recovery Time Factor (Form 1, Second Part)

The recovery time can be defined as any time in which upper limbs that are otherwise involved in performing work are virtually inactive.

As already extensively stated (Colombini et al., 2002; 2005), the following could be considered as recovery times:

a. Breaks, official or otherwise, including meal breaks (lasting at least 30 min), whether included or not as part of the paid workday.
b. Sufficiently long periods of work activity in which the muscle groups are at rest (e.g., visual inspections).
c. Periods within the cycle during which muscle groups that are otherwise active are completely at rest. For recovery periods within the cycle (visual inspections, waiting, or inactive periods) to be considered as significant, such periods must last for at least 10 consecutive seconds within the cycle and be repeated every cycle and for the entire duration of the repetitive task. Moreover, there must be a ratio of 5:1 between work time (repetitive tasks utilizing the upper limbs) and recovery time (upper limbs inactive). When there are recovery times within the cycle, the *number of hours without an adequate recovery time* will be 0. In practice, such a situation is rare.

TABLE 4.5
OCRA Checklist: Risk Scores Describing the Lack of Recovery Time Factor Corresponding to Six Scenarios, Depicting the Distribution of Pauses and Breaks During the Work Shift

Recovery Risk Factor	
Type of work interruption (with pauses or other visual inspection tasks) (max. score allowed = 10). Choose one answer. It is possible to choose intermediate values.	score
One interruption in the repetitive work lasting at least 8/10 min every hour (also count the meal break) or the recovery period is included in the cycle.	**0**
Two interruptions in the morning and two in the afternoon (plus meal break), lasting at least 8–10 min per 7–8 h shift, or at least four interruptions per shift (plus meal break), or four 8/10 min interruptions per 6-h shift.	**2**
Two pauses, lasting at least 8–10 min each, per 6-h shift (not including meal break); or, three pauses, plus meal break, per 7–8 h shift.	**3**
2 pauses, plus meal break, lasting at least 8–10 min each, per 7–8 h shift (or 3 pauses without meal break), or one pause lasting at least 8–10 min per 6-h shift;	**4**
One pause, lasting at least 10 min, per 7-h shift without meal break; or, in an 8-h shift, only a meal break (meal break is not counted in the working hours).	**6**
No actual pauses except for a few minutes (less than 5) per 7–8 h shift.	**10**

As a general rule, what is examined is whether or not there are breaks, their duration and frequency, and how the breaks are distributed throughout the shift.

The first version of the OCRA checklist provided six scenarios (Table 4.5) depicting different break distributions during the shift; each scenario corresponds to a different risk score, defining the *lack of recovery time* factor (Colombini et al., 2002).

For some time now, experienced large-scale users of the OCRA checklist believed that, unlike the OCRA index (Occhipinti, 1998), this method for calculating the recovery factor score did not adequately take into account the potential inclusion of additional breaks. This is because with the OCRA checklist method, the *recovery time* factor was merely added to the overall sum, while with the OCRA index, it acts as a multiplier of the other risk factors.

In the new version of the OCRA checklist, there is a new model for calculating the recovery factor (recovery multiplier), which corresponds more closely with the OCRA index and also better accounts for the effectiveness of corrective actions.

There are two steps to the assessment:

- The first step consists of identifying the number of work hours without adequate recovery time, which can be determined based on the six classic scenarios (Table 4.5) or, for a more accurate result, based on the exact number of hours without adequate recovery, as proposed with the OCRA index.
- The second step entails applying a specific factor, called the *recovery multiplier*, to the checklist score obtained by adding up all the scores for the various factors, such as *frequency*, *force*, *posture*, and *additional factors* (Figure 4.2).

4.1.5.1 Precise Calculation of the Number of Hours without Adequate Recovery Time

The purpose of this calculation is to determine how many hours are worked in a shift without an adequate recovery time of at least 8–10 consecutive minutes. The calculation does not include

- The hour just prior to the meal break, which is recovered during the break itself (and must last at least 30 min)
- The last hour in the shift, which is recovered by the fact that the shift ends

Figure 4.5 shows an example of calculating the number of hours without an adequate recovery time in different scenarios.

In order to easily and very accurately define the number of hours worked without adequate recovery, as per the OCRA index, it is advisable to

- Mark as *recovered* the 60 min prior to the end of the shift and before the meal break (at least 30 min).
- Working back from the last hour to the first hour of the shift, mark each individual 60-min period as *recovered* if it includes a pause and with a score of 1 if it did not include any pauses. If any periods of half an hour (or in actual fact between 20 and 40 min) appear as nonrecovered, assign a score of 0.5.

This procedure will generate the number of hours worked without an adequate recovery time.

The new Excel spreadsheet software tool ERGOepmChecklitOCRAauto-EN is designed to facilitate the determination of the new final checklist value and proposes an automatic method for calculating the number of hours without adequate recovery, using the data described in Table 4.6. Essentially, the elements listed here are

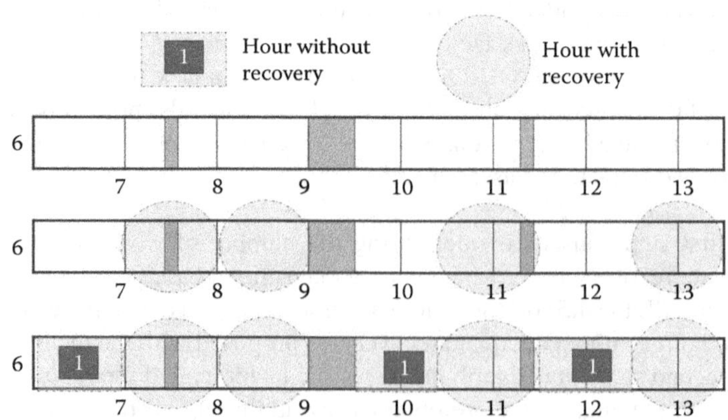

FIGURE 4.5 Example of calculation of hours without adequate recovery.

TABLE 4.6
Criteria Used for the Automatic Calculation of Hours without Adequate Recovery

		No. of Interruptions in the Workday Considered as Recovery Periods: Meal Break Lasting at Least 30 min and/or No. of (Properly Distributed) Breaks						
Duration of Shift	No. of Hours without Recovery	n.1	n.2	n.3	n.4	n.5	n.6	n.7
480	7	6	5	4	3	2	1	0
460	7	6	5	4	3	2	1	
440	6.5	5.5	4.5	3.5	2.5	1.5	0.5	
420	6	5	4	3	2.5	1.5	0	
390	5.5	4.5	3.5	2.5	1.5	0.5	0	
360	5	4	3	2	1	0		
330	4.5	3.5	2.5	1.5	0.5	0		
300	4	3	2	1	0			
270	3.5	2.5	1.5	0.5	0			
240	3	2	1	0				
210	2.5	1.5	0.5	0				
180	2	1	0					
120	1	0	0					
0	0							

subtracted from the total number of hours worked (less the last hour, which is always recovered):

- The number of adequate *pauses* (i.e., pauses lasting at least 8–10 consecutive minutes that are neither in the 60 min before the meal break nor in the last 60 min of the shift).
- The meal break (no. = 1), whether included or not in the work day.

The aforementioned software tool, ERGOepmChecklitOCRAauto-EN, can also be used to choose whether or not to adopt this automatic method (which is slightly less precise) or whether to directly enter the number of hours without adequate recovery calculated manually as described above (see Figure 4.5); if this figure is entered, the automatic estimate will not be calculated. Manual calculations are useful when job redesign plans entail improving recovery periods and times.

In fact, it is useful, if not indispensable, to specify the beginning and the end of the shift, and indicate when the pauses occur, in the specific section of the form (Figure 4.6); this will make it easier to recognize ineffective pauses (such as in the 60 min prior to the meal break or before the end of the shift), which should be counted as *adequate pauses.* If there are certain shifts featuring very different working hours or different recovery times, describe the situation and add more scores for the corresponding recovery time distribution in the various shifts.

Mark the breaks in the shift. (1 box = 1 h) First shift First hour...Last hour								
Mark the breaks in the shift. (1 box = 1 h) Second shift First hour...Last hour								
Mark the breaks in the shift. (1 box = 1 h) Third shift First hour...Last hour								

FIGURE 4.6 Schematic description of the distribution of recovery times in the shift.

4.1.5.2 Application of the New Multiplier Factor, Called the Recovery Multiplier

In the OCRA index (Occhipinti, 1998), the lack of recovery time factor is a multiplier that, when applied to the final index, adjusts the result depending on how many hours do not feature adequate recovery.

In the OCRA index, the multiplier is applied to the denominator in the equation (observed actions/recommended actions): Table 4.7 shows the multiplier and its reciprocal (which will serve for the checklist, insofar as it is applied to the numerator) corresponding to each box indicating the number of hours without adequate recovery time).

Figure 4.7 shows the resulting (exponential) trend.

By analyzing the trend for the OCRA index multipliers for the recovery factor, the following mathematical model emerges and expresses a complex exponential function:

$$\text{Original OCRA index recovery multiplier function}$$

$$= y = \text{esp}(0.0735 * \text{esp}(0.4907 * x))$$

The next logical step is to compare the two similar OCRA index and OCRA checklist scores (same level of exposure = light red) by plotting them together on a graph (Figure 4.8). The resulting behavior curves show that as the scores for recovery

TABLE 4.7
OCRA Index: Multipliers Corresponding to Number of Hours without Recovery

No. of hours without adequate recovery time	7	6	5	4	3	2	1	0
OCRA index multiplier factor (applied to the index denominator)	0.1	0.25	0.45	0.6	0.7	0.8	0.9	1
1/multiplier factor	10	4	2.22	1.66	1.43	1.25	1.11	1

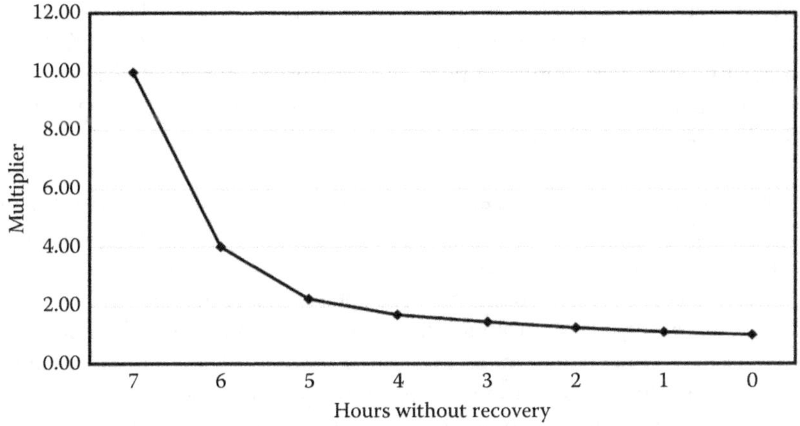

FIGURE 4.7 OCRA index: trend for recovery factor multipliers in relation to the number of hours without adequate recovery.

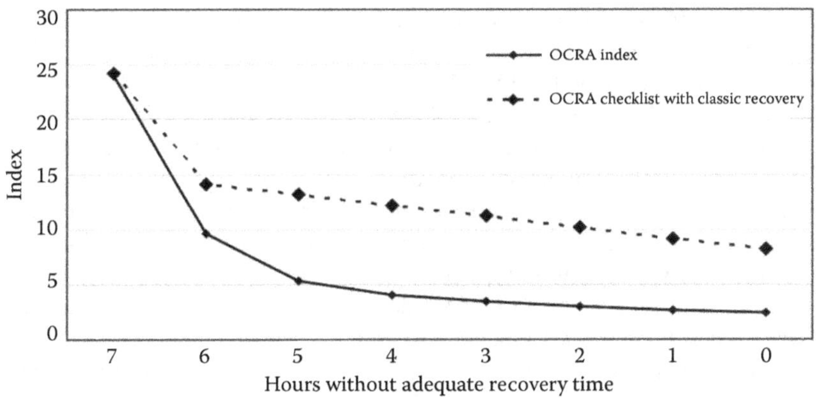

FIGURE 4.8 Comparison between two similar values for the OCRA index and the OCRA checklist (light red), depicted together in a graph showing both trends in relation to different recovery times.

change, there is an almost exponential rise for the OCRA index, while the line for the checklist score is straight. However, the straight line rises suddenly when there are no recovery times in a 7–8 h shift, and the risk score rises very quickly.

This different trend, which was chosen in the past for the checklist as it seemed easier to apply, appears to lead to an underestimation of the effectiveness of corrective actions when other recovery periods are added to the shift. The OCRA index used to be recommended for assessing job redesigns in view of its accuracy, but a growing number of operators have found it easier to test the effectiveness of organizational improvements (such as a greater number and a better distribution of pauses) earlier on in the risk mapping stage.

Therefore, the linear curve for the recovery scores in the checklist has been corrected by carefully applying the exponential curve for recovery multipliers of the OCRA index.

In order to more effectively and rationally apply this exponential curve without detracting from the significance of the scores already being utilized, two constraints needed to be taken into account:

- The sum of the risk scores deriving from the frequency, force, posture, and additional factors must remain unchanged when all of the hours are adequately recovered (i.e., the recovery multiplier must be = 1).
- When the new recovery multipliers are applied, they must be "anchored" to the recovery score of 4 (in situations of slight to medium risk). In fact, in our database of clinical data linked to exposure levels, most of the cases feature exposure levels corresponding to this recovery score, displaying remarkable predictive power at this level for upper limb work-related musculoskeletal disorders (UL-WMSDs) (see later). Therefore, it was an obvious decision to anchor the application of the exponential curve to the score of 4 and adjust the new multipliers to this point.

The recovery multipliers thus obtained are shown in Table 4.8.

Table 4.9 shows multiplier values even when the number of hours without adequate recovery is intermediate (0.5, 1.5, 2.5, etc.).

This table also describes the percentage of over- or underestimates in the final scores of the OCRA checklist in relation to the different recovery factor values, versus a score of 4 h without adequate recovery.

Figure 4.9 provides a graphic depiction of the trends that now reflect the checklist values after changing the number of hours without adequate recovery. It is worth noting that the new recovery multipliers create an exponential curve that now more closely resembles the one for the OCRA index but is not identical to it, given the need to accommodate the aforementioned constraints.

After this lengthy but necessary description of the new criteria introduced by the OCRA checklist to more accurately calculate the effects of recovery periods, the following steps can be briefly defined:

- Count the number of hours without adequate recovery (as shown in Section 4.1.5.1).
- Identify the corresponding *recovery multiplier.*
- As illustrated in Figure 4.1, apply this multiplier to the sum of the scores obtained for the four risk factors illustrated (i.e., *frequency*, *force*, *posture*, and *additional*).

TABLE 4.8
The New Recovery Multipliers for the OCRA Checklist

No. of hours without adequate recovery time	0	1	2	3	4	5	6	7	8
Recovery multiplier	1	1.050	1.120	1.200	1.330	1.480	1.700	2.000	2.500

TABLE 4.9
Multipliers Corresponding to Various Scenarios of Hours without Recovery and Positive or Negative Percentage Differences with Respect to the Condition of 4 h without Adequate Recovery Time in an 8 h Shift

No. of Hours without Adequate Recovery Time	Recovery Multiplier	Difference for 4 h without Recovery (%)
0	**1**	−24.8
0.5	**1.025**	−22.9
1	**1.05**	−21.1
1.5	**1.086**	−18.3
2	**1.12**	−15.8
2.5	**1.16**	−12.8
3	**1.2**	−9.8
3.5	**1.265**	−4.9
4	**1.33**	0.0
4.5	**1.4**	5.3
5	**1.48**	11.3
5.5	**1.58**	18.8
6	**1.7**	27.8
6.5	**1.83**	37.6
7	**2**	50.4
7.5	**2.25**	69.2
8 or more	**2.5**	88.0

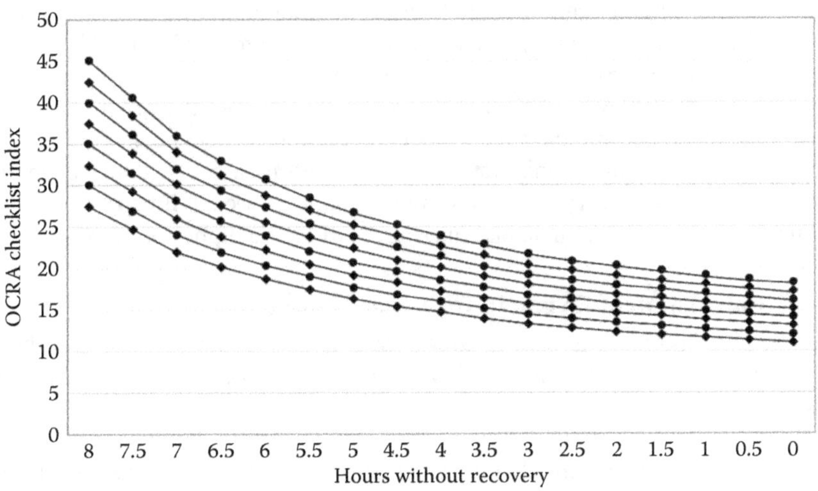

FIGURE 4.9 Example of the new curves relating to the OCRA checklist values as a function of changes in the number of hours without adequate recovery.

4.1.6 Action Frequency Factor (Form 2, First Part)

Since the mechanism that triggers tendon disorders appears to be significantly linked to frequency of movement, it follows that *frequency of action* will be an important factor for estimating the risk of biomechanical overload.

A method that can be used in the field to measure the frequency of mechanical events within the cycle involving the upper limbs is to analytically count or at least identify and estimate the number of *technical actions in a cycle* and relate them to the relevant unit of time (i.e., no. of technical actions/minute = *frequency of technical actions*).

It should be noted that *technical actions* are not the individual movements of the hand, wrist, elbow, or shoulder but rather the overall movement performed by one or more joint segments enabling a simple work element to be performed, such as *grasping, positioning, turning, pushing*, and so on.

For example, very common technical actions such as *grasping* or *positioning* often require several movements as well as involving several upper limb joints (fingers, wrist, elbow, shoulder).

Only after first undertaking a separate analysis of *awkward postures* (hence the postures and movements of each upper limb joint group) adopted to perform technical actions and then of the *force* required to complete them (along with an estimate of additional factors and organizational events) will it be possible to achieve an overall assessment of exposure risk (Colombini et al., 2002; 2005). Annex 4.2 provides an updated list of the most frequent technical actions, including some clearer definitions that will help in recognizing them. In order for the outcome of the risk assessment using the OCRA method to be reliable (making it possible to predict the likelihood of disorders developing), it is essential to strictly comply with the criteria provided. It is not possible to combine other approaches, such as the counting of *phases* as technical actions (i.e., sets of multiple technical actions serving the same purpose) or taking a number automatically extrapolated from other analytical systems with predetermined times, known as PTS, (such as motion time method [MTM], Universal Analysing System [UAS], etc.), with the OCRA method.

Technical actions may be *dynamic* (when they involve movement) or *static* (when workers hold an object in their hand for over 4 consecutive seconds).

The scores for dynamic and static technical actions are calculated differently. The final result will identify the most critical situation for each limb (based on both dynamic and static movements), and the score will be the higher of the two.

4.1.6.1 Calculation of Dynamic Technical Actions

The process for assigning scores is fairly straightforward, but to be certain that the final result is accurate, attention should be focused on the following fundamental points:

- It is worth remembering that the calculation of technical actions must keep the actions of the right limb separate from those of the left limb. It is not possible to average the two, since each limb will have a different likelihood of developing a disorder based on the relevant exposure level. It is essential to calculate the exposure of each limb separately in order to establish the causal link between risk and disease and also to ensure the safest return of the worker to the workforce.

- Depending on the aim of the analysis, it may be possible to assess only one limb, or it may be necessary to assess both. However, when carrying out the first risk mapping exercise, it is generally recommended that only the most frequently used limb be analyzed, which is usually the dominant one.
- To count the technical actions, it is important to strictly comply with the criteria for their identification (see the definitions in Annex 4.2). When using the OCRA checklist, the name of the technical action does not need to be entered, only the correct number.
- Once the number of technical actions in a cycle has been obtained (for each limb), the frequency of action must be calculated following the method illustrated; since the value has to be expressed in *technical actions/minute*, the following formula is applied:

No. Actions * 60 / T.T. Cycle

where:

No. of actions = number of technical actions present in a cycle relative to one limb
T.T. Cycle = Total Cycle Time or Rate

There are two different ways to calculate the frequency factor: the classic manual method and the automatic method. The automatic method is obviously to be preferred, but the manual method will be described first as it enables all the criteria to be illustrated as well how to calculate the score, which might otherwise be difficult to understand.

a. Manual method for calculating the frequency factor score

The first section of the OCRA checklist in hard-copy form (Table 4.10), relative to frequency, includes seven scenarios, each numbered from 0 (low frequency of action) to 10 (maximum frequency). Each line describes the speed of work actions (slow, somewhat fast, fast, very fast) via scenarios featuring increasingly frequency of action from 20 to over 70 actions per minute at intervals of about 10 actions per minute.

Once the scenario with the corresponding frequency of action has been identified, it must be ascertained if the work allows for short interruptions (at a constant or irregular rate). This second characteristic is also used to choose the score for the corresponding scenario, opting for intermediate values if a more precise result is required.

Experience has shown that the choice of an intermediate frequency score is often interpreted too subjectively by the analyst. The authors provide a guide (Table 4.11) to avoid differences in the way intermediate scores are assigned and apply them correctly. The table should be used as follows:

- Calculate the frequency of action (per individual limb) and check whether the worker is able or not to make short interruptions. For example, if it is the machine that sets the pace (e.g., a conveyor belt), the

TABLE 4.10
Classic Scenarios for Calculating the Frequency Factor Score for Dynamic and Static Technical Actions

Frequency Risk Factor			
Dynamic Technical Actions	**Score**	**R**	**L**
Arm movements are slow; frequent short interruptions are possible (20 actions per minute)	**0**		
Arm movements are not too fast; short interruptions are possible (30 actions per minute)	**1**		
Arm movements are quite fast (about 40) but short interruptions are possible	**3**		
Arm movements are quite fast; only occasional and irregular short pauses are possible (about 40 actions per minute)	**4**		
Arm movements are fast; only occasional and irregular short pauses are possible (about 50 actions per minute)	**6**		
Arm movements are very fast; the lack of interruptions makes it difficult to keep up the pace, which is about 60 actions per minute	**8**		
Very high frequencies: 70 actions per minute, or more; absolutely no interruptions are possible	**10**		
Static Technical Actions			
An object is held for at least 5 consecutive seconds, with one or more static actions maintained for less than 50% of the cycle (or observation) time	**0**		
An object is held for at least 5 consecutive seconds, with one or more static actions maintained for 2/3 of the cycle (or observation) time	**2.5**		
An object is held for at least 5 consecutive seconds, with one or more static actions maintained for 3/3 of the cycle (or observation) time	**4.5**		

worker will be unable to make short pauses (to drink a glass of water or simply stop for a few seconds); conversely, workers on a bench can always make a short pause, which they can also do if the speed of the assembly line can be modified (i.e., it is said to have a "buffer zone").

- Once the frequency has been identified, if there are short breaks the corresponding frequency score must be chosen from Section A of Table 4.11; if there are no breaks, the score must be chosen from Section B. No further intermediate scores should be chosen.

b. Automatic method for calculating the frequency factor score

Going back to Example 1 in Figure 4.4, it was determined that the *total cycle time* or *rate* was 60 s (observed total cycle time = 58 s, with an acceptable 3% difference versus the *calculated total cycle time*).

When the number of technical actions detected is entered in the appropriate boxes of the OCRA checklist Excel spreadsheet (50 actions on the right and 30 on the left), the software tool automatically calculates the frequency (Figure 4.10). In order to obtain the correct *frequency score*, mark an X in the appropriate box to indicate whether brief interruptions are possible or not. The corresponding frequency scores will appear automatically for the

TABLE 4.11
Intermediate Frequency Factor Scores *With* (Section A) or *Without* (Section B) Brief Interruptions

	Section A: frequency score when brief interruptions are possible					
Frequency	inf 22.5	From 22.5 to 27.4	From 27.5 to 32.4	From 32.5 to 37.4	From 37.5 to 42.4	From 42.5 to 47.4
Scores	0.0	0.5	1.0	2.0	3.0	4.0
Frequency	From 47.5 to 52.4	From 52.5 to 57.4	From 57.5 to 62.4	From 62.5 to 67.4	From 67.5 to 72.4	sup 72.4
Scores	5.0	6.0	7.0	8.0	9.0	9
	Section B: frequency score when brief interruptions are not possible					
Frequency	inf 22.5	From 22.5 to 27.4	From 27.5 to 32.4	From 32.5 to 37.4	From 37.5 to 42.4	From 42.5 to 47.4
Scores	0.0	0.5	1.0	2	4	5
Frequency	From 47.5 to 52.4	From 52.5 to 57.4	From 57.5 to 62.4	From 62.5 to 67.4	From 67.5 to 72.4	sup 72.4
Scores	6	7	8	9	10	10.0

Report the number of observed technical actions (right and left sides separately)	No. of actions right	Frequency	No. of actions left	Frequency
	50	50.0	30	30.0
If the technical actions are very quick and hard to count (> 70 actions/min), place an "X" in the box without counting the actions				
Short interruptions are possible (it is possible to regulate the pace)	No	Yes	Frequency scores: 6 right, 2 left	
	X			

FIGURE 4.10 Example 1: automatic calculation of frequency and respective score when the only actions present are dynamic.

left and right sides, taking into account the intermediate scores described above.

4.1.6.2 Calculation of Static Technical Actions

a. Manual method for calculating the static technical action score

The risk score for static technical actions is calculated as follows:

- Identify actions within the cycle that involve continuously holding objects or tools for periods of more than 4 consecutive seconds
- Determine the *total holding time* as the total number of seconds detected
- Compare and calculate the difference in percentage terms (%) versus the *total cycle time* (or rate)
- Determine the risk score based on the following percentage of duration intervals: 0–50% = 0; 51%–80% = 2.5; 81%–100% = 4.5

The final frequency score of static action is then entered into the relevant box.

There may be situations where both dynamic and static actions are performed simultaneously (e.g., cutting with a knife). One hand continuously holds the knife handle (static action) and, at the same time, performs cuts (dynamic actions). In a case such as this, to determine the final *frequency factor*, it is necessary to consider the representative score for frequency, which will be the higher of the two *frequency scores* (dynamic or static).

b. Automatic method for calculating the static technical action score

Consider a different situation (Example 2), featuring the same total cycle time of 60 min but where the left hand performs a static action (holding something practically all the time) but also performs dynamic actions (30 actions/min). When the numbers are entered into the appropriate boxes

Dynamic actions						
		No.	Frequency		No.	Frequency
Report the number of observed technical actions (right and left sides separately)	Right	50	50.0	Left	30	30.0
If technical actions are very quick and hard to count (> 70 actions/min), place an "X" in the box without counting the actions	Right			Left		
Short interruptions are possible (it is possible to regulate the pace)		No	Yes			
		X				

Static actions	Right: No	Right: Yes	Left: No	Left: Yes
An object is held for at least 5 consecutive seconds, incurring one or more static actions for 2/3 of the cycle (or observation) time	No		No	
An object is held for at least 5 consecutive seconds, incurring one or more static actions for 3/3 of the cycle (or observation) time	No			Yes

Frequency scores	
6	4.5
Right	Left

FIGURE 4.11 Example 2: automatic calculation of frequency and respective score when dynamic and static actions are present.

in the spreadsheet (Figure 4.11), the software tool automatically calculates the final frequency score, comparing the static with the dynamic frequencies and choosing the worst one as the representative frequency (i.e., static in the present case).

4.1.7 Force Factor (Form 2, Second Part)

To overcome the challenge of assessing the force exerted by the worker without using specific tools, use of the Borg CR-10 (Borg, 1998) scale is also recommended for compiling the OCRA checklist. The exercise entails interviewing workers and asking them to describe the effort that they subjectively perceive while performing a repetitive task (Colombini et al., 2002; 2005).

The perceived effort of the entire upper limb should be assessed for each individual technical action making up the cycle. A practical way to go about this is to disregard any technical actions requiring minimal or very slight effort (between 0.5 and 2 on the Borg scale) and to assess only those technical actions (or sets of actions) requiring at least a *moderate* effort (i.e., a score of 3 or higher on the Borg scale). To complete the assessment, the duration in percentage terms of each level of effort equal to or greater than 3 on the Borg scale is determined with respect to the *total cycle time.*

Experience has led to a few tips on how best to interview the worker in order to obtain reliable information and also avoid any uncertainties caused by the use of "subjective" data.

TABLE 4.12
Borg CR-10 Scale

Borg CR-10 Scale			
0.5	Extremely light		
1	Very light	**6**	
2	Light	**7**	Very hard
3	Moderate	**8**	
4		**9**	
5	Hard (heavy)	**10**	Extremely hard (almost max.)

Table 4.12 shows a model featuring the Borg scale used to collect information concerning perceived physical effort. The interview should be conducted as per the following steps:

- The force assessment has to be conducted after the technical actions analysis, as it requires a prior understanding of the cycle.
- The interview might be more effective if it is conducted by the same in-house technical specialist who participated in the first stage of the work analysis and was involved in describing the technical actions.
- The first question to put to the worker is whether or not there are any technical actions within the cycle that require the use of *appreciable muscular effort* of the *upper limbs.* It is important to word the question in this way because workers often mistake muscular effort with the *overall fatigue* they feel at the end of the shift.
- After extrapolating actions that require the use of force, the worker is asked to assign one of the definitions (not numerical values) indicated in the Borg CR10 scale (e.g., light and moderate). This way of asking the question, that is, using verbal descriptions of the level of force rather than scores, is important because workers often think of scores the way they think of marks in school reports. Each action thus identified will correspond to a progressive score that goes from 0 to 10 (Table 4.12). The analyst then indicates the duration of each action as a fraction of the duration of the whole cycle.
- Since the ultimate aim of the risk exposure assessment procedures is preventative, it is essential to ask the worker to explain why he or she believes that the actions reported as challenging entail "physical effort." This information is of immediate practical relevance because at times, force is required to perform certain actions because of a technical flaw in the product, inefficient tools, malfunctions, or improper mechanical aids, all issues that can be readily resolved.
- It is important that it is the actual worker who assigns a "score" to the physical exertion perceived while performing the various actions. Having an outside observer assign the score may lead to major errors. It should

be noted that it is very difficult for an outside observer to perceive the use of force, especially with regard to actions using the fingertips or smaller joints or certain positions of the joints, even when the exertion required is quite considerable. In fact, it may be helpful for the interviewer to try and perform the same action, not only to help the worker express his or her judgment on the degree of force but also to verify the results.

The results obtained through the interview conducted with the Borg scale will generally be more reliable when an adequate number of workers are involved, if possible performing the same job. This will, of course, make the results less subjective. However, in all cases, the *modal* value expressed by the workers must be used. It is also advisable for the judgments of workers suffering from work-related muscle and tendon disorders of the upper limbs to be assessed separately and for different purposes.

a. Manual method for calculating the force factor score

    In the OCRA checklist, the form for analyzing force includes three identical blocks (Table 4.13) describing actions that entail the use of force but differ based on the amount of force required.

    The blocks describe a few of the most common tasks that are associated with the use of *intense, almost maximal force*, with a score of 8 and above on the Borg scale (first section); *strong force*, with a score of 5, 6, or 7 (second section); and *moderate force*, with a score of 3 or 4 (third section). The activities to be assigned to one of these three degrees of exertion are pulling or pushing a lever, pressing a button, opening or closing, pressing or handling components, and using tools. Other activities requiring the use of force may be added.

    To fill in the form, after obtaining the previous information through interviews, the next step is to identify the duration of the actions performed during the cycle with a score for the use of force of 3 or higher. Different scores are assigned to each scenario featuring a different degree of force and duration. When there are multiple scenarios, the scores from the three blocks are added together to generate the final score.

b. Automatic method for calculating the force factor score

    Figure 4.12 shows an example of a force score estimate calculated by the aforementioned Excel software tool. By simply entering an X into the scenario or scenarios corresponding to the replies given in the interview, the software tool automatically calculates the final force score.

    To meet the needs of various users of the OCRA checklist, an alternative software tool has been created that increases the accuracy of the analysis: ERGOepmChecklitOCRAautoAP-EN.

    Instead of dividing the duration of the task into fractions of time (thirds), this tool measures and expresses the effort duration in seconds (Figure 4.13). The software automatically calculates the force scores using the intermediate scores presented in Table 4.14.

TABLE 4.13
OCRA Checklist: Assessment of Force Factor (Form 2, Second Part)

Force risk factor
More than one score can be marked and totaled to obtain the final score. IF YES:

The activity requires the use of almost maximum force: (score 8 or more on the Borg scale)

	SCORES		L	R
Pulling or pushing levers	6	2 s every 10 min		
Pushing buttons closing or opening	12	1% of the time		
Pressing or handling components using tools	24	5% of the time		
Lifting or handling objects	32	over 10% of the time		

The activity requires the use of intense force for: (scores 5–6-7 on the Borg scale)

	SCORES		L	R
Pulling or pushing levers	4	2 s every 10 min		
Pushing buttons	8	1% of the time		
Closing or opening	16	5% of the time		
Pressing or handling components using tools lifting or handling objects	24	over 10% of the time		

The activity requires the use of moderate force for: (scores 3–4 on the Borg scale)

	SCORES		L	R
Pulling or pushing levers	2	1/3 of the time		
Pushing buttons	4	about half the time		
Closing or opening	6	over half the time		
Pressing or handling components using tools lifting or handling objects	8	nearly all the time		

4.1.8 Assessment of Awkward Postures (Form 3)

An accurate description of the most common awkward postures and movements is fundamental for predicting the location of future work-related muscle and tendon disorders.

It is worth remembering that when assessing postural risk, awkward postures and movements should be described and their duration measured only if they require joints to work at angles of more than 50% of their maximum range of motion (Colombini et al., 2002; 2005). The joint segments analyzed and corresponding awkward postures are shown in a specific form contained in the hard-copy version of the OCRA checklist (Figure 4.14).

Postural risk assessments involve three main aspects:

- A description of awkward postures and/or movements broken down as follows: scapulohumeral (shoulder) joint, elbow, wrist, and hand (type of grip and finger movements) for the left and right sides, respectively.

Force-right side		Less than 1/3 of the time	About 1/3 of the time	About half of the time	About 2/3 of the time	Almost all the time	
The task requires moderate force (score 3-4 on the Borg scale)			X				
The task requires intense force (score 5-6-7 on the Borg scale)	Peaks of 1–2 s every 10 min	X	About 1% of the time	About 5% of the time		About 10% of the time or more	
The task requires almost maximal force (score 8 or more on the Borg scale)	Peaks of 1–2 s every 10 min		About 1% of the time	About 5% of the time		About 10% of the time or more	

FIGURE 4.12 Example 1: automatic calculation of force factor score using the "fraction of the time" approach.

		Right seconds	%	Score
Force-right side	The task requires moderate force (score 3-4 on the Borg scale)	20	33%	2.00
	The task requires intense force (score 5-6-7 on the Borg scale)	0.3	1%	4.00
	The task requires almost maximal force (score 8 or more on the Borg scale)		0%	

FIGURE 4.13 Example 3: automatic calculation of the force factor score using the durations in seconds of different scenarios.

- If the joint must work at an awkward angle, the duration of the task within the cycle must be measured (one-third, two-thirds, or all of the total cycle time, or of a specific observation period, or of the repetitive work). The scores for the shoulder joint take into account intervals of as little as one-tenth of the total cycle time to detect flexion or abduction of the arm at an angle of more than 80° with respect to the trunk (i.e., arm almost at shoulder height), or cases of extreme extension.
- It is possible to detect *stereotyped movements or postures*, that is, identical actions (regardless of whether or not they are associated with awkward postures or movements) by observing the following:
  - Identical technical actions or groups of identical technical actions repeated for over 50% of the cycle time or for almost the entire cycle
  - Static postures held uninterrupted for over 50% of the cycle time or for almost the entire cycle (e.g., prolonged gripping of a knife or screwdrivers)

TABLE 4.14
OCRA Checklist: Intermediate Scores Used for the Automatic Calculation of the Force Score, Using the Duration in Seconds of Different Scenarios

									Force 3-4									
Time in %	5	10	18	26	33	37	42	46	50	54	58	63	67	75	83	92	100	
Scores	**0.50**	**0.50**	**1.00**	**1.50**	**2.00**	**2.50**	**3.00**	**3.50**	**4.00**	**4.50**	**5.00**	**5.50**	**6.00**	**6.50**	**7.00**	**7.50**	**8.00**	
									Force 5-6-7									
Time in %	0.33	1.00	1.50	2.00	2.50	3.00	3.50	4.00	4.50	5.00	5,63	6.25	6.88	7.50	8.13	8.75	9.38	10.00
Scores	**4.00**	**8.00**	**9.00**	**10.00**	**11.00**	**12.00**	**13.00**	**14.00**	**15.00**	**16.00**	**17.00**	**18.00**	**19.00**	**20.00**	**21.00**	**22.00**	**23.00**	**24.00**
									Force 8-9-10									
Time in %	0.33	1.00	1.33	1.67	2.00	2,33	2,67	3.00	3,33	3,67	4.00	4,33	4,67	5.00	5,63	6.25	6.88	7.50
Scores	**6.00**	**12.00**	**13.00**	**14.00**	**15.00**	**16.00**	**17.00**	**18.00**	**19.00**	**20.00**	**21.00**	**22.00**	**23.00**	**24.00**	**25.00**	**26.00**	**27.00**	**28.00**
Time in %	8.13	8.75	9,38	10.00														
Scores	**29.00**	**30.00**	**31.00**	**32.00**														

Type of awkward posture and movement	Duration of awkward posture or movement	Score	Right	Lift
A—Shoulder: The arms are kept at about shoulder height, without support (or in other extreme postures)				
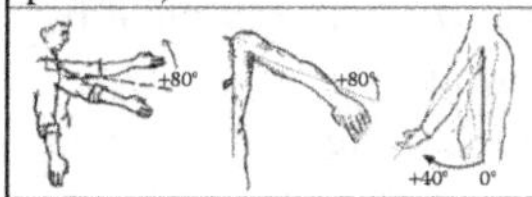	Inf. 1/3 = 10%–24% of the time 1/3 = 25%–50% of the time 2/3 = 51%–80% of the time 3/3 = more than 80% of the time	2 6 12 24		
B—Elbow: The elbow executes sudden movements (wide flexion-extension or pronosupination, jerking movements, striking movements)				
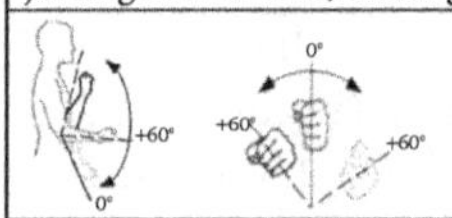	1/3 = 25%–50% of the time 2/3 = 51%–80% of the time 3/3 = more than 80% of the time	2 4 8		
C—Wrist: The wrist must bent in an extreme position, or must keep awkward postures (such as wide flexion/extension, or wide lateral deviation)				
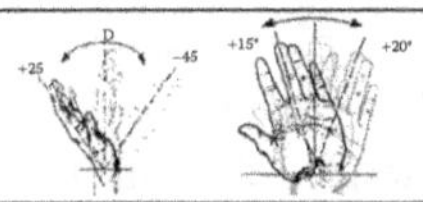	1/3 = 25%–50% of the time 2/3 = 51%–80% of the time 3/3 = more than 80% of the time	2 4 8		
D—Hand: Objects or tools are held in a pinch, palm grip, hook grip, or other kind of grip				
	1/3 = 25%–50% of the time 2/3 = 51%–80% of the time 3/3 = more than 80% of the time	2 4 8		

FIGURE 4.14 OCRA checklist: assessment of the awkward posture factor score. (Form 3, first part)

- Extremely short cycles of less than 15 s or even less than 8 s, obviously including actions that involve the upper limbs
  a. Manual method for calculating the awkward posture and movement factor score

     The questions in Sections A and D describe a separate joint segment (Figure 4.14); the last section describes the presence of stereotype.

     The following levels of stereotype may be observed:

     – *High:* When the cycle time is less than 8 s (and obviously involves the use of the upper limb); or when practically the entire task is made up of identical technical actions: the score is 3.

     – *Intermediate:* When the cycle time is between 8 and 15 s or when two-thirds of the entire task is made up of identical technical actions: the score is 1.5.

     Of all the scores calculated for the different joint segments (A—B—C—D), only the highest will be chosen, and added to the score for stereotype (E), where applicable: The total will be the score for the posture factor.

The questions relating to posture are very simple and describe the following for each joint:

- *Arms*: How long are they held at about shoulder height (i.e., compare the height of the elbow with the height of the shoulder) or in other extreme postures (arm fully extended)?
- *Wrist*: Are awkward postures required (flexion–extension above 45° and/or obvious radioulnar deviation)?
- *Elbow*: Are wide movements required entailing flexion–extension (60°, or a distance of at least 40 cm between *grasping* and *placing* an object) or in pronosupination with an angle of over 60° (almost complete rotation of objects held in the hand)?
- *Hand*: Is the grip a pinch, palm grip, or hook grip (Figure 4.15)?

With regard to the scapulohumeral joint, recent studies suggest that risk may be present even when the arm (or elbow) is held almost at or above shoulder height for more than 10% of the time (Punnet et al., 2000).

No scores are assigned to optimal power grips; however, when the grip is not optimal (e.g., when the index finger is extended forward to guide the tip of a knife or screwdriver or to press a button), an intermediate score may be added of 1 (for approximately one-third of the time), 2 (for approximately two-thirds of the time), and 3 (for approximately the entire time). It is important to note that although they do not generate awkward postures scores, identical power grips performed for two-thirds or more of the time do generate stereotype scores.

In any case, intermediate scores can be used.

b. Automatic method for calculating the awkward posture and movement factor score

Figure 4.16 shows an example of estimating the awkward posture score using the previously described Excel software tool ERGOepmChecklitOCRAauto-EN.

Simply place an X in the box next to the scenario or scenarios corresponding to those detected, and the tool will automatically calculate the final score for the posture factor.

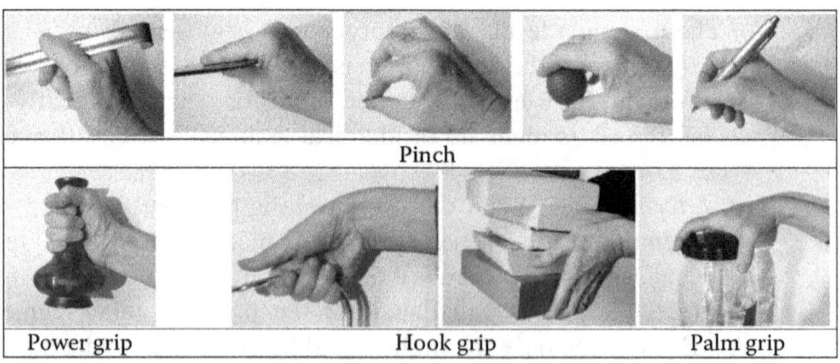

FIGURE 4.15 Main types of grips.

Awkward upper limb postures-right side		Less than 1/3 of the time	About 1/3 of the time	About half of the time	About 2/3 of the time	Almost all the time		Scores	
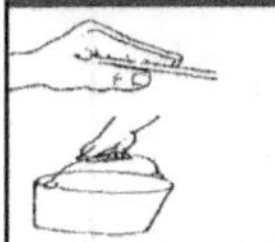	Pinch or palm or hook grip (not power grip)				X			4	
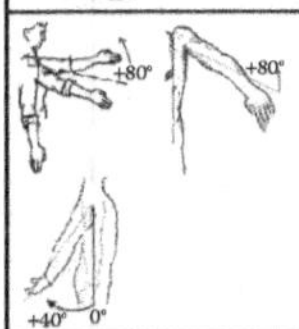	Arm more or less at shoulder height	X						2	
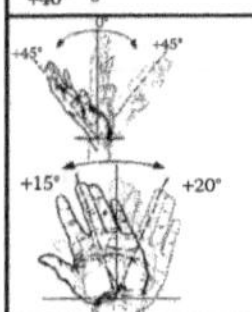	Extreme wrist deviations							0	
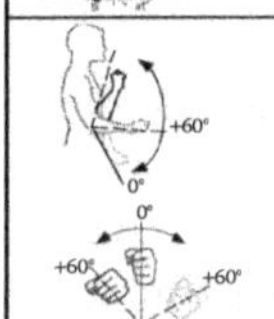	Complete object rotation (pronosupination) or wide arm-forearm (elbow) flexion-extension		X					2	
Stereotype	Duration of cycle (cycle time)	sup. 15 s		9–15 s		Equal or less than 8 s		0	
	Always repeat the same actions/gestures	About 2/3 of the time (more than 50%)			Almost all the time		X	1.5	3

FIGURE 4.16 Example 1: automatic calculation of the awkward posture score, using the "third of the time" approach.

As described in the previous paragraph related to force, an alternative software tool, ERGOepmChecklitOCRAautoAP-EN, has been created to increase the accuracy of the analysis. Instead of dividing the duration of the task into fractions of time (thirds), this tool measures and expresses the duration of different scenarios directly in seconds (Figure 4.17).

This software tool also uses intermediate scores, as shown in Table 4.15.

4.1.9 Additional Risk Factors (Form 4, First Part)

The OCRA checklist includes two sections for classifying additional risk factors (Table 4.16): The first includes scenarios featuring additional physical and mechanical factors, the second organizational factors.

Awkward upper limb postures-right		Seconds	%	Scores
	Pinch or palm or hook grip (not power grip)	30	50	4
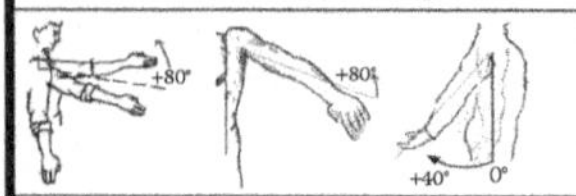	Arm more or less at shoulder height	10	17	3.5
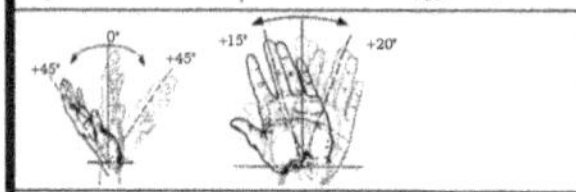	Extreme wrist deviations	30	50	4
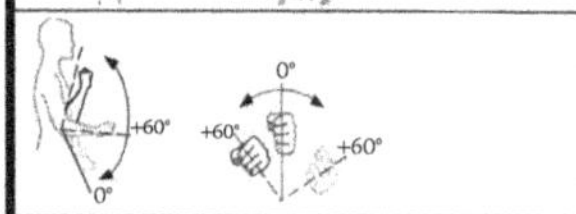	Complete object rotation (pronosupination) or wide arm-forearm (elbow) flexion-extension	30	50	4

FIGURE 4.17 Example 3: automatic calculation of the awkward posture score, using the durations in seconds of different scenarios.

a. Manual method for calculating the additional factors score

The list in Table 4.16 suggests a breakdown of additional factors into two sections:

- The *first section* lists only physical and/or mechanical factors; the score will be 2 when the action is performed for >50% of the time or the number of events per minute (frequency) is as specifically described in the table, and it will be 3 when multiple factors are present practically all the time. A higher score (4) is assigned when the task involves the use of vibrating tools or equipment (such as a pneumatic drill or grinder, etc.) for at least one-third of the time. Moreover, a score of 2 is assigned when the work entails shocks and countershocks twice or more per minute, or when the hands are used as tools to strike objects with a frequency of at least 10 hits/h.
- The *second section* lists several organizational scenarios, including two that generate scores due to the presence of the following risk factors:
  - The pace is determined by the machine but there are "buffer zones" enabling the speed to be at least partly modified (e.g., an assembly line where the operator "calls" the piece to the workstation only after the previous piece has been completed; the system allows several pieces to pile up).
  - The pace is determined by the machine but the line moves at a very slow speed.
  - The pace of the work is determined exclusively by the machine; this applies when the assembly line operator has to keep up with the speed of the machine (e.g., a conveyor belt).

Scores equal to or lower (but never higher) than those indicated may be used especially for factors that may feature different risk levels, for example, different degrees of exposure to vibrations and so on.

TABLE 4.15
Intermediate Scores Used by the Software Program to Calculate the Awkward Postures Risk Factor in Relation to the Duration of Exposure

									Hand									
Time	0.05	0.10	0.15	0.20	0.25	0.31	0.37	0.44	0.50	0.54	0.57	0.61	0.65	0.69	0.72	0.76	0.80	1.00
Score	**0.00**	**0.50**	**1.00**	**1.50**	**2.00**	**2.50**	**3.00**	**3.50**	**4.00**	**4.50**	**5.00**	**5.50**	**6.00**	**6.50**	**7.00**	**7.50**	**8.00**	**8.00**
									Shoulder									
Time	0.03	0.05	0.08	0.10	0.12	0.14	0.16	0.18	0.20	0.22	0.24	0.25	0.28	0.31	0.34	0.37	0.40	0.43
Score	**0.50**	**1.00**	**1.50**	**2.00**	**2.50**	**3.00**	**3.50**	**4.00**	**4.50**	**5.00**	**5.50**	**6.00**	**6.50**	**7.00**	**7.50**	**8.00**	**9.00**	**10.00**
Time	0.46	0.50	0.54	0.58	0.62	0.66	0.70	0.74	0.78	0.82	0.86	0.90	0.94	1.00				
Score	**11.00**	**12.00**	**13.00**	**14.00**	**15.00**	**16.00**	**17.00**	**18.00**	**19.00**	**20.00**	**21.00**	**22.00**	**23.00**	**24.00**				
									Wrist									
Time	0.05	0.10	0.15	0.20	0.25	0.31	0.37	0.44	0.50	0.54	0.57	0.61	0.65	0.69	0.72	0.76	0.80	1.00
Score	**0.00**	**0.50**	**1.00**	**1.50**	**2.00**	**2.50**	**3.00**	**3.50**	**4.00**	**4.50**	**5.00**	**5.50**	**6.00**	**6.50**	**7.00**	**7.50**	**8.00**	**8.00**
									Elbow									
Time	0.05	0.10	0.15	0.20	0.25	0.31	0.37	0.44	0.50	0.54	0.57	0.61	0.65	0.69	0.72	0.76	0.80	1.00
Score	**0.00**	**0.50**	**1.00**	**1.50**	**2.00**	**2.50**	**3.00**	**3.50**	**4.00**	**4.50**	**5.00**	**5.50**	**6.00**	**6.50**	**7.00**	**7.50**	**8.00**	**8.00**

TABLE 4.16
OCRA Checklist: Assessment of Additional Risk Factors (Form 4, First Part)

Additional Factors	
Choose one answer per section. The final score is the sum of the two partial scores.	
	Section A: Mechanical factors
2	**Inadequate** gloves (uncomfortable, too thick, wrong size) are used more than half the time for the task.
2	Presence of two or more sudden, jerky movements per minute.
2	Presence of at least 10 repeated impacts (use of hands as tools to hit) per hour.
2	Contact with cold surfaces (less than 0°C) or performance of tasks in cold chambers for more than half the time.
2	Use of vibrating tools at least one-third of the time. **Assign a score of 4** if these tools involve a high degree of vibration (e.g., pneumatic hammers).
2	Tools are used that cause compression of muscle and tendon structures (check for the presence of redness, calluses, wounds, etc., on the skin).
2	More than half the time is spent performing precision tasks (tasks on areas of less than 2 or 3 mm), requiring the worker to be physically close to see.
2	More than one additional factor (e.g., …) is present at the same time for more than half the time.
3	One or more additional factors (e.g., …) are present for almost the entire cycle.
	Section B: Organizational factors
1	The work rate is determined by the machine, but "recovery spaces" exist, allowing the rate to be sped up or slowed down.
1.5	The pace is determined by the machine, but the line moves at a very slow speed.
2	The work rate is entirely determined by the machine.

Only one answer is given in each of the two blocks (physical and/or mechanical factors and organizational factors); the sum of the partial scores entered in the two blocks generates the additional factors score.

c. Automatic method for calculating the additional risk factor score

Figure 4.18 shows an example of estimating the additional risk factor score using the usual Excel software tool (ERGOepmChecklitOCRAauto-EN). By simply entering an X into the box next to the scenario or scenarios corresponding to the situations identified in the workplace, the software tool automatically calculates the final additional risk factor score. The duration of these factors does not have to be measured; the software tool ERGOepmChecklitOCRAautoAP-EN makes the calculation automatically when an X is entered for a given scenario.

4.1.10 Calculation of the Final OCRA Checklist Score

To obtain the final OCRA checklist score, all that needs to be done is to sum the scores for all the risk factors, that is, frequency, force, posture, and additional (separately for the right and left limbs) and multiply this amount by the recovery factor and the duration factor (Figure 4.2).

Additional	Shocks and countershocks					For over half the time	
Physical factors	Repeated impacts by the hand (the hand is used as a tool)					Frequency: Almost 10 times/hour	
	Vibrating tools					For almost 1/3 of the time	
	Other: Report only for those suggested					For over half the time	
Organizational factors	The pace is set by the machine	There are "buffer areas" for slowing down the working pace		The pace is determined by the machine (the line moves very slowly)		The pace is completely determined by the machine (the line is constantly moving)	

FIGURE 4.18 Example 1: automatic calculation of the additional risk factor score using the more classic approach.

Task name	Recovery multiplier	Recovery score	Frequency	Force	Side	Shoulder	Elbow	Wrist	Hand	Srereotype	Total posture	Additional	OCRA checklist
Fan assembly	1.330	4	7	2	DX	6	0	0	4	1.5	7.5	0	20.85
Fan assembly	1.330	4	7	2	SX	6	0	0	4	1.5	7.5	0	20.85

FIGURE 4.19 Presentation of the final OCRA checklist results.

Figure 4.19 shows the final OCRA checklist results. As can be seen, the OCRA checklist not only supplies the final results for the estimated risk index but also all the scores for each individual risk factor, for the right and left limb. This representation of the results is indispensable insofar as it forms the basis for the subsequent risk management steps (such as returning the worker with UL-WMSDs to the workforce, detecting causal links, redesigning the workplace, etc.).

4.1.11 A Few Short Conclusions

These notes concerning the method adopted by the OCRA checklist refer only to its "intrinsic" content; multitask risk exposure assessments can be found in other publications (Colombini and Occhipinti, 2007; 2008; Occhipinti and Colombini, 2009; Occhipinti et al., 2009) and elsewhere in this volume.

In order to more easily apply the risk assessment method in relation to the three levels of intervention of the OCRA system (Figure 4.20), various Excel tools are now available (and can be downloaded free of charge from www.epmresearch.org), catering to specific requirements. This volume describes the software tools for performing minichecks, analyzing simple tasks (Chapter 5), and assessing a multitask scenario (Chapter 6).

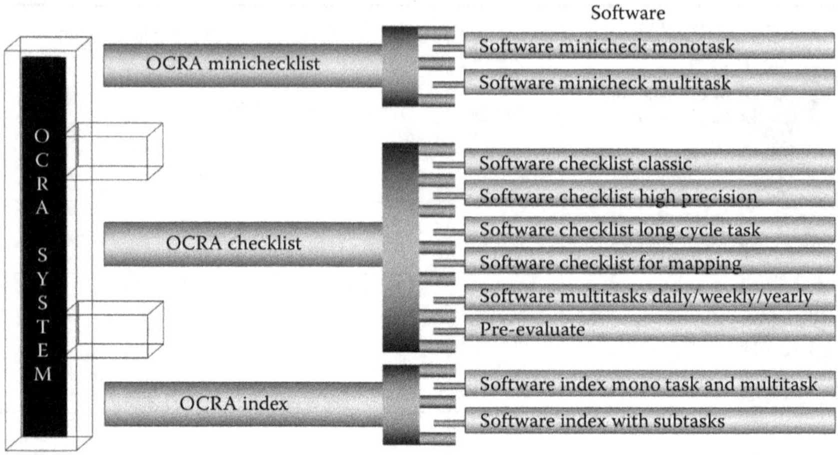

FIGURE 4.20 The OCRA system, the three OCRA tools and relevant software programs.

For the classic checklist, along with the risk mapping software (Chapter 7), other tools are now available for the automatic calculation of a classic checklist, complete with a high-precision model. The checklist also includes models for calculating exposure to prolonged tasks or for simplified calculations using the mini OCRA checklist (Chapter 5) and exposure to multiple tasks featuring a weekly or even yearly cycle (Chapter 6).

With regard to the OCRA index, a tool that has been around for some time now for calculating single and multitask scenarios, models have been created for analyzing risk by first breaking down the cycle into stages and then rebuilding it to achieve a twofold result: a better balanced workplace and more effective risk management (www.petrasoftware.it).

This volume will hopefully enhance the effectiveness of all these tools (and the system itself) during the preliminary identification (key entry) and quick assessment of biomechanical overload risk of the upper limbs (Chapter 3).

The assessment of risk due to repetitive movements has always been considered a complex exercise due to the large number and high variability of the risk factors involved. Not all of the issues associated with assessing risk have been solved, but it can safely be stated that, thanks to ongoing cooperation between many Italian and international operators, significant progress has been made toward achieving greater simplification, designing efficient preventative tools, and defining the relevant application criteria.

SOFTWARE EXERCISES

A few exercises are recommended for learning how to use the two aforementioned Excel spreadsheets, ERGOepmChecklitOCRAautoAP-EN and ERGOepmChecklitOCRAautoAP-EN.

EXERCISE 4.1

Open ERGOepmChecklitOCRAauto-EN and enter the following information:

Exercise 4.1 Overall shift duration (min)		480
Actual shift duration (min)		460
Duration of nonrepetitive tasks (e.g., cleaning and fetching supplies) in minutes		10
No. of actual breaks (recovery periods) during the shift, break duration of at least 8 min (excluding meal break)		2
Overall duration of all actual breaks (excluding meal break) in minutes		15
Actual duration of meal break if included in shift (min)		45
There are identified cycles:		
Indicate the number of units per worker per shift		780
There are identified cycles:		
Indicate the Observed cycle time (in seconds)		28
Indicate the number of observed technical actions (right and left sides separately)		30 (R) and 30 (L)
Short interruptions are possible (work pace can be regulated)		YES
Hand in pinch grip	About 2/3 of the time	Right and left
Arm more or less at shoulder height	1/3 of the time	Right and left
Repetition of same actions/movements	almost all the time	Right and left
Task requires Moderate Force (Borg score 3–4)	1/3 of the time	Right and left
Pace not determined by the machine		
Check the final result		**20.85**

EXERCISE 4.2

Open ERGOepmChecklitOCRAautoAP-EN and enter the following information. The data is the same as in the previous exercise, but for posture and force the duration is measured in seconds instead of in thirds.

Exercise 4.2 Overall shift duration (min)		480
Actual shift duration (min)		460
Duration of nonrepetitive tasks (e.g., cleaning and fetching supplies) in minutes		10
No. of actual breaks (recovery periods) during the shift, break duration of at least 8 min (excluding meal break)		2
Overall duration of all actual breaks (excluding meal break) in minutes		15
Actual duration of meal break if included in shift (min)		45
There are identified cycles:		
Indicate the number of units per worker per shift		780
There are identified cycles:		
Indicate the observed cycle time (in seconds)		28
Indicate the number of observed technical action (right and left sides separately)		30 (R) and 30 (L)
Short interruptions are possible (work pace can be regulated)		YES
Hand in pinch grip	20 s	Right and left
Arm more or less at shoulder height	8	Right and left
Repetition of same actions/movements	almost all the time	Right and left
Task requires moderate force (Borg score 3–4)	10	Right and left
Pace not determined by the machine		
Check the final result		**20.85**

4.2 SIGNIFICANCE OF THE FINAL OCRA CHECKLIST SCORE AND OCRA INDEX: RISK CLASSIFICATION AND PREDICTIVITY

4.2.1 Introduction

Over the last 20 years, the authors collected, processed, and published much data from numerous studies on risk and injury due to biomechanical overload of the upper limbs (Colombini and Occhipinti, 1996; Occhipinti and Colombini, 2004b and 2007; Colombini et al., 2005), concerning collective exposure indicators on the one hand (expressed in terms of both the OCRA index and the OCRA checklist score) and effect indicators on the other (based on clinically proven cases of UL-WMSDs among various groups of exposed workers).

Thanks to the resulting database, it has been possible to generate

- *Critical values* for the OCRA index and OCRA checklist final score for the purposes of classifying risk associated with tasks entailing repetitive movements and exertion of the upper limbs.
- Reliable models for predicting the occurrence of UL-WMSDs in exposed worker populations, based on exposure indicators.

The resulting data, methods, criteria, and outcomes have been published by international scientific journals and incorporated into international standards (ISO 11228-3, EN 1005-5, and ISO TR 12295).

Therefore, this section will first summarize the main elements needed for the practical implementation of results that have already been published and embedded in the standards. Since new data has emerged in the meantime based on the specific use of the OCRA checklist with regard to the occurrence of UL-WMSDs, it has been possible (as will be indicated) to fine-tune the models for predicting upper limb musculoskeletal disorders based directly on the OCRA checklist scores.

4.2.2 2004 Database

Table 4.17 summarizes the salient information contained in the database (Occhipinti and Colombini, 2004b; 2007). It considers 22 groups of exposed workers (total number: 4624, of whom 1879 were male and 2745 female), their composition (total and by gender), the average OCRA index and OCRA checklist score, and the reported prevalence of the selected effect variable (PA) corresponding to the number of individuals suffering from one or more UL-WMSDs diagnosed for every 100 exposed workers.

The table also indicates similar data for a "reference" group (total number: 749, of whom 310 were male and 349 female) composed of workers who had never been exposed to tasks entailing any risk of upper limb biomechanical overload.

The exposure indicators were assigned *nominally* to this group (i.e., OCRA index = 0.5; checklist score = 1.5) as for almost no exposure.

In analyzing the exposure variables (OCRA index and OCRA checklist score), the OCRA index was often available, but the checklist score was only occasionally

TABLE 4.17
Principal Characteristics of Groups Examined in 2004: Total Composition (Number and Gender), Exposure Indexes (OCRA Index and OCRA Checklist Score), Prevalence of Workers Suffering from One or More UL-WMSDs (PA: Prevalence of Affected Individuals)

Type of Job/Task	Total No.	No. Males	No. Females	OCRA Checklist (score)	OCRA index	% Affected (PA)
Electric motor assembly 1	431	126	305	15.2	4.7	11.4
Electric motor assembly 2	288	173	115	12.0	3.4	8.7
Freezer assembly	374	264	110	11.5	3.2	8.6
Refrigerator assembly A	350	270	80	14.7	4.5	15.4
Refrigerator assembly B	42	32	10	13.0	3.8	14.3
Refrigerator assembly C	31	31	0	14.4	4.3	19.4
Refrigerator assembly D	118	63	55	15.0	4.6	15.3
Refrigerator assembly and cabling	42	22	20	19.4	7.2	31.0
Oven assembly	650	150	500	10.2	2.8	13.2
Shock-absorber assembly	242	159	83	19.5	7.3	24.0
Meat processing (chickens)	943	0	943	20.0	7.7	22.4
Ceramics finishing	22	0	22	24.0	21.0	63.6
Sandpapering vehicle parts	121	55	66	21.0	13.0	17.4
Sandpapering wooden doors	25	0	25	34.0	24.7	72.0
Supermarket cashiers	100	0	100	17.0	7.0	26.0
Packaging pickled food	29	0	29	29.0	21.0	72.4
Upholstering car seats	59	33	26	32.0	41.7	79.7
Pork slaughtering	86	67	19	28.0	23.8	47.7
Tile sorting	46	0	46	30.0	41.0	93.5
Motor assembly 1	467	355	112	10.0	3.4	3.9
Motor assembly 2	53	37	16	12.0	3.9	7.5
Stator assembly	105	42	63	17.0	5.8	13.3
Control group	749	310	439	1.5	0.5	4.4

available; in such cases, the missing score was calculated based on the ratio, validated some time ago (Occhipinti and Colombini, 2004b), between the two parameters according to the following cubic regression equation:

$$\text{Checklist score} = 3.7\left[\text{OCRA}\right] - 0.16\left[\text{OCRA}\right]^2 + 0.0021\left[\text{OCRA}\right]^3$$

For all groups, the study of the effect variable took into account only clinically proven cases (based on patient record + specialist visit + appropriate instrumental exams) of work-related upper limb disorders. Cases diagnosed exclusively on the basis of reported symptoms or other syndromes or pathologies with uncertain disease classification were excluded.

4.2.3 Methods and Criteria for Identifying Limit Values for the OCRA Index and OCRA Checklist Score

For the OCRA index, the critical values required for identifying different risk levels (absent, acceptable, uncertain, present) were calculated by means of the model outlined in Figure 4.21, using

a. The regression function (derived from the database) between the OCRA index and the prevalence of workers affected by one or more UL-WMSDs (PA) and the relevant 95% confidence intervals.
b. PA calculated for the reference group and relevant 95% confidence intervals (given the relative sample size).

The best simple regression function between the OCRA index and PA (% affected subjects) is linear and can be expressed by the following general equation:

$$PA = 2.39(\pm 0.14) \times OCRA$$

This function shows a fairly high association between the two variables (adjusted $R^2 = .92$) and is statistically very significant ($p < .00001$). This regression function assumes that for an exposure index of 0, the prevalence of the analyzed effects is nil, and therefore the best function was sought without a constant. Consideration was also given to the *weight* of the groups in the comparison related to their relative numerical size.

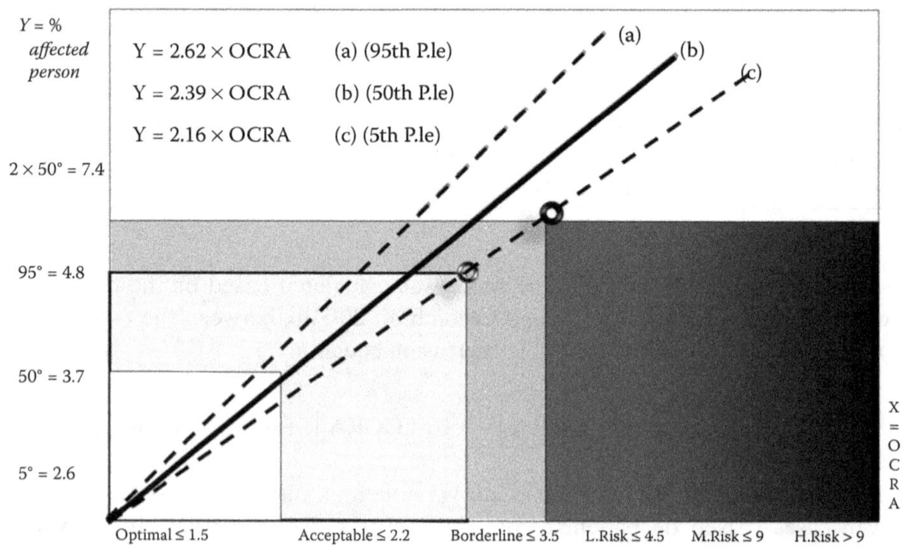

FIGURE 4.21 Model for identifying critical OCRA values based on selected PA (i.e., prevalence of people affected by one or more UL-WMSDs) in the reference group, and using the regression equation OCRA/PA and relevant confidence intervals (90%).

With regard to the specific aspect mentioned in paragraph (b), where the prevalence of the effect in question (PA) is estimated in a generic *nonexposed* population, it seemed useful not to directly use the rough rate emerging from the sample under examination but to calculate a standardized rate, taking into account the structure for four subgroups broken down by age (<35 years; ≥35 years) and gender (males; females) in the Italian population as of 31 December 1999. The resulting standardized PA rate is 3.7%. Using routine statistical methods for estimating the sampling variability of proportions, the 90% confidence intervals were also calculated for the standardized rate thus obtained, taking into account the numerical size of this particular sample. Assuming a normal distribution of the character, these intervals correspond to the 5° and 95° distribution percentile and equate to 2.6% (5° percentile) and 4.8% (95° percentile), respectively.

Going back to the general model, the decision was taken to begin by looking for the critical OCRA values by utilizing the OCRA/PA regression formula (with the relevant confidence intervals) and the data for the various percentiles of the standardized PA rate obtained in the reference group according to the following criteria:

a. *Optimal OCRA index limit* corresponding to the point where the regression function (mid-value) intersects with the mid-value of the standardized PA rate in the reference group.
b. *Still acceptable OCRA index limit* corresponding to the point where the 90% lower confidence interval of the regression equation intersects with the point corresponding to the 95° percentile of standardized rate distribution in the reference group (4.8%). In this scenario, the minimal PA (lower limit of the regression equation at the 5° percentile), based on the corresponding OCRA Index value, is still similar to the maximum PA (95° percentile), hypothetically detectable in a nonexposed population.
c. The lower limit of the *OCRA index values,* which can be "reasonably assumed" to represent overt risk, is situated at the intersection between the regression function (lower limit of the equation at the 5° percentile) and double the PA value in the nonexposed population (7.4%). This key value (double the PA in the nonexposed population) has been chosen as it is utilized in numerous guidelines and international standards. Above this level, the predicted minimum PA value (5° percentile) based on the OCRA index in exposed individuals is very likely to be significantly higher than in the population of nonexposed individuals. The resulting range of OCRA values between the value of point (b) and the value of point (c) describes an *uncertain or borderline area* between acceptability and the presence of overt risk.
d. Additional critical OCRA index values can be found at the intersection between the regression function (mid-value) and points corresponding to *n. times PA* (mid-value) reported in the reference population. In this case, the values of three times and six times were chosen to identify the limits for *medium* and *high* risk.

Thus obtained, the critical OCRA index values were used to reaggregate the data in the database into groups with different exposure levels and to calculate the odds

TABLE 4.18
Odds Ratio (and Relevant 95% Confidence Interval) Considering PA (Prevalence of Individuals Affected by One or More UL-WMSDs), in a Comparison between Groups with Different Exposure Levels according to the OCRA Index and a Low-Exposure Group (OCRA ≤2.2), as Reaggregated in the Database

OCRA Index Class	ODDS Ratio Lower 95% Limit	ODDS Ratio Mid-Value	ODDS Ratio Upper 95% Limit	p-Value
Up to 2.2	#	#	#	#
2.3–3.5	1.45	2.16	3.23	.00006
3.6–4.5	2.38	3.74	5.89	.00000
4.6–9.0	3.63	5.30	7.78	.00000
Over 9.0	15.99	24.31	37.11	.00000

ratios (and relevant confidence intervals) resulting from the comparison between the PA of each reaggregated group and that of the low-exposure group.

The results are shown in Table 4.18. It should be noted that in reaggregating the groups for this specific purpose, the low-exposure group (OCRA ≤ 2.2) is represented only by the reference group.

Table 4.18 is of critical importance for confirming the identification of the critical OCRA values. The table shows that when the new critical OCRA values are applied to the available database, there is a steady increase in the odds ratios as the exposure level moves from low to slight, medium, and high. According to the X^2 tests using the Mantel–Haenszel method for single-layer groups, the comparisons are highly significant. The odds ratios (and relevant confidence intervals) for each range of OCRA values also suggest "overestimates" in the prevalence of affected workers (suffering from one or more UL-WMSDs) that can be expected with respect to those with negligible exposure levels.

So far, the calculations regarding exposure have been largely based on the OCRA index. However, it should be borne in mind that, as already stated, there is an extremely high association between the OCRA index values and the OCRA checklist scores; therefore, it is possible to link the critical OCRA index values with similar OCRA checklist values.

With that in mind, and also considering the results obtained with the 2004 database, a classification was devised that divides risk into three main categories, as per the relevant international standards: *green* = risk absent or acceptable; *yellow* = risk uncertain or very slight; *red* = risk present.

These three main exposure categories may be further divided into six different subcategories for a more "coherent" interpretation of both the OCRA index values and the OCRA checklist score, as specified in Table 4.19.

The table shows each of the six subcategories along with possible actions to be taken following the risk assessment using the OCRA checklist method presented in this and other chapters.

TABLE 4.19
OCRA Index and OCRA Checklist Score, at Incremental Levels of Risk and brief Description of Corrective Actions

Zone	OCRA Index Values	OCRA Checklist Values	Risk Classification	Suggested Actions
Green	Up to 1.5	Up to 5	Optimal	None
Green	1.6–2.2	5.1–7.5	Acceptable	None
Yellow	2.3–3.5	7.6–11	Borderline or very slight	Recheck; if possible improve working conditions
Red-low	3.6–4.5	11.1–14	Slight	Improve working conditions; health surveillance; training (**)
Red-medium	4.6–9.0	14.1–22.5	Medium	Improve working conditions; health surveillance; training (***)
Red—high	More than 9.0	More than 22.5	High	Improve working conditions; health surveillance; training (****)

A number of important points need to be made regarding the use of the classification shown in Table 4.19, especially if applied to the final OCRA checklist score:

- While the classification of exposure to different levels or categories of risk as discussed in this chapter may represent a useful contribution toward providing a framework for assessing risk and guiding preventative actions, if improperly used it could also determine misinterpretations of the results obtained using the OCRA method. This applies especially to the most critical risk indexes.

  For example, although it is arguably true that an OCRA value of 3.5 indicates uncertain (or very slight) risk, and that an OCRA value of 3.6 indicates the presence of risk, it is equally true that there is no substantial difference between these two values; therefore the user must factor in the "continuity" expressed by the OCRA results (also using the predictive models provided) rather than simply dividing the results into *risk* and *no risk*.
- The final OCRA checklist scores should be used as "ranges" rather than hard and fast cut-off values. Since the checklist deals with certain variables in a more simplified way than the OCRA index method, it is advisable to avoid making hard distinctions between values included in the same range.

  For example, if the checklist generates a score of 12.0, it is not possible to make a definite distinction between that figure and a score of 13.0; both values belong to the same risk range (light red; slight risk).

  The problem, of course, remains for results straddling critical values, where the aforementioned considerations are even more important.

  In short, since the OCRA method is now employed in many different scenarios, including for supervisory and medicolegal purposes, the authors wish to stress that the proposed classification system should be regarded largely as a guide for interpreting the assessment analysis, with a view to steering the resulting preventative actions, rather than as an inflexible standard for setting limits to be used indiscriminately for purposes other than those stated herein.

4.2.4 Predictive Models

With regard to identifying models for predicting the prevalence of UL-WMSDs in exposed workers (i.e., predicting health outcomes) based on exposure indicators and using the 2004 database, mention has already been made of the result of the analysis into the best model for linking OCRA index and PA.

The relative regression equation has also been used to identify the critical OCRA index values and, consequently, the critical OCRA checklist scores.

Therefore, the same equation can be used to predict the prevalence of PA—individuals who will be affected by one or more clinically diagnosed UL-WMSDs—based on the OCRA values for the various scenarios in which the method is applied.

In other words, the OCRA values and the straight line regression curve, with its relevant confidence intervals, can be used to suggest the likely prevalence of individuals affected by one or more UL-WMSDs (PA) in a given work environment.

Table 4.20 shows an example in which various precalculated critical OCRA index values are used to predict the expected PA, calculating both the mid-value and the confidence intervals (in this instance 90%) for that particular estimate.

As stated previously, in the case of the OCRA checklist, the final scores should be used as ranges rather than hard and fast cut-off values; therefore, predictions of the possible future prevalence of affected workers (PA), based on the checklist, must be "anchored" to the average predictive values for the respective range (from the lowest to the highest value in the range) as shown in detail in Table 4.21.

The models for predicting health effects on exposed workers (i.e., prevalence of UL-WMSDs) based on the OCRA exposure indexes and OCRA checklist were designed primarily to provide users with estimates of the consequences of risk

TABLE 4.20
Estimate of Expected PA (Mid-Value and 90% Confidence Intervals) in Relation to Key OCRA Index Values Using the Simple Linear Regression Function OCRA/PA

	% Affected (PA)		
OCRA	**5° Percentile Minimum**	**50° Percentile Mid-Value**	**95° Percentile Maximum**
1.0	2.15	2.39	2.60
2.2	4.73	5.26	5.72
3.5	7.52	8.36	9.10
4.5	9.67	10.75	11.70
9.0	19.35	21.51	23.40

TABLE 4.21
Estimated Expected PA Range based on the Final OCRA Checklist Score and in Relation to the Key OCRA Index Values

Area	OCRA Values	Checklist Values	% Affected (PA)
Green	Up to 2.2	Up to 7.59	Up to 5.26
Yellow	2.3–3.5	7.6–11	5.27–8.35
Red-low	3.6–4.5	11.1–14	8.36–10.75
Red-medium	4.6–9.0	14.1–22.5	10.76–21.51
Red-high	More than 9.0	More than 22.5	More than 21.51

assessments capable of leading themselves and their dialog partners (such as company management) to make more realistic priorities for corrective action plans, considering the technical and financial costs associated with prevention as well as the costs associated with nonprevention.

In practice, experience has shown that predictive models are an extremely useful tool for communicating with decision-makers, especially corporate decision-makers, when establishing corrective actions following risk assessments and bearing in mind that predictions are now based on case series of officially recognized occupational diseases in most developed countries.

Moreover, these predictive models are also recommended for simply classifying risk in the ranges of *green*, *yellow*, or *red*.

For example, if a workplace has an index of *purple* (checklist >22.5), which, after corrective action, drops to medium *red* (checklist = 14.5), formally it will still be "at risk," but in actual fact the expected number of workers who are expected to have one (or more) occupational diseases has decreased from approximately 21% to approximately 11%; this is what matters most.

Lastly, from the purely technical, scientific, and statistical standpoint, the forecasting models presented here can only provide estimations of expected UL-WMSDs over a multiyear time frame of 7–8 years on average, given a certain level of exposure to repetitive upper limb movements and exertions. Nonetheless, subject to further explanations that shall follow in this chapter, in terms of risk management procedures, these models are a fundamental tool for occupational health operators and decision-makers and for estimating and measuring health effects over the medium and long term.

4.2.5 Results of Incidence Studies

The results are presented here of a retrospective cohort study published by *La Medicina del Lavoro* (Nicoletti et al., 2008b), involving three large upholstered furniture factories with over 5000 employees (including unskilled workers and administrative staff). The study also assessed the annual incidence of UL-WMSDs in relation to exposure according to the OCRA index method.

The study included all employees on the payroll of the three furniture companies as of 1 January 2000 and new hires up till 31 December 2004. The individuals/

years at risk were calculated starting from 1 January 2000 or as of when they were hired, if after that date, until the first diagnosis of a UL-WMSD or, if none were diagnosed, until the end of the study (31 December 2004 or termination of employment). Overall, 21,484 man/years were calculated, and 493 incident cases were reported.

Upper limb disorder rates were analyzed using Poisson multiple regression models to calculate the ratios (rate ratio [RR]) and the relevant 95% confidence intervals (CI 95%). The same co-variates (for exposure, adjustment, or stratification) were used throughout: gender, age (six categories: <30, 30–34, 35–39, 40–44, 45+), calendar year (five categories, from 2000 to 2004), company (three categories). To describe the degree of overload, the OCRA index was used with six categories: <2.2 (risk absent or acceptable), from 2.2 to 3.5 (uncertain or very slight risk), and from 3.6 to 4.5 (slight risk; in this study, no risk group fell within this interval), from 4.6 to 9.0 (medium risk), from 9.1 to 11.0 (high risk), >11.9 (very high risk, a risk class considered similar to the previous one in other studies, but specified here only due to the large size of the groups of workers exposed to this condition).

The reference group chosen for this study was composed of quality control workers with low exposure (OCRA 2.2–3.5) to biomechanical overload. The various disease rate trends, over time and/or based on the OCRA index in the three factories, were assessed by entering the appropriate interaction terms (categories) into the model. The principal analysis considered the clinical diagnosis of any UL-WMSD.

Annual incidence rates (i.e., number of cases per 1000 workers/year) in any OCRA index exposure class (or with OCRA checklist score) are shown in Figure 4.22.

Table 4.22 shows the ratios between incidence rates (and relative confidence intervals) in the transition between incremental OCRA index categories. As previously stated, reference was made (IRR = 1) to the group in the *yellow* exposure class (OCRA 2.3–3.5).

The relationship between the OCRA index and the incidence of UL-WMSDs was in line with the results of other cross-sectional epidemiological studies, which enabled the OCRA risk categories to be defined, with the exception of the lowest risk class (administrative staff), whose incidence rates were probably lower than

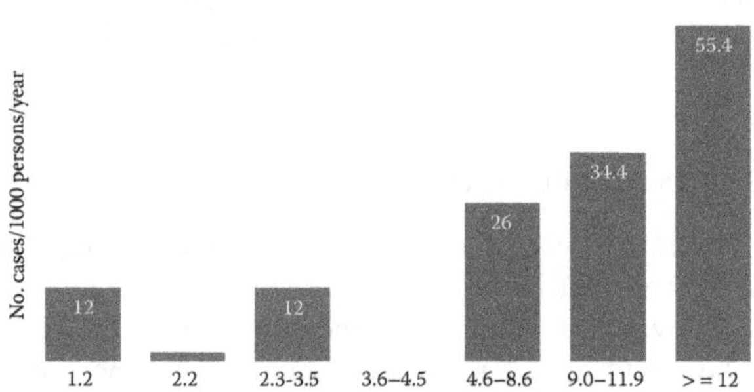

FIGURE 4.22 Annual incidence rate (no. of cases × 1000 people/year) per OCRA Index risk class, according to a cohort study in the upholstered furniture industry.

TABLE 4.22
IRR (Incidence of Relative Risk) for UL-WMSDs: Analysis according to OCRA Risk Class

			Confidence Interval (95%)	
OCRA Index Range	**OCRA Checklist Range**	**IRR Mid-Value**	**Min**	**Max**
<2.2	< 7.6	0.12	0.05	0.30
Between 2.3 and 3.5	Between 7.6 and 11.0	1.00	=	=
Between 4.6 and 8.9	Between 14.1 and 22.5	2.17	1.62	2.91
Between 9.0 and 11.9	Between 22.5 and 25.0	2.90	2.23	3.77
≥12.0	≥25.0	4.58	3.51	5.97

expected. The other risk categories showed incidence rates in line with the aforementioned estimates:

- The transition from *yellow* (uncertain or very slight risk, equal to OCRA index 2.3–3.5) to *red* (medium risk, OCRA index 4.6–9) increases the IRR by 2.2, and the forecasting models based on prevalence studies suggest a similarly higher relative risk (more specifically, from a risk of less than two times that of the control population to a risk of between three and six times that of the control population).
- For the *high risk* class (OCRA index above 9), the models indicate a risk level six times higher than that of the reference population (hence, more than three times higher than the yellow class), and the incidence rates detected in this study confirm a similar increase: IRR of 3 for the class with OCRA index above 9 and below 12 and of 4.6 for the even higher risk class.

Data in Figure 4.22 and Table 4.22 can be used to make further forecasts, in this case of the expected incidence of UL-WMSDs, based on exposure indicators developed using the OCRA method. The authors of this briefly described paper also add that the average incidence rate for the "reference" workers (1.2 cases per 1000 individuals/year) could represent a threshold above which exposure to risk might be significant and therefore deserves an in-depth analysis by the occupational health and safety staff.

4.2.6 Other Forecasting Models for Calculating the Prevalence of UL-WMSDs Based on OCRA Checklist Scores

From 2004, more data was added to the existing database used for calculating critical OCRA index values and OCRA checklist scores; the data was generally acquired using the OCRA checklist method and included matched results for prevalence among workers with one or more clinically diagnosed UL-WMSDs (Nicoletti et al., 2008a; Meroni et al., 2010).

Therefore, while confirming the previous OCRA checklist thresholds (i.e., critical values) in order to better frame and classify risk, the pre- and post-2004 data was combined to define more accurate forecasting models for the expected prevalence of UL-WMSDs based directly on the OCRA checklist scores.

The new database amounted to over 11,000 workers broken down into 30 groups, featuring different exposure levels (including the original reference group). Table 4.23 shows the most significant data in the database used for this purpose.

In Table 4.24, the database shows how the SPSS© software found the best associations (regression models) between the independent variable (for exposure),

TABLE 4.23
Main Features of Groups (Total Workers: 11,774) Included in the Study of the OCRA Checklist Forecasting Models: Breakdown by Gender, OCRA Checklist Score and Prevalence of Workers affected by One or More UL-WMSDS (PA)

Job/Task	Total	No. Males	No. Females	Checklist	% PA
Electric motors assembly 1	431	126	305	15.2	11.37
Electric motors assembly 2	288	173	115	12.0	8.68
Freezer assembly	374	264	110	11.5	8.56
Refrigerator assembly A	350	270	80	14.7	15.43
Refrigerator assembly B	42	32	10	13.0	14.29
Refrigerator assembly C	31	31	0	14.4	19.35
Refrigerator assembly D	118	63	55	15.0	15.25
Refrigerator assembly and cabling	42	22	20	19.4	30.95
Oven assembly	650	150	500	10.2	13.23
Shock-absorber assembly	242	158	83	19.5	23.97
Meat processing (chickens)	943	0	943	20.0	22.38
Assembly motor 1	467	355	112	10.0	3.85
Assembly motor 2	53	37	16	12.0	7.55
Assembly motor 3	105	42	63	17.0	13.33
Upholsterers A	783	783	0	25.0	18.60
Hide cutters A	514	488	26	21.7	8.20
Stitchers A	840	4	836	23.2	11.30
Preparers A	205	196	9	20.6	13.20
Upholsterers B	85	85	0	24.9	20.00
Hide cutters B	54	50	4	20.4	10.00
Stitchers B	143	0	143	24.3	8.40
Preparers B	56	56	0	20.0	7.10
Upholsterers C	76	76	0	23.0	28.90
Hide cutters C	25	24	1	15.2	16.00
Stitchers C	75	1	74	20.9	9.30
Preparers C	33	33	0	17.7	15.20
Nonexposed to repetitive tasks					
First reference group	1383	1306	77	7.4	6.10
VDU 20–30 h	577	329	248	6.2	4.33
VDU > 30 h	1440	792	648	7.4	3.13
Second reference group	749	310	439	1.5	4.41

TABLE 4.24
Linear and Exponential Regression Equations between Checklist and Prevalence of Affected Workers; Relative Association and Statistical Significance

Exposure Indicator (CK)	Effect Indicator (PA)	Function Type	Equation	E.S. (b)	R^2 ADJ	p
Checklist	Prevalence of workers with UL-WMSD	Linear	PA = 0.742 * CK	0.055	.856	.00001
Checklist	Prevalence of workers with UL-WMSD	Exponential	Ln(PA) = 0.138 * CK	0.008	.90	.00001

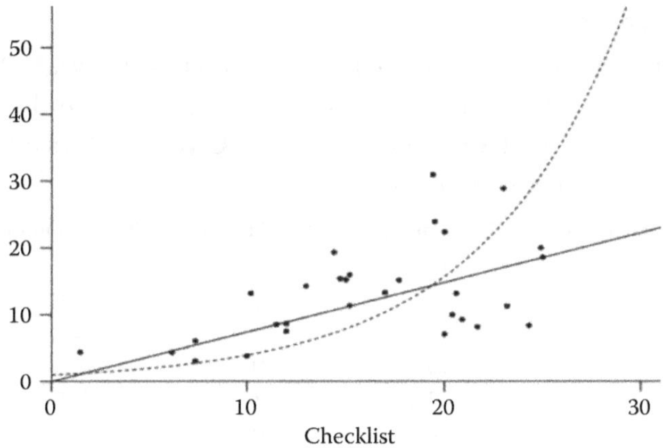

FIGURE 4.23 Scatter plot comparing checklist score data pairs and PA (% of affected workers) and linear and exponential regression curves.

that is, *OCRA checklist score* and the dependent variable *% of workers with UL-WMSDs* (PA).

Without considering the constant in the models (i.e., checklist = 0; PA (%) ≈ 0) and weighting the compared groups based on their numerical size, at least two adequate associations emerged: linear and exponential.

These two models are depicted graphically in Figure 4.23 with respect to the database.

The analytical results of the two regression models are shown in Table 4.24; here we see the regression equations (with relative standard error), the magnitude of the association through R^2 adjusted for random error, and the consequent statistical significance.

For forecasting purposes, both models could also be used for the specific values of the OCRA checklist score; however, if one examines the graph in Figure 4.23, it appears clear that for OCRA checklist scores of up to 20, the "linear" model is closer to the actual data collected, while above that level, the "exponential" model more

TABLE 4.25
Expected Percentage of Workers Affected by UL-WMSDs Based on OCRA Checklist Scores of 20 or Less, According to a Linear Regression Model (Mid-Values and 95% Confidence Interval)

OCRA Checklist Score	PA Minimum	PA Mid-Value	PA Maximum
7.5	4.7	5.6	6.4
11	6.9	8.2	9.4
14	8.8	10.4	12.0
17	10.7	12.6	14.6
20	12.6	14.8	17.1

TABLE 4.26
Expected Percentage of Workers Affected by UL-WMSDs Based on OCRA Checklist Scores of Over 20, According to an Exponential Regression Model (Central Values and 95% Confidence Interval)

Checklist Value	PA Minimum	PA Mid-Value	PA Maximum
20.5	11.9	16.9	23.5
22.5	15.2	22.3	32.0
23	16.2	23.9	34.5
25	20.6	31.5	47.0
26	23.2	36.2	54.8
27	26.2	41.5	63.9
28	29.6	47.7	74.6
29	33.4	54.7	87.0
30	37.7	62.8	~100

accurately depicts the potential growth of the expected prevalence of individuals with UL-WMSDs, at least for OCRA checklist scores of up to 30.

Hence the suggestion that the linear model be used for forecasting purposes and the relative equation (with 95% confidence intervals) with OCRA checklist scores of 20 or less, and that the exponential model equation be used for checklist scores of over 20. For OCRA checklist scores above 30, the expected *affected person prevalence* values should be used resulting from the application, in the exponential model, of the OCRA checklist score of 30.

The linear regression equation shown in Table 4.24 will be considered with its 95% confidence interval; in this case, the upper, middle, and lower levels of the forecasting range are determined by the following equations:

- MINIMUM → PA = 0.629*CK
- CENTRAL → PA = 0.742*CK
- MAXIMUM → PA = 0.856*CK

Table 4.25 shows what happens when these linear equations are applied for selected checklist scores (up to 20).

Similarly, the exponential regression equation, with a 95% confidence interval, will generate the following values:

- MINIMUM → Ln(PA) = 0.121*CK
- CENTRAL → Ln(PA) = 0.138*CK
- MAXIMUM → Ln(PA) = 0.154*CK

Table 4.26 shows what happens when these equations are applied for selected checklist scores (over 20 and below 30).

ANNEX 4.1

The OCRA Checklist

ANNEX 4.1: OCRA CHECKLIST MODEL
Shortened procedure for the identification of upper limb overload in repetitive tasks by the EPM International School of Ergonomics

Page 1

Company	Area
Compiled by	date

Name of workstation/department	
Brief description of task	
–number of identical or very similar workstations	
–number of shifts per day	
–number of workers in identical/very similar workstations/ departments per day	

Organizational data	Description	Actual duration (min.)
Shift duration (for duration consider ACTUAL duration)	Official	
	Actual	
Official pauses	Contractual (hour and number)	
Actual breaks (for duration and numbers consider ACTUAL duration and ACTUAL numbers)	ACTUAL (hour and number)	
Lunch break (for duration consider ACTUAL duration). Consider NET DURATION OF REPETITIVE TASKS only if included in shift duration	Official (hour)	
	Actual	
Nonrepetitive tasks (e.g., cleaning and fetching supplies.) (for duration consider ACTUAL duration)	Official	
	Actual	
No. of units (or cycles) (consider ACTUAL numbers)	Planned	
	Actual	
Net cycle time (s)		
Observed cycle time		

Note

	Duration Multiplier

Type of work interruption (with pauses or other visual inspection tasks)

(max. score allowed = 10).

Choose one answer. It is possible to choose intermediate values.

0	One interruption in the repetitive work lasting at least 8/10 min every hour (also count the meal break) or the recovery period is included in the cycle.
2	Two interruptions in the morning and two in the afternoon (plus meal break), lasting at least 8–10 min per 7–8 h shift, or at least four interruptions per shift (plus meal break), or four 8/10 min interruptions per 6-h shift.
3	Two pauses, lasting at least 8–10 min each, per 6-h shift (without meal break); or three pauses, plus meal break, per 7–8 h shift.
4	Two pauses, plus lunch break, lasting at least 8–10 min each, per 7–8 h shift (or three pauses without lunch break), or one pause lasting at least 8–10 min per 6-h shift.
6	One pause, lasting at least 10 min, per 7-h shift without meal break; or, in an 8-h shift, only a meal break (meal break not counted in the working hours).
10	– No actual pauses except for a few minutes (less than 5) per 7 –8 h shift.

First hour								Last hour
Shift duration in minutes (mark breaks in the shift)								

	Recovery Multiplier

PAGE 2

ARM ACTIVITY AND FREQUENCY WITH WHICH WORK CYCLES ARE PERFORMED

(max. score possible = 10)

Choose one answer for each upper limb. Do NOT use intermediate scores.
If both static and dynamic actions are present, consider both static and dynamic actions.
•For the most representative task, choose the one with the highest risk value.

DYNAMIC TECHNICAL ACTIONS

Score	Description
0	Arm movements are slow; frequent short interruptions are possible (20 actions per minute)
1	Arm movements are not too fast; short interruptions are possible (30 actions per minute)
3	Arm movements are quite fast (about 40), but short interruptions are possible
4	Arm movements are quite fast; only occasional and irregular short pauses are possible (about 40 actions per minute)
6	Arm movements are fast; only occasional and irregular short pauses are possible (about 50 actions per minute)
8	Arm movements are very fast; the lack of interruptions makes it difficult to keep up the pace (about 60 actions per minute)
10	Very high frequency; 70 or more actions per minute; absolutely no interruptions are possible

STATIC TECHNICAL ACTIONS

Score	Description
2.5	An object is held for more than 4 consecutive seconds, with one or more static actions maintained for 2/3 of the cycle (or observation) time
4.5	An object is held for more than 4 consecutive seconds, with one or more static actions maintained for 3/3 of the cycle (or observation) time

	R	L
Number of actions in the cycle		
Frequency of actions per minute		
Possibility of short interruptions		

R	L
FREQUENCY SCORE	

PRESENCE OF ACTIVITIES INVOLVING THE REPEATED USE OF FORCE BY THE HANDS-ARMS (AT LEAST ONCE EVERY FEW CYCLES DURING THE ENTIRE TASK ANALYZED: ☐ YES ☐ NO

More than one score can be marked and sum to obtain the final score. IF YES:

			Score	Duration
THE ACTIVITY REQUIRES THE USE OF ALMOST MAXIMUM FORCE: (8 points or more on the Borg scale)				
	Pulling or pushing levers			
	Pushing buttons		6	2 s every 10 min
	Closing or opening		12	1% of the time
	Pressing or handling components		24	5% of the time
	Using tools		32	Over 10% of the time (*)
	Lifting or handlingobjects			
THE ACTIVITY REQUIRES THE USE OF INTENSE FORCE: (5–6–7 points on the Borg scale)				
	Pulling or pushing levers			
	Pushing buttons		4	2 s every 10 min
	Closing or opening		8	1% of the time
	Pressing or handling components		16	5% of the time
	Using tools		24	Over 10% of the time (*)
	Lifting or handling objects			
THE ACTIVITY REQUIRES THE USE OF MODERATE FORCE: (3–4 points on the Borg scale)				
	Pulling or pushing levers			
	Pushing buttons		2	1/3 of the time
	Closing or opening		4	About half the time
	Pressing or handling components		6	Over half the time
	Using tools		8	Nearly all the time
	Lifting or handling objects			

(*)THIS CONDITION IS ABSOLUTELY UNACCEPTABLE

R	L
FORCE SCORE	

PRESENCE OF ACTIVITIES INVOLVING AWKWARD POSTURES AND MOVEMENTS OF THE UPPER LIMBS DURING THE ENTIRE TASK ANALYZED: ☐ YES ☐ NO More than one score: add the highest of the four posture scores to the stereotype score to obtain the final score.

TYPE OF AWKWARD POSTURE AND MOVEMENT	Duration of awkward posture or movement	Score	RIGHT	LEFT
A: Shoulder—the arms are kept at about shoulder height, without support, (or in other extreme postures)				
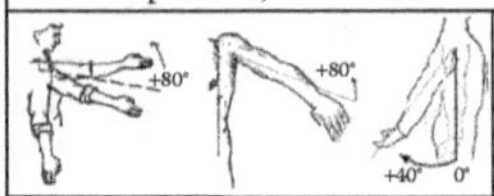	Inf. 1/3 =10%–24% of the time 1/3 = 25%–50% of the time 2/3 = 51%–80% of the time 3/3 = more than 80% of the time	2 6 12 24		
B: Elbow—the elbow executes sudden movements (wide flexion–extension or pronosupination, jerking movements, striking movements)				
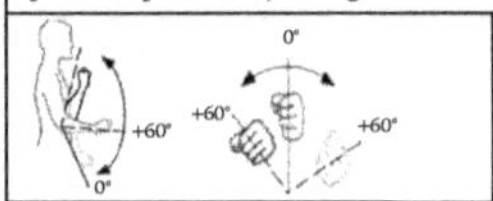	1/3 = 25%–50% of the time 2/3 = 51%–80% of the time 3/3 = more than 80% of the time	2 4 8		
C: Wrist—the wrist is bent in an extreme position, or adopts an awkward posture (such as wide flexion/extension, or wide lateral deviation)				
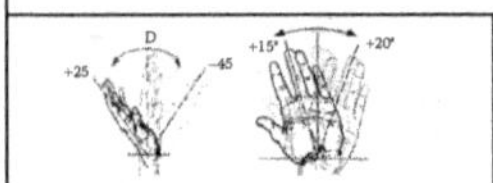	1/3 = 25%–50% of the time 2/3 = 51%–80% of the time 3/3 = more than 80% of the time	2 4 8		
D: Hand—the hand grasps objects or tools in pinch, hook grip, or other kinds of grasp				
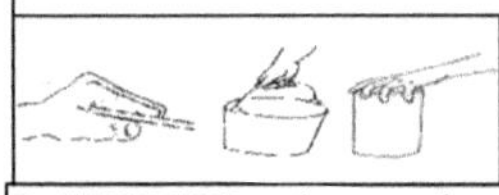	1/3 = 25%–50% of the time 2/3 = 51%–80% of the time 3/3 = more than 80% of the time	2 4 8		
E - STEREOTYP can be assessed at two levels			R	L
High level: a score of 3 is assigned when the cycle time is less than 8s (and, obviously, involves use of the upper limb) or when identical technical actions are performed almost the entire time.				
Intermediate level: a score of 1.5 is assigned when the cycle time is between 8 and 15s or when identical technical actions are performed for 2/3 of the time.				

The overall score for the posture factor is the sum of the highest value calculated for a joint segment plus the stereotypy value, where applicable.

R	L
AWKWARD POSTURE SCORE	

PRESENCE OF ADDITIONAL FACTORS *Choose one answer per section. The final score is the sum of the two partial scores.*

Section A: physical-mechanical factors

Score	Factor
2	Inadequate gloves (uncomfortable, too thick, wrong size) are used more than half the time for the task.
2	Presence of two or more sudden, jerky movements per minute.
2	Presence of at least 10 repeated impacts (use of hands as tools for striking) per hour.
2	Contact with cold surfaces (below 0°C) or tasks performed in cold chambers for more than half the time.
2	Use of vibrating tools at least one-third of the time. **Assign a score of 4** if these tools involve a high degree of vibration (e.g., pneumatic drill).
2	Tools are used that cause compression of muscle and tendon structures (check for the presence of redness, calluses, wounds, etc., on the skin).
2	More than half the time is spent performing precision tasks (tasks on areas of less than 2 or 3 mm), requiring the worker to be physically close to see.
2	More than one additional factor (e.g., ...) is present at the same time for more than half the time.

3	One or more additional factors (e.g., ...) are present almost the entire cycle.
	Section B: organizational factors
1	The work rate is determined by the machine, but 'recovery spaces' exist allowing the rate to be sped up or slowed down.
1.5	The pace is determined by the machine but where the line is moving at a very slow velocity)
2	The work rate is entirely determined by the machine.

R	L
ADDITIONAL FACTORS	

PAGE 4

Multipliers for total duration of repetitve task(s) in shift

60–120 min = 0.5 121–180 min = 0.65 181–240 min = 0.75	241–300 min = 0.85 301–360 min = 0.925 361–420 min = 0.95	421–480 min = 1 Sup 480 min = 1.5

Recovery multiplier

No. of hours without recovery	0	1	2	3	4	5	6	7	8
RECOVERY MULTIPLIER	**1**	**1.050**	**1.120**	**1.200**	**1.330**	**1.480**	**1.700**	**2.000**	**2.500**

EVALUATION OF FINAL CHECKLIST SCORE FOR TASK/ACTIVITY

(frequency + force+posture + additional) × "Net duration of repetitive task multiplier" × recovery multiplier)	**R**	**L**
OCRA checklist final score		

Link between OCRA index and OCRA checklist final score

Color code	*OCRA value*	*Check-list score*	*Risk classification*	*Suggested actions*
Green	**UP TO 1.5**	**UP TO 5**	**Optimal**	**None**
Green	**1.6–2.2**	**5.1–7.5**	**Acceptable**	**None**
Yellow	**2.3–3.5**	**7.6–11**	**Borderline or Very slight**	**Recheck or improve**
Red-low	**3.6–4.5**	**11.1–14**	**Slight**	**Improve + health Surv. +training (2°)**
Red-medium	**4.6–9.0**	**14.1–22.5**	**Medium**	**Improve + health Surv. +training (3°)**
Red high	**More than 9.0**	**More than 22.5**	**High**	**Improve + health Surv. + training (4°)**

ANNEX 4.2
List of Technical Actions

Action	Description	Synonyms	Specifications
Take	The act of grasping an object with the hand or fingers, to perform a task.	Grip	The act of grasping with the right hand and regrasping with the left hand must be counted as individual actions and attributed to the limb that actually performed them. Do not use the term *move the object to the other hand* because it is difficult to determine which limb performed the action.
Position	The act of placing an object or a tool in a predetermined place.	Support, place, straighten, return to the sampling point, as to reposition, replace, etc.	The technical actions TAKE and POSITION are almost always present before any other action that marks the beginning of the technical process.
Retake	The act of taking the object a second time with the same hand.	Regrip	Count each retaking/regripping action as a new technical action. Note: for screws, also refer to the "Rules for sets of actions"
Reach	The act of reaching for an object placed beyond the length of the arm, at a point not reached by walking. The operator moves the trunk and shoulder to reach the object.	Get	The operator moves the trunk (bending and/or tilting and/or rotating) to reach the object. If the object is placed at a suitable distance (less than 42 cm or as per UNI EN ISO 14738, in any direction, i.e., up, down, sideways, etc. …), the REACH action must not be counted, but only the TAKE action.
Replace	The act of replacing a tool, previously used, in its original position (at rest)	Replace, put back	Remember to use the REPLACE action for storing tools after use (even if the position is not specified).
Assist	The act of assisting a suspended tool to return to its original position, when it does not pull back on its own	Support	A typical example is a hanging screwdriver 'or welding gun with a faulty spring' that prevents it from pulling back properly.
Hold	The act of holding an object in the hand, between TAKING and POSITIONING, for a longer than 5 consecutive seconds. This is a static action.	Keep in the hand, keep in prehension	The action should not be counted when the object is held for less than 5 consecutive seconds.

Prop up	The act of propping up an object or the body, using the upper limbs, without taking, for longer than 5 consecutive seconds; this is a static action.	Bear	The action should not be counted if the body or object is propped up for less than 5 s.
Insert	The act of fitting an object into a deep (25 mm), narrow cavity (the space between the walls of the cavity and the object must be less than 5 mm).	Slot in, fit in	When fitting a component into a cavity measuring 25 mm or more in length, the action is INSERT (if less than 25 mm the action is PLACE). If the two components fit together tightly (the object will not fit if inserted upside down) the actions are POSITION + INSERT
Extract	The act of removing an object or a tool from a deep (at least 25 mm), narrow cavity or support (with the space between the walls of the cavity and the object less than 5 mm).	Remove	If an object or tool is extracted or removed from a cavity (or support) measuring less than 25 mm in length, the action is not EXTRACT.
Thread	The act of passing an object through a round hole or ring		This is the action of passing a needle, rope, or cable through rings, or placing a bolt through a washer, etc.
Pull	The act of dragging an object in a specific direction	Tug, yank	Any act of pulling continuously must be counted as a technical action. Every pull and/or change of direction counts as a new technical action. Each pulling action must be counted even if it does not involve the use of force.
Push	The act of moving an object in a forward direction	Thrust, drive, ram	Any act of pushing continuously must be counted as a technical action. Every push and/or change of direction counts as a new technical action. Each pushing action must be counted even if it does not involve the use of force.
Stretch	The act of grasping the ends of a cable (or similar object) with both hands to straighten	Extend, spread	Typically, when untangling a cable with repeated grasping and stretching. Every time the object is regrasped it should count as a single action ("scrolling" is part of the STRETCHING action)

(*Continued*)

ANNEX 4.2 (CONTINUED)
List of Technical Actions

Action	Description	Synonyms	Specifications
Press	The act of applying force to an object using a tool (screwdriver, drill) without causing the object to move	Depress, compress	PRESSING should be counted as a technical action only if the applied force is more than "slight," i.e., least 3 on the BORG scale.
Block	The act of pressing or pushing one limb against the another	Resist, oppose	
Power	The act of operating a machine or tool by pressing a button or moving a lever with parts of the hand, or one or more fingers	Activate, operate	If the action is carried out many times without moving the tool, each individual action is counted. If the action also entails holding the tool (or lever or other tool), also count the TAKE (before the POWER action).
Carry weight	The act of manually transferring a load (using the upper limbs), walking a distance of at least one meter (two steps)		If the load does not have the minimum requirements outlined, no action CARRY WEIGHT action is counted between the two TAKE and POSITION actions.
Drag (both static and dynamic)	The act of pulling or pushing an object (not on castors) along the ground, while walking	Tow	Given that generally this action lasts more than 5 s, it must be counted as a static action.
Move weight	The act of returning after 'REACH' and 'TAKE' actions		If the load does not have the minimum requirements outlined, there is no MOVE WEIGHT action between TAKE and POSITION.
Raise weight	The act of transferring a load from a lower to a higher position, over a vertical distance of at least 50 cm.		If the load does not have the minimum requirements described, the RAISE WEIGHT action will not be counted between the two TAKE and POSITION actions.
Lower weight	The act of transferring a load from a higher to a lower position, over a vertical distance of at least 50 cm.		If the load does not have the minimum requirements described, the LOWER WEIGHT action will not be counted between the two TAKE and POSITION actions.

Rotate	The act of rotating is when an object is grasped and repositioned in a different direction: the change of direction must be more than 90 °, otherwise the action is POSITION.	Revolve	Every change of direction should be counted as one ROTATE action. For tools, consider only TAKE and POSITION: (postural changes).
Screw-unscrew	The act of manually turning a screwdriver or other hand tool to tighten or loosen a threaded component.		Count every complete rotation as a technical action before a new grasp. Remember that the TAKE action almost always precedes screwing/unscrewing (using a screwdriver, tool). Every time the tool is regrasped, a new screwing/unscrewing action should be counted. When the screwdriver is rotated using the tips of fingers, count a SCREW action without any TAKE action.
Turn	The act of manually rotating a bolt, cap, or other threaded object, or rotating an object around its axis.	Rotate the wheel, unscrew a screw-cap	Count every complete rotation as a technical action before a new grasp. Remember that the TAKE action almost always precedes screwing/unscrewing (using a screwdriver, tool). Every time the tool is regrasped, a new screwing/unscrewing action should be counted. When the screwdriver is rotated using the tips of fingers, count a SCREW action without any TAKE action.
Roll	The act of tightening a cable (or similar component) around a post.		Count as a technical action every complete turnaround the post
Open	The act of opening the front of a tool designed to cut or hold the object to be worked. The act of opening the object that rotates on a hinge (e.g., a door).		If the tool does not have an opening spring, count an OPEN action before a CLOSE action (in this case the same as POSITION). Conversely, if the tool has a well-functioning spring, the action should not be counted.
Close	The act of closing the front of a tool designed to grasp the object to be worked, or a rotating door hinge (e.g., a door).		The CLOSE action, in the case of closing a gear to grasp an object to be worked, is the same as the POSITION action. In case of closing of a tool such as scissors, use the term CUT.

(Continued)

ANNEX 4.2 (CONTINUED)
List of Technical Actions

Action	Description	Synonyms	Specifications
Cut with scissors	The act of using sharp blades (scissors, knives, cutters, scalpels, or similar), to divide an object into two parts.		SCISSORS Count each cut as one technical action. If the cutting action only uses the first two thirds of the blade of a pair of scissors, count the action as CUT and not OPEN. After TAKE, count the action as OPEN (if necessary), POSITION (only the first cut-off point), and CUT and continue to count OPEN and CUT for each consecutive cut until finished cutting along the same line. When starting to cut at another point by repositioning the scissors, count another POSITION action. If the cut is performed by holding the scissors open and sliding the blades, after the TAKE action, count OPEN (if necessary), POSITION (only the first cut-off point), then a single cut called 'SPREADING CUT', until the next change of direction or repositioning of the scissors.
Cut with knife	The act of using cutting a blade or knife to divide an object into two parts		KNIFE Any cut (or repositioning of the blade) and any change in the direction of the cut count as one technical action. After the TAKE action, count the CUT action (without counting the blade POSITION action). If the knife is used for boning and the tip is used before cutting, also count a POSITION action.
Rip	The act of dividing an object into two parts using the hands	Rip, tear apart	Count one technical action for every rip.
Hit	The act of striking a spot using a tool or the upper limb to obtain a technical result	Hammer, beat, strike	Count each stroke or blow as a technical action.
Brush	The act of passing a tool (brush, file, sandpaper, cloth, etc.) over a surface	Paint, coat	Count each single "pass" over the surface as a technical action.
Sand		Scrape	Count each "pass" over the surface as a technical action.

Clean		Rub	Count each "pass" over the surface (one circular and/or linear movement) as a technical action.
Mark		Mark	Count each "pass" of the pen (chalk, pencil, pen, etc.) over the surface (one circular movement and/or linear movement) as a technical action.
Write	The act of writing using a specific instrument or tool.	Draw, mark, trace	This should be regarded as a predominantly static action. If the lines are longer than 2 cm, count each change of direction as 1 technical action.
Smooth	The act of holding the hand flat against a surface to level it.	Smooth, spread out	Count each "pass" over the surface (one circular and/or linear movement) as a technical action.
Spread	The act of applying pressure with the fingers to a surface to obtain a technical result.	Press together	The typical action for joining two parts together, or spreading a surface (such as spreading dough, modeling). Each pressing/spreading action using one or more fingers should be counted as a technical action.
Scroll	The act of dragging an object with the fingers over a surface.		Count each time the hand scrolls up to take the object as a technical action.
Throw	The act of imparting a parabolic trajectory to an object, so that it reaches destination.		This action is different from RELEASE, where the object is passively released and falls vertically to destination. RELEASE must not be counted as a technical action.
Shake	The act of moving an object quickly to and fro or up and down to obtain a result (e.g., mixing the contents of a container, etc.).		Count every shake as an action.
Curve	The act of curving an object.		
Bend	The act of bending an object.		
Straighten	The act of changing the shape of an object from deformed to straight.		
Align	The act of arranging an object to obtain a technical result (e.g., locate, extract, embed, etc.). The action is characterized by short and rapid movements.	Adjust, arrange	The action is performed when, after positioning, micromovements are required to position the object properly, or to bring two objects together before joining or splitting. Count every micromovement as a single technical action. If the alignment takes longer than 5 s, count it as a static action.
Embed	The act of enveloping or enclosing one object snugly into another.	Match, fit	This action is often preceded by the ALIGN action.

THE FOLLOWING DO NOT COUNT AS TECHNICAL ACTIONS		
WALK		Without carrying a load
PASS		An object from one hand to the other
Release	A tool or object	The action should not be counted as a technical action if the object or tool is not placed in a specific spot after use, but is "released" by opening the hand or fingers (i.e., by passively returning or dropping it).

5 Variants of the OCRA Checklist

5.1 INTRODUCTION

As stated in Chapter 4, in addition to the *classic* version of the Occupational Repetitive Actions (OCRA) checklist (which uses fractions of time to determine the duration of awkward postures and force used), it is also possible to use the *high-precision* version (which uses actual time in seconds to determine the duration of awkward postures and force used).

However, the checklist now includes other models and procedures that have been developed and will be presented in this chapter:

- The OCRA minichecklist monotask, a model entailing simplified calculations (see Chapter 6 for minicheck multitask)
- The model for calculating exposure to tasks with prolonged cycles

The models and procedures for estimating exposure to daily, weekly, or even annual multiple tasks, including those referring to the OCRA minichecklist, will be described in Chapter 6.

All the relevant software can be downloaded free from www.epmresearch.org.

5.2 ASSESSMENT OF RISK DUE TO BIOMECHANICAL OVERLOAD OF THE UPPER LIMBS USING SIMPLIFIED TOOLS: THE OCRA MINICHECKLIST. CONTENT, APPLICATION, AND VALIDATION

5.2.1 Introduction and Aims

To meet the need for risk assessment tools that even less experienced staff can apply, the aim has been to create a lean and manageable tool for measuring the risk determined by repetitive movements of the upper limbs: the OCRA minichecklist. This tool, which will be described in detail in this chapter, may be useful for conducting a simple assessment of upper limb exposure to the risk of biomechanical overload. The minichecklist is not an alternative to the classic OCRA checklist or index but rather a simplified offshoot of it, which nonetheless allows a reliable evaluation to be made, albeit with a certain margin of error; it is simpler than the classic OCRA checklist and less precise than the OCRA index. A detailed description of the simplifications, as well as the relevant criteria, is shown. This simple tool is also available online as an Excel spreadsheet, ERGOepmMINIcheckOCRAmonotask-EN, which can also be downloaded from the website www.epmresearch.org.

Although simpler, the tool is not too far removed from the classic OCRA checklist assessment, and has proved to be particularly well suited to the following situations:

- Small and very small businesses where the work organization is often less rigid than in larger organizations that produce larger volumes of standard products.
- As a quick tool for checking the validity of a company's risk assessment document or for making a quick preliminary inspection of potentially risky situations that might require a more in-depth assessment and urgent preventive actions.
- As an initial screening tool for a large organization, after completing the pre-mapping exercise, applying key enters, and conducting a quick assessment. The result will be a ballpark estimate but a very speedy one, especially if there is an urgent need to map risk and put an action plan in place.
- In special environments such as hospitals, where work schedules may not be rigid and staff perform multiple tasks.
- The simplified tool could also be useful in agriculture, to carry out a rough but quick preliminary risk map.
- For medical specialists or insurance companies who need to double check risk assessments or investigate patient work histories. This aspect will be further explored in a later section.
- For health and safety representatives, in order to monitor problem situations in the workplace and request the necessary corrective actions.
- In all those cases where a rough risk exposure analysis is better than nothing.

Of course, expert users of the OCRA method will find this tool so simple that no explanations are required. However, newcomers to the field will need some basic training, which the Ergonomics of Posture and Movement (EPM) School organizes periodically.

5.2.2 OCRA Minichecklist for Tasks Featuring Only One Repetitive Task (Monotask)

With regard to biomechanical overload, the pre-mapping questionnaire will already have shown whether or not it is necessary to tackle second-level actions, that is, whether or not the conditions call for a real risk estimate.

There should be one pre-mapping form for each worker or group of workers (homogeneous group) performing the same job or set of tasks in the shift. The term *job* refers to a set of tasks performed during a shift or even over a longer period of time.

Two different types of exposure need to be considered in order to carry out an effective risk assessment:

- The worker or homogeneous group of workers carries out only one type of repetitive task.

- The worker or homogeneous group of workers carries out several repetitive tasks.

This first section offers a stepwise illustration of the approach toward analyzing exposure to a single repetitive task: the *monotask model*.

5.2.2.1 Description of the Company and Identification of the Repetitive Task (Figure 5.1)

In this section, the name of the company is recorded, together with the department and the name of the repetitive task performed by the worker or homogeneous group. The number of workers is also indicated, with a breakdown by gender.

5.2.2.2 Description of the Shift for Calculating Recovery Time, Net Duration of Repetitive Work, and Respective Multipliers

This part of the organizational analysis is similar to the approach adopted by the classic OCRA method, especially with regard to calculating the *net duration of repetitive work*.

The details to be entered into the form, after the duration of the shift, are

- Duration of nonrepetitive tasks (cleaning, fetching, etc.) in minutes.
- Actual number of pauses during the shift lasting at least 8 min (excluding meal break); only "good pauses" must be counted, that is, not within 1 h of the meal break or during the last hour of the shift.
- The actual total duration of pauses (excluding the meal break) in minutes. This duration should include all the pauses.
- The actual duration of the meal break if included in the shift (and thus paid for) in minutes.
- If the meal break is more than 30 min long (outside work time) or there are other interruptions (such as transfers to other premises lasting more than 30 min), these should be counted.

This initial section has been simplified; once this basic organizational data has been entered, the program generates the first three risk-related scores:

- Number of hours without adequate recovery
- *Duration multiplier (correction factor)* for repetitive work
- *Lack of recovery time multiplier (correction factor)*

Company		Area				
Assembly line/workstation/task performed by the worker or a homogeneous group of workers		No. of workers		No. of males		No. of females
Brief description of the task						

FIGURE 5.1 Description of organizational data and identification of the repetitive task performed by a homogeneous group of workers in monotask jobs.

For this last risk factor, as explained in Chapter 4 with regard to the classic OCRA checklist, the number of hours without adequate recovery corresponds to a multiplier to be applied to the sum of the scores deriving from the other risk factors. The procedure for calculating the *recovery* and *duration multipliers* for the OCRA minichecklist is the same as the classic checklist.

For the OCRA minichecklist, it is not necessary to know how many pieces are produced in the shift in order to work out the *total calculated cycle time* compared to the *observed cycle time*; the cycle time is determined by *observational assessments* of the task.

All that needs to be ascertained (see Figure 5.2) is whether

- The repetitive task is composed of actual cycles (if so, a representative duration in seconds must be found).
- The task consists of the same movements for virtually the whole time (e.g., nothing but rasping, or tightening a screw, or gluing).

A brief explanation is also in order of how shifts and pauses are distributed. The rule remains, in any event, that for pauses to be counted as actual recovery time, they must last for at least 8 consecutive minutes and must not occur within 1 h of the meal break or in the last hour of the shift. Otherwise, pauses are not counted as recovery time but only as time to deduct from the duration of the shift in order to obtain the *net duration of repetitive work.* In considering pauses, the breaks actually taken by workers must be assessed (i.e., the pause duration that most accurately reflects the real-life situation), not "official" pauses.

Before evaluating the various risk factors, it is first essential to decide which limb is to be analyzed; the limb used most extensively to perform the task is advisable (in fact, the OCRA minichecklist analyzes only one limb). If both limbs are used equally, the acronym BIL (bilateral) must be indicated.

5.2.2.3 Assessment of Risk Factors: Frequency of Technical Actions

Technical actions are defined as actions that involve the joints, muscles, or tendons of the upper limbs. Such actions should not be confused with individual joint movements (e.g., flexion or extension of the wrist, elbow, or shoulder); they refer to the overall movement that enables a simple task to be performed, for instance, grasping a part, positioning it, turning it, and so on.

The classic way of measuring frequency is to count how many technical actions are performed in a given period of time (Colombini et al., 2002, 2005):

- When repetitive tasks are characterized by cycles, count (estimate) the number of technical actions performed by one limb (left or right) in a cycle (a).
- When identical repetitive tasks are repeated virtually all the time, count (estimate) the number of technical actions performed by one limb (left or right) in a representative period of 1 or 2 min (b).

Overall shift duration (minutes): actual shift duration (minutes)			460	4
				Multiplier. Rec
Duration of nonrepetitive tasks (e.g., cleaning, fetching supplies, etc.) in minutes			10	1.330
No. of actual breaks (recovery periods) during the shift, with a duration of at least 8 min (except meal break) that can be considered as recovery periods			2	
Overall duration of all actual breaks (excluding meal break) in minutes			15	
Actual duration of the meal break if included in shift duration (minutes)			45	
No. of other breaks (i.e., meal break out of working time; travel time to/from different company locations). Mark one number only when these breaks last at least 30 min.				
Description of repetitive work				Multiplier Dur
There are no identified cycles but the same actions are repeated all the time		Net duration of repetitive tasks in the shift (in minutes)	390	0.950
There are identified cycles	Yes	Report duration (in seconds) of representative observation.	28	

FIGURE 5.2 Description of the shift for calculating recovery times, net duration of repetitive work, and respective multipliers.

Once the cycle time (or observation period) has been established, the software calculates the action frequency per minute, where the time is represented by either the *observed cycle time* or the *observed representative time*.

Several examples of the most common technical actions are also provided in Chapter 4.

Figure 5.3 shows the simplified assessment of the frequency factor using the OCRA minichecklist.

It is worth reiterating that these assessments are made by *observing the worker*, and that the observation period based on which the frequency of technical actions is estimated is a cycle or a representative period of 1 or 2 min.

The OCRA minichecklist includes only three scenarios for assessing frequency:

- Scenario A: The frequency of the observed actions is very low: less than one every 2 s (less than 30 actions a minute).
- Scenario B: The frequency of the observed actions is so high (many more than one per second) that it is difficult to calculate them.
- Scenario C: An alternative scenario between scenario A and scenario B, or when a part or tool is held in a prolonged grip (static actions).

These criteria help to speed up the preliminary part of the risk assessment (which is as important as it may often be complex). The two extreme scenarios are selected, which do not require a great deal of experience, and an intermediate score is assigned that is representative of the middle scenario, the one that is most commonly encountered.

Each scenario is preassigned a score (which is not visible in the software) that reflects the criteria of the classic OCRA checklist: The score is 1 for scenario A and 9 for scenario B. The intermediate scenario is assigned a fixed score of 5 (the score generally used for a frequency of 50 actions a minute). This figure was chosen because it is representative of the most common frequencies.

Figure 5.3 (Example 5.1, first section) shows how an intermediate scenario is reported (the chosen scenarios must be marked with an X). The frequency factor score will automatically appear only for the chosen option (in the example, the score is 5).

Alternatively, the technical actions in the cycle (or observed representative period) can be counted directly and the number (for the limb being analyzed) entered into the appropriate box (Figure 5.3, Example 5.2, second section). This will automatically produce the frequency of action and relative score; the result will be more accurate only when describing intermediate frequencies (i.e., between the lowest and the highest).

Table 5.1 shows the scores used by the software corresponding to each frequency reported. Compared to the classic OCRA checklist, the only scores used are those referring to cases in which the worker can modify the pace of the task, thus meeting the need for short interruptions in performing the actions. This is obviously the most common situation among craftsmen and in workplaces where the pace of production is more flexible.

Example 5.1: First section							
Frequency	Calculation of observed frequency of technical actions (number of technical actions per minute)	A: Low frequency (less than one action every 2 s)		Between A and C or holding an object in the hand for most of the time (STATIC)	X	C: Very fast technical actions; unable to count (at least one per second)	
N.B.: To increase the accuracy of the frequency score, indicate in this box the number of technical actions actually counted in the cycle or in a representative period						Frequency per minute	
Frequency score = 5							

Example 5.2: Second section							
Frequency	Calculation of observed frequency of technical actions (number of technical actions per minute)	A: Low frequency (less than one action every 2 s)		Between A and C or holding an object in the hand for most of the time (STATIC)		C: Very fast technical actions; unable to count (at least one per second)	
N.B.: To increase the accuracy of the frequency score, indicate in this box the number of technical actions actually counted in the cycle or in a representative period					28	Frequency per minute	60
Frequency score = 7							

FIGURE 5.3 Determination of action frequency using predefined scenarios (Example 5.1, first section) or by entering the action number of observed technical actions performed by the test limb (Example 5.2, second section).

TABLE 5.1

Scores Used by the Software and Corresponding to Each Frequency, If the Actual Number of Technical Actions Is Entered into the OCRA Minichecklist

Section A: Frequency Factor Score When Brief Interruptions Are Possible

Frequency	Below 22.5	From 22.5 to 27.4	From 27.5 to 32.4	From 32.5 to 37.4	From 37.5 to 42.4	From 42.5 to 47.4
Scores	0.0	0.5	1.0	2.0	3.0	4.0
Frequency	From 47.5 to 52.4	From 52.5 to 57.4	From 57.5 to 62.4	From 62.5 to 67.4	From 67.5 to 72.4	Above 72.4
Scores	5.0	6.0	7.0	8.0	9.0	9.0

5.2.2.4 Assessment of Risk Factors: Awkward Postures

The postures and movements performed by different segments of the upper limb while performing repetitive tasks are crucial for determining the risk of developing musculoskeletal disorders and diseases.

Postures are described and assessed as for frequency (based on cycles or representative periods) and refer to the four main anatomical segments of the upper limb, these being the most prone to work-related overload. Awkward postures must be reported in the following cases:

- *Shoulder*: When the task requires the elbow to reach or exceed shoulder height
- *Elbow*: When the task requires the elbow to be frequently and significantly bent and stretched (flexion/extension) or when parts are rotated frequently (pronosupination of the forearm)
- *Wrist*: When there is significant flexion or extension of the wrist or radial and ulnar deviation
- *Type of grip*: When the hand is actually not in a grip position (such as gripping a handle measuring 1–5 cm) but the fingers are pinching or in a palmar grasp, hook grip, and so on

The classic OCRA checklist defines an awkward posture or movement as one in which the task requires the joint segment to exceed its normal angular excursion by more than 50%. To classify the effort, the scores (one for each of the aforementioned joint segments) increase based on the duration of the awkward position (i.e., one-third, two-thirds, or all of the repetitive work).

Figure 5.4 describes the simplified approach proposed with the OCRA mini-checklist: It also includes pictures illustrating the most common awkward postures and how to recognize them.

For each joint segment and for the durations indicated, an *X* must be entered as the event occurs. For the wrist and elbow, awkward postures must be reported only when the event occurs at least two-thirds of the time (i.e., cycle time or representative period).

Awkward postures of the upper limbs		Less than 1/3 of the time	About 1/3 of the time	About half of the time	About 2/3 of time	Almost all the time	Scores
Hand in pinch	Objects or tools are held in a pinch, palm grip, hook (not in grip) for				X		4
Shoulder	The arms are kept at about shoulder height, without support (or in other extreme postures) for		X				6
Wrist	Extreme wrist deviations for						0
Elbow	Completed object rotation (pronosupination) or significant arm-forearm (elbow) flexion-extension for						0
Stereotype	Duration of cycle (cycle time)	Over 15 s		9–15 s		Equal to or less than 8 s	3
	Same actions/gestures repeated			About 2/3 of the time (more than 50%)	X	Almost all the time	1.5

FIGURE 5.4 Detecting awkward postures and stereotypy for the purposes of calculating the relevant scores.

TABLE 5.2
Scores and Durations Used to Determine the *Awkward Postures* Risk Factor for Each Joint Group

Awkward Upper Limb Postures and Movement Scores	Less than 1/3 of the Time	About 1/3 of the Time	About Half of the Time	About 2/3 of the Time	Almost All the Time
Pinch or palm or hook grip (not power grip)	0	2	3	4	8
Arm more or less at shoulder height	2	6	8	12	24
Extreme wrist deviations	—	—	—	4	8
Complete object rotation (pronosupination) or wide arm-forearm (elbow) flexion–extension	—	—	—	4	8

The stereotypy scores, as for the awkward postures, are the same ones adopted for the classic OCRA checklist. Stereotypy must be reported when the following situations are detected:

- Extremely short cycles of less than 15 s, or even less than 8 s, obviously including actions that involve the upper limbs
- Actions of the same type (identical technical actions performed with the same posture, including static) repeated for over 50% of the cycle time or for most of the cycle

It is worth remembering that there may be stereotypy even without awkward postures; for instance, identical technical actions repeated for a good part of the time, even if entailing "power" grips, still generate stereotypy scores.

Table 5.2 shows the values used for the *awkward postures factor* score (the same as those used in the classic OCRA checklist) that vary depending on the joint group and duration. The scores for stereotypy are the same as those used in the classic OCRA checklist and are shown in Figure 5.4. The final score for awkward postures of the upper limb is automatically generated by the software. This score is based on the worst (highest) score of the four joint segments analyzed plus the stereotypy score, if present.

5.2.2.5 Assessment of Risk Factors: Force (Figure 5.5)

Force represents the biomechanical effort required to perform a certain technical action (or set of actions).

It is always difficult to quantify the level of exertion, or force, in a real workplace. Here are some suggestions that might help to overcome these difficulties:

		Less than 1/3 of time	About 1/3 of time	About half of time	About 2/3 of time	Almost all the time	Scores
Force	The task requires moderate force (borg score 3-4)		X				2
	The task requires intense force using tools or during technical actions	Peak force about 1% of the time		Peak force lasting 2%–9% of the time		Peak force lasting 10% or more of the time	0

FIGURE 5.5 Detecting the risk factor for force for the purposes of calculating the relevant score.

- Interview workers and try to perform the task in question with them.
- Use the Borg scale (10-point category scale for rating of perceived exertion) to describe the muscular force subjectively perceived while performing certain tasks.

The simplified version of the classic OCRA checklist employs the same procedure for acquiring accurate information; workers still have to be interviewed! For an effective interview to detect force levels, the worker has to be asked if the actions performed in the course of a repetitive task (Figure 5.5)

- Require *moderate force* (i.e., more than *slight*, because if only *slight* force is required, the information is negligible and is not counted for the purposes of generating scores) and for how long they are performed with respect to the duration of the cycle (or representative period). For one-third of the time, the score = 2, for half = 4, for two-thirds = 6, and for the whole time = 8.
- Require *intense* or *very intense force*, described as *peak force*; and for how long.

The use of moderate force and peak force may be present in the same action and must therefore be reported simultaneously.

5.2.2.6 Assessment of Additional Risk Factors (Figure 5.6)

Alongside the aforementioned risk factors, the literature highlights various other work-related factors that must be taken into consideration when assessing exposure to risk. Here they are described as additional, not because they are less important but because each of them could be present or absent from time to time in the workplace.

Additional factors may include, but are not limited to

- Using vibrating tools (for even only a part of the actions)
- High-precision work (tolerance of approx.1–2 mm in positioning a part)
- Localized compression of anatomical structures of the hand or forearm by tools, objects, or surrounding structures

Additional	Shocks and countershocks	For over half the time			
Physical factors	Repeated impacts with the hand (the hand is used as a tool)	Frequency: Almost 10 times/hour			
	Vibrating tools	For almost 1/3 of the time			
	Other: Report only those suggested			For over half the time	
Organizational factors	The pace is set by the machine	There are "buffer areas" for slowing down the working pace		The pace is completely determined by the machine (the line is constantly moving)	

FIGURE 5.6 Detecting additional risk factors for the purposes of calculating the relevant score.

- Exposure to cooling or refrigeration
- Use of gloves that hamper the dexterity required by the task
- Slippery surfaces of parts
- Execution of abrupt, tearing, or extremely quick movements
- Execution of actions with shocks and counter shocks (e.g., hammering or striking hard surfaces using the hand like a tool)

These factors must only be reported if they occupy a good part of the time (at least two-thirds of the cycle or representative period) or if their frequency is high (e.g., hammering).

The OCRA minichecklist model lists only the most common additional factors. However, there is space for reporting other additional factors; all the additional factors are listed on a comments page that opens by clicking on "other ADDITIONAL."

The score for an additional factor is 2; if there are several factors, the maximum score that will automatically appear is 3 or 4.

5.2.2.7 OCRA Minichecklist Final Score

The final score of the new OCRA minichecklist and amended OCRA checklist is determined thus:

- The sum of the various risk factor scores: frequency, force, awkward postures, additional.
- The sum total of the scores is then multiplied by the correction factor for recovery and duration.

Table 5.3 shows the final exposure levels according to the OCRA method and their significance; for the OCRA minichecklist, the indexes are the same as those used in the classic OCRA checklist.

A comparison was made between the results obtained using the classic OCRA checklist and the OCRA minichecklist in order to calculate the margin of error.

The comparison included 74 completed OCRA checklists picked randomly from the files of four different experts who were asked to apply the OCRA minichecklist method to their assessments.

The results are depicted graphically in Figure 5.7, showing a high degree of correlation ($p = .000$). The standard deviation was 1.39.

In order to enable this simplified analytical model to more reliably predict exposure risk, the values of the straight lines defining the areas within which points around the regression line are clustered with a 95% confidence interval have been calculated, as illustrated in Figure 5.7.

Thus, taking the value obtained using the OCRA minichecklist, corrected with the formulas shown here, that is, Equations 5.1 and 5.2 for the lower and upper limits, respectively, it is possible to calculate the upper and lower exposure values within which the final risk exposure value is most likely to lie, even using the simplified version of the model.

$$\text{Lower} = 0.31 + (0.93^{*}\ \text{minichecklist score}) \tag{5.1}$$

$$\text{Upper} = 1.8 + (1.07^{*}\ \text{minichecklist score}) \tag{5.2}$$

TABLE 5.3
Classification of Final Values Obtained Using the OCRA Method (Index and Checklist)

Color Code	OCRA Index Value	OCRA Checklist Value	Risk Classification
Green	Up to 1.5	Up to A 5	Optimal
Green	1.6–2.2	5.1–7.5	Acceptable
Yellow	2.3–3.5	7.6–11	Borderline or very slight
Red–Low	3.6–4.5	11.1–14	Slight
Red–Medium	4.6–9.0	14.1–22.5	Medium
Red–High	More than 9.0	More than 22.5	High

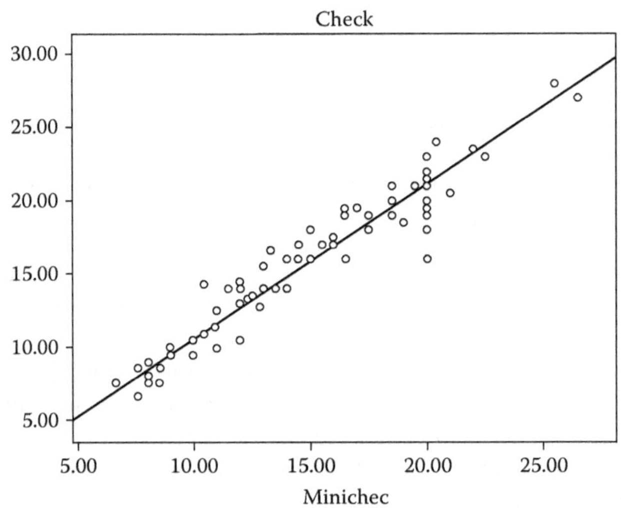

FIGURE 5.7 Scattergram showing the classic OCRA checklist scores compared with the minichecklist in 74 paired analyses.

The following data will be generated automatically in the last section of the OCRA minichecklist model:

- The score determined by each individual risk factor; this is useful for observing the extent to which risk is affected by the two organizational factors *recovery* and *duration of repetitive work.*
- The final indexes: the central value and the highest and lowest values that define the risk area (Figure 5.8).

Final score weighted for recovery and net duration Example 5.1. Using an approximate estimation of the fequency		
Minimum	Central	Maximum
16.73	18.32	21.4
Final score weighted for recovery and net duration Example 5.2. Using precise number of technical actions to calculate fequency		
Minimum	Central	Maximum
19.08	20.85	24.1

FIGURE 5.8 Final scores (Examples 5.1 and 5.2).

EXERCISE 5.1

Open the software (ERGOepmMINIcheckOCRAmonotask-EN) and enter the data listed in the table. Perform two assessments, once using the exact frequency of 28 actions and once placing an X in one of the three boxes describing frequency (for the reader's information, the frequency is 60 actions a minute).

EXERCISE 5.1 OVERALL SHIFT DURATION (min)		**480**
Actual shift duration (min)		460
Duration of nonrepetitive tasks (e.g., cleaning, fetching supplies) in minutes		10
No. of actual breaks (recovery periods) during the shift, break duration of at least 8 min (excluding meal break)		2
Overall duration of all actual breaks (excluding meal break) in minutes		15
Actual duration of meal break if included in shift (min)		45
There are identified cycles: Indicate the observed cycle time (in seconds)		28
Indicate the number of observed technical actions (right and left sides separately)		28 (R) and 28 (L)
Short interruptions are possible (work pace can be regulated)		Yes
Hand in pinch grip	About 2/3 of the time	Right and left
Arm more or less at shoulder height	1/3 of the time	Right and left
Repetition of same actions/ movements	almost all the time	Right and left
Task requires moderate force (Borg score 3–4)	1/3 of the time	Right and left
Pace not determined by the machine		
Check the final result : see figure 5.8		

5.3 CRITERIA AND ANALYTICAL MODELS FOR EVALUATING TASKS WITH A LONG CYCLE TIME

It is not unusual to come across situations where risk must be calculated for repetitive tasks with a long and complex cycle time, that is, tasks lasting several minutes and entailing several different operations. In such cases, counting technical actions and calculating the duration of awkward postures and the use of force may be challenging, and, in the case of longer tasks, downright impossible. However, since risk needs to be assessed even in such situations, a specific software program has been developed, again using Excel: ERGOepmchecklistOCRAlong-recAP-EN. The analytical strategy will be illustrated in this chapter through a few examples of its application.

The definition of a *long* task is not set in stone. Whether or not to use this model depends more on the structural characteristics of the task than its actual duration. Structural characteristics may include

- Definite duration in excess of 2 min.
- "Variable" task components, including several obvious micro-phases (see Chapter 2). Tasks with a long cycle time, in which the same actions are performed the entire time, are in fact very easy to analyze and do not require this approach.

As usual, the task analysis begins with the organizational analysis, to identify the *net duration of the repetitive work* in the shift and the distribution and duration of recovery times. This important initial section is the same for all the OCRA checklist variants (Figure 5.9).

Figure 5.10 shows the second fundamental step, which is the calculation of the *total duration of the cycle* (or pace) estimated by dividing the *net duration of the repetitive task* by the *number of parts handled or worked* in the shift.

The approach is the same one adopted for the classic OCRA checklist. In tasks with a longer cycle time, one of the most complicated steps is counting the number of technical actions, which may be innumerable. In such cases, it is advisable to break down the repetitive work cycle into micro-phases (also called *phases* or *simple groups of elements*) as illustrated in the example in Figure 5.11a. When a micro-phase is homogeneous but very long (i.e., micro-phase 3 = 240 s), it is necessary to identify and indicate a long enough time to carry out the analysis (i.e., 60 s): The software automatically estimates the correct times based on the percentage duration of examination of the subperiod (in this case 25% of the whole micro-phase), as shown in Figure 5.11b.

There are no hard and fast rules about how to break a repetitive work cycle into micro-phases. Different approaches toward task analysis, such as the *MTM* system (motion time measurement) or the *chronometric* method, have suggested the same type of breakdown as a basis for organizing new tasks but using different terminology (*phases* or *elements*). Experts will find these breakdown criteria easy to work with. Less experienced users will use their common sense to break the cycle down into *standard/homogeneous operating times*; either way, the end result will be the same.

Overall shift duration (minutes)	480		Actual shift duration (minutes)	460
Duration of nonrepetitive tasks (e.g., cleaning, fetching supplies, etc.) in minutes				20
No. of actual breaks (recovery periods) during the shift, with a duration of at least 8 min (except meal break) that can be considered as recovery periods				2
Overall duration of all actual breaks (excluding meal break) in minutes				30
Actual duration of meal break if included in shift duration (minutes)				45
No. of other breaks (i.e., meal break not included in working time; travel time to/from different company locations). Enter a number only when these breaks last **at least** 30 min.				
Description of repetitive task				
There are identified cycles: Report the number of units per worker per shift			Net duration of repetitive tasks in shift (in minutes)	365

FIGURE 5.9 Organizational analysis for detecting the net duration of repetitive work and studying the distribution and duration of recovery time.

Repetitive work description			
There are identified cycles: Report the number of units per worker per shift	103.3	Net duration of repetitive tasks in shift (in minutes)	365
There are identified cycles: Report the observed cycle time (in seconds)	205	Average NET duration of repetitive work in shift, excluding any recovery time within the cycle (in minutes)	0
There are identified cycles but the same actions are repeated all the time: Report the time (in seconds) of your representative observation		NET duration of calculated total cycle time (or takt time) (in seconds)	212.0

FIGURE 5.10 Number of parts worked or handled and calculation of total cycle time (or pace).

Phase duration (seconds)	Name of micro phase (or phase or element)	Technical actions (right)	No. of dynamic actions	Duration of static actions (seconds)	Awkward postures and movements (duration in seconds)				Force (level in Borg scale and duration in seconds)		
					Hand (pinch)	ARM (at about shoulder height)	WRIST (flexion/extension. more than 45° or radioulnar deviation)	ELBOW (wide flexion-extension or pronosupination)	Moderate force	Intense force 5–6–7	Peak force 8–9–10
40	1: Place 10 Components	Take, place, turn, fit 10 components	40		30		10	5			
20	2: Place 10 Screws	Take, place, turn, fit 10 screws	20		20		10	2			
22	3: Place fan	Reach, take fan wheel twice, place, rotate five times, push	11			10			5	1	
40	4: Hold fan (while adding component X with left hand)			40			40				
40	4: Place 10 Screws	Take, turn twice, position 10 screws	40		35		10	10			
50	5: Screws	Take screwdriver, place and tighten 10 screws, and replace screwdriver	22			25			25		
212	Total duration of risk factors		133	40	85	35	70	17	30	1	0
				19%	% Static actions						

FIGURE 5.11 (a) Example of a repetitive work cycle broken down into phases.

Right upper limb							Awkward postures and movements (duration in seconds)				Force (level in Borg scale and duration in seconds)		
Phase duration (seconds)	Observed time in phase (indicate only if differenet from phase duration)	% Time observed in the phase	Name of microphase (or phase or element)	Technical actions list (right)	No. of dynamic actions	Duration of static actions (seconds)	Hand (pinch)	ARM (at about shoulder height)	WRIST (flexion/extension. more than 45° or radioulnar deviation)	ELBOW (wide flexion-extension or pronosupination)	Moderate force 3–4	Intense force 5–6–7	Peak force 8–9–10
40		100%	1: Place	Take, place, turn, fit 10 components	40		30		10	5			
20		100%	2: Place 10 components	Take and place 10 screws	20		20		10	2			
240	60	25%	3: Manually tighten screws for 3 min	Take, place, tighten screws	52			10			10	1	
300													
					268	0	50	40	20	7	40	4	0
						0%	% Static actions						

FIGURE 5.11 (CONTINUED) (b) Example of a repetitive work cycle broken down into phases. When a micro-phase is homogeneous but prolonged (i.e., micro-phase 3 = 240 s), it is necessary to identify and indicate a long enough time to carry out the analysis (i.e., 60 s); the software automatically estimates the correct times based on the percentage duration of examination of the subperiod (in this case, 25% of the whole micro-phase).

Every phase is assigned a time, the sum of which must add up to the *total cycle duration* as shown in Figure 5.10 and not the observed cycle time.

Going back to Figure 5.11 (the lower part of which appears in the software), the following procedure is recommended for evaluating frequency, awkward postures, and force.

These three risk factors are now analyzed within each micro-phase, each one being treated as if it were a cycle, according to the following steps (for each micro-phase and each limb):

- Count the *number* of dynamic technical actions.
- Identify the static actions and enter their duration in seconds.
- Assign a time for awkward postures (duration in seconds).
- Assign a duration to each of the three levels of force.

At this point, the software automatically calculates both the sum of the technical actions and the total duration of each of the risk factors shown in the table, which represent the overall objective risk present in the work cycle.

This data, which all together can be described as the *total risk factors*, can now be entered into the first section of the Excel file (ERGOepmchecklistOCRAlong-recAP EN) that is the number of technical actions for calculating frequency (Figure 5.12), the duration of awkward postures (Figure 5.13), the duration of the tasks performed using force, and the level of force (Figure 5.14).

When a micro-phase is homogeneous but prolonged (Figure 5.11b, phase 3 = 240 s), identify and indicate a long enough time to carry out the analysis (60 s). In the phase, indicate only the number of actions or time spent in awkward postures in the representative period; the software automatically estimates the correct times.

The section on additional risk factors is the same as in the classic OCRA checklist (Chapter 4). Figure 5.15 shows the final results.

EXERCISE 5.3.1

Having acquired the data described in the figures shown in Section 5.3, run the software (ERGOepmchecklistOCRAlong-recAP EN), and enter all the data; check the final result.

5.4 RECOVERY PERIODS WITHIN THE CYCLE

Recovery periods within the cycle are as rare as they are overestimated. In short, recovery periods are constant, compulsory, and unchangeable periods of inactivity of the upper limbs during a repetitive work cycle. This section will address the definitions and criteria for accurately estimating recovery periods within the cycle, with examples and automatic calculation models (ERGOepmchecklistOCRAlong-recAP EN). There is a 60-s cycle, 50 s of which are spent performing technical actions featuring repetitive movements of the upper limbs (50 technical actions in 50 s: 60 actions/min), and for 10 consecutive seconds the upper limbs are at rest (e.g., waiting for the machine to process a part).

Dynamic actions						
Report the number of observed technical actions (right and left sides separately)	Right	133	37.6	Left		
If the technical actions are very quick and hard to count (>70 actions/min), place an "X" in the box without counting the actions	Right			Left		
Short interruptions are possible (it is possible to regulate the pace)		No	Yes			
			x			
Static actions		Right		Left		
		No	Yes	No	Yes	
An object is held for at least 5 consecutive seconds, and one or more static actions are performed for 2/3 of the cycle (or observation time)		No				
An object is held for at least 5 consecutive seconds, and one or more static actions are performed for the entire the cycle (or observation time)		No				

FIGURE 5.12 Dynamic actions and calculation of frequency (right limb).

Awkward postures and movements of the upper limb		Seconds	%	Score
	Hand pinch or palm or hook grip (not power grip)	85	40%	3
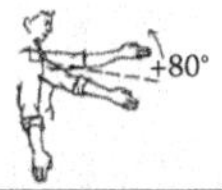	Arm more or less at shoulder height	35	17%	3.5
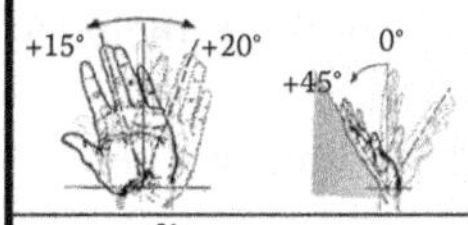	Extreme wrist deviations	70	33%	2.5
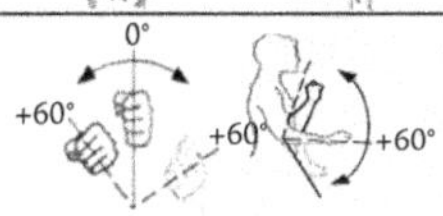	Elbow: Complete rotation of object; (forearm pronosupination) or wide arm-forearm flexion-extension	17	8%	0

FIGURE 5.13 Overall duration of awkward postures (right limb).

		Right (seconds)	%
Force right	The task requires moderate force (Borg score 3-4)	30	14%
	The task requires intense force (Borg score 5-6-7)	1	0.5%
	The task requires almost peak force (Borg score 8 or more)		0%

FIGURE 5.14 Levels of force and duration (right limb).

Task name	Recovery multiplier	Recovery score	Frequency	Force	Side	Shoulder	Elbow	Wrist	Hand	Srereotype	Total posture	Additional	Ocra checklist
Fan assembly	1.330	4	3	4.5	R	3.5	0	2.5	3	0	3.5	0	13.9
Fan assembly	1.330	4	0	L	SX					0	0	0	0.00

FIGURE 5.15 Final results.

One cycle	
Repetitive work	Rest
50 s	10 s

FIGURE 5.16 Example of a *minimum* cycle with recovery period: The recovery time must last at least 10 consecutive seconds and recur constantly and consecutively within a 60-s cycle.

In this case, the ratio of work time to recovery time in the cycle is 5:1. This is an example of the minimum condition that would qualify as a cycle with an included recovery period (Figure 5.16).

Therefore, in summary, there is a recovery period within the cycle only when

- The ratio of repetitive work to recovery time is 5:1.
- The recovery time is more than 10 s.
- Inactive periods within the cycle must be consecutive and dictated by the pace of the machine (i.e., the pauses are compulsory and constant); the worker cannot modify the duration of the recovery period.

To calculate the *lack of recovery time* score, it is important to note that for micropauses within the cycle to be qualified as recovery periods, the breaks must be consecutive and repeated constantly throughout the entire task. Therefore, working hours featuring adequate micropauses within the cycle will be counted as *hours at risk*=0 (recovery multiplier: 1), for the “recovery times” risk factor.

Furthermore, with regard to the *duration multiplier*, it is important to note that consecutive periods within the cycle used as recovery times must be deducted

- From the net duration of repetitive work
- From the total cycle time, which is thus made up only of the *active part of the cycle*

This is basically the same example as the one shown in the Figures in Section 5.3 but with a cycle of 212 s, of which

- 170 are active (83%).
- 35 are consecutive, constant, and compulsory inactive waiting times (17%) that recur throughout the shift.

To determine whether a recovery period time is present (adequate) within the cycle, the ratio of active to inactive times (here, 170:35) must be equal to or less than 6.5%. In the example, the ratio is 4.9; therefore, recovery period is present.

If each cycle provides 36 s of inactivity for the upper limbs, multiplying this amount by the 103 cycles performed in the shift produces approximately 62 min of inactivity during the shift (Figure 5.17).

There are identified cycles: Report the number of units per worker per shift		103	NET DURATION OF REPETITIVE TASK IN THE SHIFT (in minutes)	365
There are identified cycles: Report the observed cycle time (in seconds)		205	Average net duration of repetitive work in the shift excluding recovery time within the cycle (in minutes)	303
There are no idenitified cycles but the same actions are repeated all the time: report the time (in seconds) of your representative observation			CALCULATED NET TOTAL CYCLE TIME DURATION (TAKT TIME) (in seconds)	212.0
There are recovery periods within the cycle (cross if "yes")		Yes	Duration of active time present in the cycle (in seconds): For use as cycle time only when there is a recovery period in the cycle	176
Total duration of active time observed in the cycle (in seconds)	83%	170		
Total duration of passive time observed in the cycle (consecutive and constant)	17.1%	35	% difference between observed and calculated cycle time (accepted limit 5%)	3%
Sum of TOTAL RECOVERY MINUTES WITHIN THE CYCLE during the shift		60		
N.B.: To qualify as recovery time within the cycle, this number must be equal to or less than 6.5		4.9		

FIGURE 5.17 Example of calculation of net repetitive work duration and cycle time when there are adequate recovery periods within the cycle.

This amount must be deducted from the *net duration of repetitive work*, calculated by deducting breaks/pauses and/or nonrepetitive work from the duration of the shift. In the example, the *net duration of repetitive work* is now 303 min rather than 365, with a consequent reduction of the duration multiplier.

The cycle time used for calculating the various subsequent factors will thus be

- For *frequency*, the *active* cycle time
- For *use of force*, *awkward postures*, *and additional factors* (physical/mechanical), the *total cycle time* (active + passive)

6 Assessing Task Rotation

6.1 INTRODUCTION AND GENERAL STRUCTURE OF A MULTITASK ANALYSIS

Task rotation is when a worker alternates between two or more tasks during a certain period of time; this situation occurs quite often in modern work organizations and, if properly designed, can represent one of the most effective strategies for reducing the risk of biomechanical overload.

If the aim is preventive, it is essential to conduct a risk assessment analysis according to the specific procedures and methods described in this chapter.

In special situations, such as when the worker has to perform a large number of tasks and the tasks are distributed "asymmetrically" throughout the shift, risk assessments can become extremely complex. This is why it is necessary to carry out a thorough preliminary study of how the work is organized.

At any rate, the risk analysis process involves a number of steps, listed here, which are described in greater detail in later chapters (Table 6.1).

The first step consists in defining the time required to complete the task rotation schedule; this is the *cycle time*, which may be:

- Daily
- Weekly
- Monthly
- Yearly

The risk analysis process involves the following steps, which are described in further detail, with examples (Table 6.1).

Step 2 involves identifying tasks (as described in Chapter 2, which focuses entirely on work organization analyses). Certain types of jobs (e.g., cleaning, cooking, and farming), which will be explored in a specific chapter, may involve an extremely large number of tasks, even more than 100!

Therefore, each task must be analyzed using the Occupational Repetitive Actions (OCRA) checklist (Step 3) as if it was the only task performed throughout the entire period; this entails calculating *intrinsic* risk indexes.

Once that information has been acquired, the proper risk exposure analysis can be started (Step 4), which involves identifying one or more homogeneous groups or workers who are exposed to the same risks and therefore perform the same tasks with the same duration of exposure (i.e., daily exposure duration and possible duration of tasks in the various cycle times).

Next comes the application of computational mathematical models and lastly the interpretation of the final results.

TABLE 6.1
Procedures for Stepwise Multitask Exposure Risk Analysis

Step 1—Determine Cycle Time
Determine how long it takes to complete a full task rotation: daily, weekly, monthly, yearly
Step 2—Identify Tasks
Analyze the work organization to identify repetitive tasks performed in the period
Step 3—Calculate Intrinsic Risk Indexes of Repetitive Tasks Detected in the Period
Use the OCRA checklist to calculate the intrinsic risk of repetitive tasks, analyzing each task as if it were performed for the entire shift
Step 4—Analyze Risk with Respect to Real-Time Exposure
4.1 Identify a homogeneous group; collect personal data
4.2 Describe duration of real exposure of the homogeneous group: Typical work day, current allocation of tasks in the period(s), proportional duration per day or representative period (week/month/year), also weighted in relation to time constants
4.3 Apply computational mathematical models based on exposure constants
4.4 Interpret results

Since the calculations are highly complex, Excel spreadsheets have been developed, which will be illustrated in later chapters, with examples.

6.2 MULTITASK EXPOSURE ANALYSIS WITH DAILY TASK ROTATION

6.2.1 Introduction

The OCRA method for assessing risk associated with repetitive movements of the upper limbs consists of two tools, the OCRA checklist and the OCRA index. The tools feature different analytical details and purposes, although both are inspired by the same conceptual model (Colombini et al., 2002, 2005; Occhipinti and Colombini, 2004a).

The OCRA checklist is the simpler of the two tools and contains fewer analytical details; it is used for the initial screening of workstations and repetitive manual tasks, and the end result is an estimation of risk exposure.

The OCRA index was chosen as the preferred risk assessment method by two international standards relating to high-frequency repetitive manual work (ISO, 2007b; CEN, 2007); it is a more complex tool that goes into greater analytical detail. It should be used when a more thorough assessment of existing repetitive tasks is needed or for designing new ergonomically sound workstations entailing manual tasks. The latest versions of both tools, as reported in the references section and cited by international standards, include specific computational procedures for jobs

TABLE 6.2
Procedures for Daily Stepwise Multitask Exposure Risk Analysis

Identify Tasks

Analyze the work organization to identify repetitive tasks performed in the period:

- *At different workstations*
- *At the same workstation*

Calculate Intrinsic Risk Indexes of Repetitive Tasks in the Period

Use the OCRA checklist to calculate the intrinsic risk of repetitive tasks, analyzing each task as if it were performed for the entire shift

Analyze Risk with Respect to Real Time Exposure

Analysis of the type of task rotation:

- *Task rotation takes place within a period of less than 90 consecutive minutes for each task performed*
- *Task rotation takes place within a period of more than 90 consecutive minutes for each task performed*

Application of computational mathematical models based on exposure constants

- *Time-weighted average for duration (first case)*
- *Multitask complex (second case)*

where multiple repetitive tasks are performed on a rotating basis by the same group of workers (multitask analysis).

Very briefly, to calculate the OCRA index for multiple repetitive tasks in a shift, a traditional procedure has already been proposed that is based on the total number of technical actions actually performed by a single limb (ATA) during all the repetitive tasks present in the shift, and the total number of all technical actions recommended to be performed (RTA) for each task (RPAi) weighted by the multipliers for recovery periods and the total daily duration of the repetitive tasks.

Going back to the OCRA checklist, if it is necessary to estimate the exposure index for a worker or a homogeneous group of workers exposed to several tasks in a daily cycle, the procedures indicated in Table 6.2 (Occhipinti et al., 2009) must be followed.

Essentially, there are two main ways of organizing worker turnover and consequently two different calculation methods:

1. Task rotation takes place within a period of less than 90 consecutive minutes for each task performed: The *weighted average* mathematical model is used.
2. Task rotation takes place within a period of more than 90 consecutive minutes for each task performed: The *multitask complex* mathematical model is used.

Examples of how to apply these two mathematical models are provided.

6.2.2 Identification of Tasks and Turnover Duration, Criteria, and General Definitions for Calculating the Intrinsic Risk Indexes of Each Task Present in Multitask Exposure

The first step is to identify the repetitive tasks performed by each worker exposed to several tasks. Whether the tasks are carried out at multiple workstations or (at different times) at the same workstation, the calculation remains the same.

The following organizational aspects must first be estimated for each task:

- The *net duration of each task in the shift* (net duration of repetitive work in the shift for each task)
- The duration of all repetitive work in each task is added up to obtain the *total duration of repetitive work in the shift*
- The proportional *duration* or *time fraction* (FT) *of each task* is obtained with respect to the total duration of the repetitive work in the shift (FT_A, FT_B,.., FT_N)

Therefore, the following values are calculated for each separate task:

- IRi = *intrinsic risk index*: The risk index for each task, as if each task
  - Lasted for an entire 480-min shift (*duration multiplier* = 1)
  - Included two 10-min breaks plus a meal break (lack of recovery time multiplier = 1.33)
- IRir = *intrinsic risk index with actual recovery* for each task, calculated using the *lack of recovery time multiplier* actually present in the task
- IRic = *adjusted intrinsic risk indexes*, that is, calculated using both the actual *lack of recovery time multiplier* (of the shift in which tasks are being analyzed), and the *total duration multiplier* of the repetitive tasks in the shift (Dm_{tot})

6.2.3 Calculating the Time-Weighted Average: IR MultiMP

This approach and calculation model is only suitable when the task rotation rate is fairly high, for instance once every 90 min or less. In such cases, it can be assumed that *higher risk exposure* is somewhat offset by *lower risk exposure*, with the worker alternating between the two within a relatively short time frame. Accordingly, rotating tasks serves to reduce risk proportionally with respect to the risk level and duration of each task identified in the turnover.

The mathematical model used to calculate the final exposure index in work featuring several repetitive tasks involves weighting the final risk indexes of the individual checklist scores for the different tasks under examination based on

- The total duration of repetitive tasks in the shift.
- Their corresponding specific duration in the shift (expressed in time fractions).

Duration of repetitive tasks (minutes)	60–120	121–180	181–240	241–300	301–360	361–420	421–480	>480
Duration multipliers (Dm)	*0.5*	*0.65*	*0.75*	*0.85*	*0.925*	*0.95*	*1*	*1.5*
Duration of repetitive tasks (minutes)	30–59	15–29	7.5–14	3.75–7.4	1.87–3.74	Inf 1.87		
Duration multipliers (Dm)	*0.35*	*0.2*	*0.1*	*0.05*	*0.018*	*0.007*		

FIGURE 6.1 OCRA checklist duration multipliers as a function of the duration of repetitive tasks in the shift.

In other words, to obtain the IR risk score for a worker performing a job entailing repetitive tasks at several workstations using the multitask analysis with time-weighted average method (IR MultiMP), the following formula must be applied:

$$\text{IR MultiMP} = \left[\left(\text{IRir}_A \times \text{FT}_A\right) + \left(\text{IRir}_B \times \text{FT}_B\right) + \cdots + \left(\text{IRir}_N \times \text{FT}_N\right)\right] * \text{Dm}_{\text{tot}} \tag{6.1a}$$

where:

$\text{IRir}_{A,B,..,}$ N are the intrinsic scores with recovery obtained from the intrinsic OCRA checklists for the various workstations that the worker uses, calculated using the recovery multiplier corresponding to the actual distribution and duration of recovery times in the shift (the duration multiplier is the same = 1)

FT_A, FT_B,.., FT_N represent the duration in time fractions of the various repetitive tasks versus the total duration of repetitive work

Dm_{tot} = *total duration multiplier*, relative to the net duration of all repetitive tasks in the shift (Section 4.1 and Figure 6.1)

Alternatively, the following formula can be applied to obtain the same result:

$$\text{IR MultiMP} = \left[\left(\text{IRic}_A \times \text{FT}_A\right) + \left(\text{IRic}_B \times \text{FT}_B\right) + \cdots + \left(\text{IRic}_N \times \text{FT}_N\right)\right]$$
$$= \sum \text{IRic}_J * \text{FT}_j \tag{6.1b}$$

where:

$\text{IRic}_{A,B,..,}$ N are the *intrinsic scores* obtained using the checklist for the various workstations used by the worker, adjusted for the real duration of the recovery times and total duration of repetitive tasks in the shift Dm_{tot}

FT_A, FT_B,.., FT_N represent the duration in time fractions of the various repetitive tasks versus the total duration of repetitive work

The approach illustrated briefly here for the OCRA checklist provides results that can be defined as *time-weighted averages* (IR MultiMP).

6.2.4 Application of the Multitask Complex Mathematical Model: IR MultiComp

Conversely, if the repetitive task rotation occurs more than once every 90 min, the time-weighted average approach could underestimate the actual exposure level (by flattening the exposure peaks).

This problem is particularly acute in the study of certain jobs where tasks featuring high intrinsic risk indexes are alternated with lighter tasks.

In such cases, it is more realistic to adopt an approach based on the *task generating the highest overload as minimum.* With this approach, the result will be at the least equivalent to the OCRA indicator for the most overloading task in terms of its duration, and at the most equal to the OCRA indicator for the same task applied, however, (only theoretically) to the overall duration of all the repetitive tasks examined. A special procedure can be used to estimate the actual indicator resulting within the hypothetical upper and lower values of the range.

The indicators calculated using this procedure are defined as the multitask complex risk index (IR MultiComp).

The IR MultiComp computation methods also take into account the duration multipliers that adjust the exposure level as a function of the total time spent performing repetitive tasks within a routine work shift. Figure 6.1 shows the duration multipliers to use as a function of both the overall duration (in minutes) of the repetitive work (the sum of the duration of each of the repetitive tasks present in the shift and included in the rotation) and of the continuous intrinsic durations of each task.

In this case, the procedure is based on the following formula:

$$\text{IR MultiComp} = \text{IRic}_{1(\text{Dm}1)} + (\Delta\text{IRic}_1 \times \text{K}) \tag{6.2}$$

$$\text{K} = \frac{(\text{IRic}_{1(\text{Dm max})} * \text{FT}_1) + (\text{IRic}_{2\,(\text{Dm max})} * \text{FT}_2) + \cdots + (\text{IRic}_{\text{N}(\text{Dm max})} * \text{FT}_\text{N}}{\text{IRic}_{1(\text{Dm max})}} \tag{6.3}$$

where:

- $1,2,3,J,\ldots,N$ = repetitive tasks listed according to the IR of the OCRA checklist (task 1 = the task with the highest IR; task N = the task with the lowest IR)
- Dm_j = duration multiplier according to the actual duration of each task $_j$ in the shift
- Dm_{tot} = duration multiplier for the total duration of all repetitive tasks in the shift
- $\text{IRic}_{1(\text{Dm}1)}$ = the IRi of each task, calculated with the lack of recovery time multiplier actually present in the tasks and considering Dm_1 (it is worth remembering that task_1 is the highest risk task) $\text{IRic}_{1\,(\text{Dm max})}$ = the IR of task_1 (the highest risk task) considering Dm_{tot}
- $\Delta\,\text{IRic}_1$ = $\text{IRic}_{1\,(\text{Dm max})} - \text{IRic}_{1(\text{Dm}1)}$

FT_j = time fraction (between 0 and 1) of each task with respect to the total repetitive work time in the shift

In practice, in order to calculate IR MultiComp, the following steps are necessary:

- Calculate a traditional IR for each task performed by the worker in the shift, considering its intrinsic continuous (real) duration, using the relative Dm_j as the duration multiplier. Repeat the calculation with the recovery multiplier for the entire shift. Select the worst task, which will correspond to $IRir_{1(Dm1)}$.
- Calculate the same IR for each task keeping all the parameters identical except for the duration multiplier, which in this case will be considered in relation to the total duration of all the repetitive tasks in the shift (Dm_{tot}). Thus, the respective $IRic_{J(Dm\ max)}$ will be obtained for each task. Find the highest value (i.e., the risk index for the worst task): $IRic_{1(Dm\ max)}$.
- Calculate $\Delta IRic_1$ for task 1 (the highest risk task): $IRic_{1\ (Dm\ max)} - IRic_{1\ (Dm1)}$.
- Calculate the time fraction of tasks 1, 2, 3, and so on. (FT_j), dividing their respective duration (in minutes) by the total repetitive work time in the shift.
- Calculate K using the formula [3]; in practice:
    - *Multiply each* $IRic_{j(Dm\ max)}$ *by its respective duration* FT_J *and add up the resulting values.*
    - *Divide this amount by* $IRic_{1(Dm\ max)}$.

K will be within the range of 0 and 1.

6.2.5 Examples of Calculations Using Both of the Mathematical Models Presented Here

Take three repetitive workstations and their relative intrinsic checklist IR (for the whole shift); consider one worker rotation in the shift with the corresponding durations:

- Intrinsic checklist values (as if each task lasted for an entire 480 min shift, with duration multiplier = 1):
    - Task A = IRi checklist = 25; duration in the shift = 100 min
    - Task B = IRi checklist = 13.5; duration in the shift = 140 min
    - Task C = IRi checklist = 8.5; duration in the shift = 160 min
- Intrinsic checklist values for real duration (as if each task lasted for the real 400 min in the shift, with duration multiplier = 0.95 (Dm_{tot}) and with the real recovery multiplier)
    - Task A = IRir checklist = 23.7; duration in the shift = 100 min
    - Task B = IRir checklist = 12.83; duration in the shift = 140 min
    - Task C = IRir checklist = 8.08; duration in the shift = 160 min

The total time assigned to repetitive work is 400 min (Dm_{tot} = 0.95) and the various time fractions for these minutes are

- Task A: $FT_A = 25\%$ (0.25)
- Task B: $FT_B = 35\%$ (0.35)
- Task C: $FT_C = 40\%$ (0.4)

The two rotation schemes alternating three tasks shown in Figure 6.2 should also be considered, where the first (Scheme A) features a task rotation duration of more than 90 min and the second (Scheme B) a task rotation duration of less than 90 min.

Example—Scheme A: Calculation of exposure risk for task distribution in the shift with task rotations of more than 90 min.

If the rotations are less frequent (for instance, if the tasks are performed consecutively with each lasting more than 90 min) then IR MultiComp will be calculated using the relevant formula [2].

For this calculation, refer to the summary data for applying the final formula shown in Table 6.3.

Based on the data shown in Table 6.3 and performing the necessary calculation, the result will be

$$IRic_{1\,(Dm1)} = IR\left(\text{risk index}\right)\text{for task}_1 \text{ considering } Dm_1$$

$$\left(\text{task}_1 \text{ being the highest risk task}\right) = 12.5$$

Scheme A: Shift with task rotation duration of more than 90 min							
A	A	B	B		C	C	C
Scheme B: Shift with task rotation duration of less than 90 min							
A	C	B	C		B	C	A

FIGURE 6.2 Example of two different rotation schemes alternating three tasks in the shift.

TABLE 6.3
Summary Data for Calculating IR MultiComp as Required by the Task Distribution Shown in Scheme A of Figure 6.2

	Task A	Task B	Task C
$IRic_{j\,(Dmj)}$ (considering Dm_j)	12.5	8.8	5.5
$IRic_{j\,(Dm\,max)}$ (considering Dm_{tot})	23.8	12.8	8.1
Dm_i	0.5	0.65	0.65
FT_i	0.25 (25%)	0.35 (35%)	0.4 (40%)

$$\text{IRic}_{1\ (\text{Dm max})} = \text{IR for task}_1 \text{ considering } \text{Dm}_{\text{tot}} = 23.8$$

$$\text{IRic}_1 = (23.8 - 12.5) = 11.3$$

$$\text{K} = \left[(23.8 * 0.25) + (12.8 * 0.35) + (8.1 * 0.4)\right] / 23.8 = 0.57$$

$$\text{IR MultiCom} = 12.5 + (11.3 * 0.57) = 18,9$$

Example—Scheme B: Calculation of exposure risk for task distribution in the shift with task rotations of less than 90 min.

Applying the IR MultiMP and taking into account the total duration of repetitive work ($\text{Dm}_{\text{tot}}=0.95$), the formula [1] results in

$$\text{IR MultiMP} = \left[(25 * 0.25) + (13.5 * 0.35) + (8.5 * 0.4)\right] * 0.95 = 13.01$$

or

$$\text{IR MultiMP} = \left[(25 * 0.95 * 0.25) + (13.5 * 0.95 * 0.35) + (8.5 * 0.95 * 0.4)\right] = 13.01$$

This value represents the exposure level for a worker alternating between three workstations based on the durations indicated but with rotation frequencies of less than 90 min.

6.2.6 Discussion and Conclusions

When using the OCRA index and checklist to assess exposure to several repetitive manual tasks featuring potential biomechanical overload of the upper limbs, the traditional approach has been to refer to calculation models based on the time-weighted average concept. However, this approach is unsuitable for certain applications, if not actually misleading, such as when there is high continuous exposure for approximately half the shift followed by slight exposure for the rest of the shift. In such cases, the weighted average does not reflect the continuous peak exposure of the first half.

This emerges clearly from the example in which the value obtained by applying the weighted average (IR = 13.01) is significantly lower than the value obtained using the multitask complex model (IR = 18.9).

Based on multiple task load-lifting analyses already reported in the literature (Waters et al., 2007) and tested in practice, calculation models of both the OCRA index and OCRA checklist have been borrowed and tested for analyzing multiple repetitive tasks deriving from the *task generating the highest overload as minimum* concept (Occhipinti et al., 2009).

Various applications have been examined, leading to the assumption that the calculation model based on the weighted average value is still valid even if the task rotations are fairly frequent (at least every 90 min).

Conversely, if exposure to repetitive tasks is based on less frequent rotations (i.e., more than 90 min) the IR MultiComp calculation method should be used.

This also translates into an invitation to rotate tasks (requiring different levels of biomechanical effort) more frequently as a preventative strategy for minimizing exposure risk.

Since it is clearly difficult to apply the aforementioned mathematical models manually, an Excel spreadsheet tool has been developed (ERGOepmCHECKlist OCRAmultiDAY-EN), which can be downloaded free of charge from the www.epmresearch.org website.

SOFTWARE EXERCISES

Readers are urged to carry out a few exercises on multitask exposure risk analysis using the aforesaid software.

EXERCISE 6.1

Example 6.2.1—Step 1 (Sheet 1. ORGANIZATIONAL DATA): Open the software tool and enter the following data concerning the company.

Example 6.2.1—Step 2 (Sheet 1. ORGANIZATIONAL DATA): Enter the duration of the shift, distribution of breaks, and duration of nonrepetitive work in order to obtain the net duration of total repetitive work in the shift, regardless of whether the worker (or homogeneous group of workers) performs more than one task.

Example 6.2.1—Step 3 (Sheet 1. ORGANIZATIONAL DATA): Enter the main data concerning the identified tasks. It should be recalled that in this exercise, all five tasks are performed at the same workstation. Since the net duration of repetitive

EXERCISE 6.1: EXAMPLE 6.2.1—STEP 1 (SHEET 1: ORGANIZATIONAL DATA)

Company	XXXXXX	Area	YYYYYYYYYYY			
*Name of task being performed by homogeneous group	Workplace “Z” with more models to assembly	No. of workers	2	M	2	F
*Short description of working hours, breaks, and so on	There may be several models at the same workstation involving various tasks with different exposure levels: five tasks are rotated as described					
*Presence of repetitive tasks: the checklist should be applied when the task is organized in cycles, regardless of their duration, or when the task is characterized by the repetition of the same working gestures					Yes	X
					No	

EXERCISE 6.1: EXAMPLE 6.2.1—STEP 2 (SHEET 1: ORGANIZATIONAL DATA)

Summary of Net Duration of Repetitive Work on a Representative Day

Shift Duration (min)	480
*Duration of nonrepetitive tasks (e.g., cleaning and fetching supplies) in minutes	40
*No. of actual breaks (recovery periods) during the shift, with a duration of at least 8 min (except meal break) that can be considered as recovery periods	3
*Overall duration of all actual breaks (excluding meal break) in minutes	30
*Actual duration of meal break if included in shift duration (minutes)	50
*No. of other breaks (i.e., meal break out of working time; travel time to/from different company locations). Mark one number only when these breaks last at least 30 min.	
Net Duration of Repetitive Tasks in the Shift (in minutes)	360

	Daily	Weekly	Monthly	Yearly
*Type of Rotation of Repetitive Work	x			

EXERCISE 6.1: EXAMPLE 6.2.1—STEP 3 (SHEET 1: ORGANIZATIONAL DATA)

Name of the Repetitive Tasks Performed by a Homogeneous Group	There Are No Real Cycles but the Same Actions Are Always Repeated	There are Real Cycles	Cycle Time (in seconds) (Time per Piece)	Duration in Minutes of Each Repetitive Task in the Shift	% Duration of Repetitive Tasks in the Period Considered	Brief Description of the Task
D		X	30	50	14%	
E		X	30	60	17%	
F		X	30	30	8%	
G		X	30	200	56%	
H		X	30	20	6%	

work for the five tasks is 360 min, these minutes must be divided by the five tasks. The software tool automatically generates the corresponding duration ratios.

It is possible to specify the type of cycle and duration of each task, even if the data is merely descriptive and will not be used in subsequent calculations.

Example 6.2.1—Step 4 (Sheet 2. Exposure calculation): For each task in the study, the software will automatically generate the total duration multipliers (which consider the total duration of repetitive tasks in the shift as equal to 360 min), along with the intrinsic duration multipliers for each task in the shift.

Example 6.2.1—Step 5 (Sheet 2. Exposure calculation): Enter the OCRA checklist results for each task:

EXERCISE 6.1: EXAMPLE 6.2.1—STEP 4 (SHEET 2: EXPOSURE EVALUATION)

% Duration	Net Repetitive Tasks in the Shift (in minutes)	Duration Multipliers for the Intrinsic Duration of Each Repetitive Task	Total Net Duration of Repetitive Tasks in the Shift (in minutes)	Duration Multiplier for the Total Duration of Repetitive Tasks in the Shift	Name of the Repetitive Tasks Performed by a Homogeneous Group
14	50	0.35	360	0.925	D
17	60	0.500	360	0.925	E
8	30	0.35	360	0.925	F
56	200	0.750	360	0.925	G
6	20	0.2	360	0.925	H

- Enter the lack of recovery time multiplier as obtained from Sheet 1, Organizational Data.
- Enter the number of hours without adequate recovery time as obtained from Sheet 1, Organizational Data.
- Copy all the scores obtained for the various risk factors; the posture score and the final score both appear automatically.

Example 6.2.1—Step 6 (Sheet 2. Exposure Calculation): The risk indexes calculated using the two methods will appear. Use the one that best suits the setting at hand; the difference is obvious.

6.3 ANNUAL MULTITASK ANALYSIS IN AGRICULTURE: TASK DISTRIBUTION AND DURATION OVER THE YEAR. INTRINSIC RISK INDEXES AND FINAL EXPOSURE INDEX CALCULATION

6.3.1 Scope of Application and General Outline of the Analytical Process

Having explained how to deal with task rotation on a daily basis, the next step is to define a set of procedures and criteria for estimating risk in more complex situations, such as agriculture, where workers perform multiple tasks variously distributed in qualitative and quantitative terms over the year (annual cycle).

Since it is quite challenging to apply the mathematical models involved, a free Excel spreadsheet tool has been developed for this purpose, which can be downloaded from the www.epmresearch.org website. The program (ERGOepmCHECKOCRAmultiIYEAR-EN) will be described and an explanation provided of the methods and criteria for calculating the final exposure index; along with the rationale, readers will also be given a simple tool for estimating final risk.

EXERCISE 6.1: EXAMPLE 6.2.1—STEP 5 (SHEET 2: EXPOSURE EVALUATION)

Name of Repetitive Tasks Performed by a Homogeneous Group	Recovery Multiplier	Recovery Score	Frequency	Force	Side	Shoulder	Elbow	Wrist	Hand	Stereotype	Total Posture	Additional	OCRA Checklist
D	1.2	3	9	0	R	6	0	0	8	3	11	2	24.42
E	1.2	3	2	1	R	1	0	2	0	0	2	1	6.66
F	1.2	3	4	3	R	1	0	2	6	3	9	1	18.87
G	1.2	3	3	0	R	2	0	0	4	1.5	5.5	2	11.66
H	1.2	3	1	2	R	1	0	2	0	3	5	1	9.99

EXERCISE 6.1: EXAMPLE 6.2.1—STEP 6 (SHEET 2: EXPOSURE EVALUATION)

Result of the Exposure Evaluation of Repetitive Work Using the OCRA Minichecklist and Multitask Calculation Model

Time-weighted average	13.1	Task rotation takes place within a period of less than 90 consecutive minutes for each task performed
Multitask complex	17.39	Task rotation takes place within a period of more than 90 consecutive minutes for each task performed

It should be noted that one of the most recent activities undertaken by the World Health Organization (WHO) with a view to preventing work-related disorders and diseases consists in creating toolkits for the quick general identification of possible risk factors; these simple tools can be used by accident prevention officers, occupational health specialists, workers and trade union representatives, business owners and occupational health and safety staff, and so on.

This contribution offers a preliminary operational tool for investigating multitask exposure risk within an annual cycle (already available in a software package); one of its aims is to meet the need expressed by the International Ergonomics Association (IEA) to focus on neglected sectors such as agriculture.

The general risk evaluation process entails steps that shall be listed here (Table 6.4) and explained further on with examples. The examples will refer primarily to farm jobs, but the same analytical structure and calculation can be applied to any work that involves task rotations on a yearly basis.

TABLE 6.4
Procedures for Stepwise Multitask Annual Exposure Risk Analysis

	(A) Preliminary Step
A1 (6.2.1. INTRODUCTION)	Analysis of work on a farm in order to identify tasks performed in the period; qualitative definition of work during each month of the year
A2	Analysis of each repetitive task, using the OCRA checklist, to calculate intrinsic risk
A3	Definition of time constants (related to exposure time, pre-entered as constants)
	(B) Real Risk Analysis
B1	Identification of a homogeneous group: Personal data
B2	Description of time exposure for a homogeneous group: Typical day, task allocation, calculation of duration per month and year, weighted in relation to time constants
B3	Application of mathematical models
B4	Analysis of models and application for exposure to several crops per year
B5	Interpretation of results

6.3.2 Analysis of Work Organization for One Crop, in Order to Identify the Tasks Performed in the Period, with a Qualitative Description of Each Task Every Month

It is anything but a simple matter to identify farming tasks, which may be very numerous and performed by different workers or groups of workers. At the outset, therefore, it is necessary (as shown in Table 6.5) to

- Identify a specific crop.
- Break down the crop-growing activities work into *macrophases* and *phases*; all of the relevant tasks must be identified.
- List all the tasks required annually to grow and harvest the crop, regardless of who performs them. The allocation of tasks to workers (either on an individual basis or as a group) and the evaluation of exposure risk for each homogeneous group will be dealt with later.

Obviously, the same operation can be carried out in several different ways; each method should be viewed as a separate task and listed accordingly.

It is important to note that all the tasks performed on the farm over the year must be evidenced, including preparing the soil, applying fertilizers and disinfectants, and other seemingly ancillary activities, regardless of who performs them.

TABLE 6.5
Identification of Macrophases, Phases, and Tasks Performed throughout the Year for the Entire Crop, Regardless of Who Performs Them

Macrophase	Phase	Film	Task Description
Soil preparation	Soil preparation		• Plowing (tractor)
			• Installing irrigation system
	Sow seeds/plant seedlings		• Planting (manual)
			• Planting (mechanical)
Pruning	Dry pruning		• Pruning large branches with chainsaws
			• Pruning with manual shears
			• Pruning with pneumatic shears
	Green pruning		• Pruning with manual shears
			• Pruning with pneumatic shears
Harvesting	Harvesting		• Manual harvesting on ground
			• Manual harvesting on ladder
			• Automatic harvesting
Treatments	Soil fertilizing		• Preparing machine to apply fertilizer
			• Driving tractor
			• Composting (manual)
	Crop treatment		• Disinfection (manual)
			• Disinfection (tractor)

It can be very useful to film workers performing tasks and to store the most representative activities directly in the analytical software.

Table 6.5 also illustrates the preliminary breakdown of a hypothetical but indicative classification into micro-phases, phases, and tasks in a fruit-growing farm.

Figure 6.3 shows the *annual work schedule* listing the various tasks and when they are performed during the year. As an initial approach toward describing the tasks, it may be useful to use different colors to highlight the boxes corresponding to the various periods in which the tasks must be performed. This simple description will be a handy guide for locating specific monthly and yearly tasks relating to the farm work in question and therefore also for preparing an *on-site inspection calendar* so that monitoring activities can be performed at the right time. It is worth bearing in mind that if a task is missed, it may be necessary to wait a whole year before repeating the analysis!

6.3.3 Analysis of Each Individual Task Using the OCRA Checklist to Calculate the Intrinsic Score and Generate the "Basic Software" for Each Crop

Calculating the *intrinsic risk score* for a certain task means evaluating the task as if it is the only one performed by the worker all the time (i.e., for the whole shift and the whole year). To estimate the IRi, reference is made to a *shift constant* featuring

- 430/460 net minutes of repetitive work (modal value 440, where $Dm_{tot} = 1$)
- One 30-min meal break
- Two 10-min pauses

The IRi is calculated by applying the OCRA checklist to all the tasks performed for the specific crop and using both the right and the left arm.

MACRO-PHASE	PHASE	TASK DESCRIPTION	JAN	FEB	MAR	APR	MAY	JUN	JUL	AUG	SEP	OCT	NOV	DEC
SOIL PREPARATION	SOIL PREPARATION	Plowing (tractor)	■											
		Installing irrigation system	■											
	SOW SEEDS/PLANT SEEDLINGS	Planting (manual)											■	■
		Planting (mechanical)											■	■
PRUNING	DRY PRUNING	Pruning large branches with chainsaws										■	■	
		Pruning wth manual shears										■	■	
		Pruning with pneumaticl shears										■	■	
	GREEN PRUNING	Pruning wth manual shears		■	■	■	■							
		Pruning with pneumaticl shears		■	■	■	■							
HARVESTING	HARVESTING	Manual harvesting on ground						■	■	■				
		Manual harvesting on ladder						■	■	■				
		Automatic harvesting						■	■	■				
TREATMENTS	SOIL FERTILIZING	Preparing machine to apply fertilizer		■	■	■							■	■
		Driving tractor		■	■	■							■	■
		Composting (manual)		■	■	■							■	■
	CROP TREATMENT	Disinfection (manual)	■	■	■	■	■							
		Disinfection (tractor)	■	■	■	■	■							

FIGURE 6.3 Qualitative distribution of tasks over the year, regardless of who performs them: Annual schedule of phases relative to the crop under examination.

The final indexes are entered into the appropriate areas of the software, along with all the scores for the various risk factors required by the OCRA method and checklist (Figure 6.4):

- *No. of hours without adequate recovery* (which is always equal to 4, as this is the predefined constant for intrinsic indexes).
- *Recovery multiplier* (which is always equal to 1.33, as this is another predefined constant for intrinsic indexes).
- *Frequency score.*
- *Use of force score.*
- *Final posture score* (including respective scores for shoulder, elbow, wrist, hand, and stereotypy). Stereotypy scores may be assigned to each individual task according to the criteria set forth in the OCRA method. However, if the task rotation is very frequent, the stereotypy score may be excluded from the calculation of the final indexes; the software, in fact, allows this information to be supplied for either retention or exclusion from the final risk indexes.
- *Additional risk factor score.*

Having acquired the organizational data indicated here (macrophases, phases, and tasks relating to the specific crop-growing activities) and the intrinsic risk index assessment with the OCRA checklist for all the tasks in the

Phase Risk calculation for the right side	Task name	Recovery multiplier	**Recovery score**	**Frequency**	**Force**	Side	Shoulder	Elbow	Wrist	Hand	Srereotype	**Total posture**	**Additional**	**OCRA checklist intrinsic indexes**	Posture score with stereotypy	Posture score without stereotypy
Soil preparation	A	1.33	4	6	0	R	12	0	1	8	3	15		27.93	15	12
	B	1.33	4	3	0	R	2	0	1	4	3	7		13.30	7	4
Sow seeds/plant seedlings	C	1.33	4	2	1	R	1	2	3	6	3	9		15.96	9	6
	D	1.33	4	4	1	R	1	2	3	8	3	11		21.28	11	8
Dry pruning	E	1.33	4	5	0	R	0	0	0	2	2	4		11.31	3.5	2
	F	1.33	4	9	0	R	12	0	0	8	0	12		27.93	12	12
Green pruning	G	1.33	4	5	0	R	0	0	0	2	0	2		9.31	2	2
	H	1.33	4	9	0	R	12	0	0	8	0	12		27.93	12	12
Harvesting	I	1.33	4	6	0	R	8	0	0	8	0	8		18.62	8	8
	L	1.33	4	3	0	R	0	0	0	4	0	4		9.31	4	4
Soil fertilizing	M	1.33	4	3	0	R	0	0	0	2	0	2		6.65	2	2
	N	1.33	4	4	0	R	0	0	0	3	0	3		9.31	3	3
Crop treatment	O	1.33	4	3	0	R	0	0	0	5	0	5		10.64	5	5
	P	1.33	4	4	0	R	0	0	0	3	0	3		9.31	3	3

FIGURE 6.4 Calculation of intrinsic risk indexes (as if the task lasted for 8 continuous hours with one meal break and two 10-min pauses) for all the tasks identified for the specific farming activities, regardless of who performs them. Calculate one index per limb.

study (for both the left and the right arm), the software will calculate the final indexes; this will be called ERGOepmCHECKOCRAmultiYEAR-EN: "*BASIC FOR A SPECIFIC CULTIVATION*." Based on this preliminary data, and having identified one or several homogeneous groups, it will be possible to assess risk exposure.

With regard to the intrinsic checklist data, the software includes a box for indicating whether the various tasks assigned to and performed by a homogeneous group are alternated very quickly (such as farm cleaning). If the response is affirmative, the software deletes the stereotypy calculation from all tasks.

6.3.4 Exposure Time Constants

Before going on to the organizational analysis of the risk-exposed worker or homogeneous group of workers, listed here are the exposure constants to which reference is made for calculating exposure time prevalences to various tasks and also for reconstructing the *fictitious working day* that will be representative of the whole year. In agriculture, it is quite common for work schedules to be irregular, depending on the season, and for tasks to be assigned haphazardly.

It has been found useful to adopt several exposure constants representing the typical exposure level for the industry:

- Eight hours per day
- Five days a week
- Four weeks a month
- Eleven months a year

Table 6.6 lists all the constants used; although expressed in different ways, that is, hours, days, and so on, they all reflect the aforesaid criteria.

6.3.5 Identification of a Homogeneous Group

The next step is to assign tasks to an individual worker or group of workers exposed to the same risk, that is, to identify *homogeneous groups*. For each type of crop, tasks will be assigned to different groups of workers. When tasks of the same nature

TABLE 6.6
Exposure Time Constants

Hours/day constant	8	Hours/month constant	160
Minutes/day constant	440	Days/month constant	20
Days/week constant	5	Months/year constant	11
Minutes/week (440 min * 5 days) constant	2200	Days/year constant	220
Weeks/month constant	4	Hours/year constant	1760

and duration are assigned to the same group of workers, we may speak of a *homogeneous group in terms of risk exposure*. A homogeneous group may sometimes be made up of just one person if no other workers perform the same tasks qualitatively and quantitatively.

For instance, typically, a single group of workers may be assigned the job of actually growing a crop (tasks may include pruning, harvesting, etc.), while other workers prepare and disinfect the soil, apply fertilizers, and so on.

The analysis starts by opening the software ERGOepmCHECKOCR AmultiYEAR-EN: "BASIC" FOR A SPECIFIC CULTIVATION—and assigning one *file* to each homogeneous group assigned to the specific crop; each homogeneous group will thus have a *BASIC CULTIVATION FILE* for assessing the specific exposure risk.

The Excel software proposes templates for an initial description of the personal details of the group (Tables 6.7 and 6.8).

TABLE 6.7
Definition of Homogeneous Group of Workers with Regard to a Specific Crop (i.e., Name of Group, Number of Workers, Brief Description of Work) and Main Personal Data

Sheet 1: Farm and Homogeneous Group

Name of farm	XXXXXXXX
Address	YYYYYYY
Name of employer	ZZZZZZZZ
Total number of employees	15
Name of homogeneous group (job assigned to homogeneous group or even individual person)	Fruit tree growing, pruning, and harvesting workers
No. of workers performing the same tasks (if analysis involves a homogeneous group)	10
Short description of job	This group of workers performs all the tree pruning and fruit harvesting work but does not prepare the soil or apply fertilizers or pesticides

Form 2: Names of Workers in Homogeneous Group

Personal Data		Date of Birth	Gender	Age	Joined the Group in	Left the Group in	Employed by the Company since
Surname Name	XX						
Surname Name	YY						
Surname Name	ZZ						
Surname Name	HH						
Surname Name	etc.						

TABLE 6.8
Description of a *Typical Representative Working Day* (for Each Month) with Respect to Annual Exposure: Duration of Shift and Distribution of Pauses

Sheet 3: Working Hours	
Description of a Typical Working Day	
Overall shift duration (minutes)	420
No. of official breaks (recovery periods) during the shift, with a duration of *at least* 8 min (except meal break)	
No. of actual breaks (recovery periods) during the shift, with a duration of at least 8 min (except meal break) that can be considered as recovery periods	3
Overall duration of all actual breaks (excluding meal break) in minutes	30
Actual duration of meal break if included in shift duration (minutes)	
The shift is performed consecutively?	NO
No. of other breaks (i.e., meal break out of working time; travel time to/from different company locations). Mark one number only when these breaks last **at least** 30 min.	1
Put On/Take Off Protective Clothes and Equipment	10
Cleaning	
Time to reach the workstation	20
Other	
Total duration of nonrepetitive tasks (e.g., cleaning, fetching supplies) in minutes	30
Estimated net duration of repetitive work in minutes	360

6.3.6 Description of Homogeneous Group Risk Exposure: Typical Day, Task Assignment, Calculation of Proportional Task Duration by Month and Year Using Constants

Now begins the actual risk assessment for each individual worker or homogeneous group. This first section, focusing on the organizational analysis, needs to be broken down into its constituent parts.

6.3.6.1 Phase A: Description of a Typical Working Day

The description of a representative *typical* or *modal* working day then follows (Table 6.8). It is obvious that in farming, the net duration of a shift may vary depending on the weather. This is why the organizational structure of a modal day, which is the one that appears most frequently during the year, will be illustrated.

As for a normal OCRA checklist, the analysis will discover the shift duration, number and duration of pauses, and duration of nonrepetitive tasks in order to obtain the net duration of repetitive work and the first automatically generated score for the lack of recovery times (recovery multiplier).

The example shows a score of 1.12, reflecting the presence of three pauses plus a meal break in a 420-min shift. The modal day is detected primarily to adjust the IRi for each task (as listed in the *BASIC CULTIVATION FILE*) with the actual organizational data.

6.3.6.2 Phase B: Estimation of the Total Number of Hours Worked Every Month of the Year

In order to make this estimation, two pieces of information must be entered in the software (Figure 6.5):

- Number of hours worked every month by the entire homogeneous group for the total number of tasks. This information is readily available on the farm as it is required to calculate the cost of the active workforce. These hours often also include time worked by casual workers hired on a seasonal basis (seasonal or temporary workers).

Since the aim is to obtain the number of hours worked by each member of a homogeneous group, the presence of such workers must also be described:

- The presence of workers not operating on the farm during certain months of the year, possibly in addition to the homogeneous group (number of seasonal or temporary workers)

The software will automatically calculate:

- The number of hours/worker per month
- The percentage with respect to the constant of 160 h/month

Figure 6.6 provides a graphic depiction of the hourly distribution of work over the various months of the year.

In the first graph, when the threshold of 160 h/worker is passed, the ratios enter the critical zone (100% of the constant is exceeded).

The example shows an uneven distribution of working hours over the various months of the year, with some months where the hours are below the constant and others where they are above; these months often correspond to the harvest.

Harvesting requires speed, and the addition of seasonal workers is often required.

Figure 6.5 shows that in the summer months, from April to September, seasonal workers are added to the workforce, with their numbers peaking in August. It should be borne in mind that with these additional employees, who perform the same tasks as the homogeneous group included in the study, the *hours worked per month* must include the total number of hours worked by both regular and casual workers.

The software will automatically calculate the hours/person worked each month.

6.3.6.3 Phase C: Assignment of Tasks to a Homogeneous Group (Or Individual Worker) and Calculation of Proportional Duration in Each Individual Month

This step entails assigning typical crop-growing tasks to the homogeneous group (or also to an individual worker) in both qualitative and quantitative terms.

HOMOGENEOUS GROUP	JAN	FEB	MAR	APR	MAY	JUN	JUL	AUG	SEP	OCT	NOV	DEC
TOTAL HOURS WORKED PER MONTH by the whole group (fixed, seasonal, casual)		900	1600	1920	1920	2300	3800	3400	2080	1600	1300	900
TOTAL HOURS WORKED PER MONTH /EMPLOYEE		90	160	160	160	164	211	213	160	160	130	90
TOTAL MONTHS WORKED /YEAR		1	1	1	1	1	1	1	1	1	1	1
% OF CONSTANT 160 HOUR MONTH		56%	100%	100%	100%	103%	132%	133%	100%	100%	81%	56%
Total active employees per month/ year		10	10	12	12	14	18	16	13	10	10	10
NR. PERMANENT EMPLOYEES in the homogeneous group		10	10	10	10	10	10	10	10	10	10	10
NR. SEASONAL OR TEMPORARY EMPLOYEES, added to the group				2	2	4	8	6	3			
TOTAL HOURS WORKED / YEAR PER EMPLOYEE	1698											
CONSTANT OF HOURS WORKED PER YEAR (reference)	1760											

FIGURE 6.5 Estimation of number of hours worked per month and per worker in the homogeneous group.

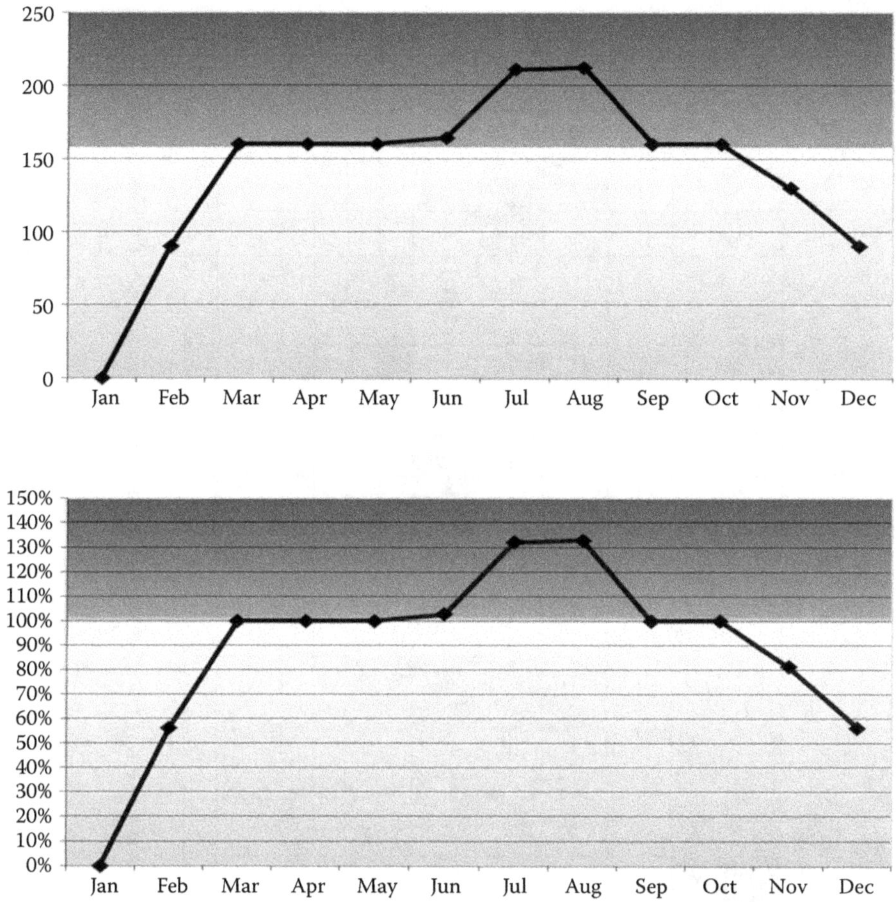

FIGURE 6.6 Graph depicting the number of hours worked per month and per worker, and ratios with respect to a constant 160 h/month for the homogeneous group of workers assigned to pruning and harvesting.

While the assignment is relatively easy to define qualitatively, it may be more complex from the quantitative standpoint, especially if the tasks are very numerous (for instance, in a greenhouse or vineyard, etc.).

The following tasks must be entered into the software for the *qualitative* definition (Figure 6.7):

- Tasks actually performed by the homogeneous group during the entire year—not yet broken down into months (placing an *X* in the box corresponding to active tasks)
- Tasks actually performed by the homogeneous group on a monthly basis (e.g., using different colors for the boxes corresponding to the tasks performed over the various months of the year)

MACRO-PHASE	PHASE	TASK	ACTIVE TASK	JAN	FEB	MAR	APR	MAY	JUN	JUL	AUG	SEP	OCT	NOV	DEC
SOIL PREPARATION	SOIL PREPARATION	A													
		B													
	SOW SEEDS/PLANT SEEDLINGS	C	X												
		D	X												
PRUNING	DRY PRUNING	E	X												
		F	X												
	GREEN PRUNING	G	X												
		H	X												
HARVESTING	HARVESTING	I	X												
		L	X												
TREATMENTS	SOIL FERTILIZING	M													
		N													
	CROP TREATMENT	O													
		P													
TOTAL /MONTH															

FIGURE 6.7 Example of qualitative description of pruning and harvesting tasks per month among a homogeneous group of workers.

To obtain a *quantitative description* of the *active* tasks, enter the percentage duration per month of each task into the software. The sum of the percentages per month in the column must always add up to 100% (Figure 6.8).

The proportional description of the tasks does not require extreme precision; the employer, or the members of the homogeneous group, can usually provide the information quite easily.

Once the quantitative data has been provided as described, the software automatically estimates the number of hours worked on each task every month by matching the proportional task descriptions shown in Figure 6.8 to the total number of hours worked per month/worker (Figure 6.5). An example is shown in Figure 6.9.

The software will now calculate the critical figure enabling the final risk to be evaluated: the *total number of hours worked per year on each task* by each member of the homogeneous group and the proportion of these hours to both the total number of hours worked and to the constant 1760 h/year (Figure 6.10).

It is worth noting that in this last table and in the example shown, the total proportion of hours worked versus the constant is 96%, that is, less than 100%; thus, all the percentage durations of each task are less than the intrinsic durations calculated versus the total hours actually worked over the year.

This brings us to the main findings that can be used to convert a year into a fictitious day. Once all the preliminary organizational information has been acquired, along with the data concerning risk, it is possible to apply the formulas for calculating the final risk indexes.

6.3.7 Application of Mathematical Models: Preliminary Preparation of "Fictitious Working Day" Representative of the Whole Year and of Every Month of the Same Year

Two models are proposed for calculating the final exposure risk index: one based on the time-weighted average and the other on the multitask complex, which is based on the most overloading task (calculated with respect to its actual duration) as the minimum exposure score that must be increased versus the score of the other tasks, taking their relative durations into account.

These two models have already been used for calculating the daily turnover, that is, when the task rotation occurs within or outside the 90-min time frame.

Both mathematical models have been used and adjusted for calculating annual and monthly multitask exposure risk, as illustrated.

In order to apply them to annual and monthly exposure, as mentioned before, it has been necessary to convert the data relative both to the individual months and to the year into a fictitious working day, representative first of each month of the year and then of the full year, respectively. The data for performing this conversion are summarized (Figures 6.11 and 6.12):

- Active tasks in each month and over the entire year
- Their duration in hours relative to each month and over the entire year

MACRO-PHASE	PHASE	TASK	ACTIVE TASK	JAN	FEB	MAR	APR	MAY	JUN	JUL	AUG	SEP	OCT	NOV	DEC
SOIL PREPARATION	SOIL PREPARATION	A													
		B													
	SOW SEEDS/PLANT SEEDLINGS	C	X											40%	10%
		D	X											20%	90%
PRUNING	DRY PRUNING	E	X										20%	10%	
		F	X										80%	30%	
	GREEN PRUNING	G	X		70%	60%	60%	50%	40%						
		H	X		30%	40%	40%	50%	60%						
HARVESTING	HARVESTING	I	X							50%	50%	50%			
		L	X							50%	50%	50%			
TREATMENTS	SOIL FERTILIZING	M													
		N													
	CROP TREATMENT	O													
		P													
Total month					100	100	100	100	100	100	100	100	100	100	100
TOTAL HOURS WORKED/MONTH (Table 6.10)					90	160	160	160	164	211	213	160	160	130	90

FIGURE 6.8 Example of quantitative description of pruning and harvesting tasks per month among a homogeneous group of workers.

MACRO-PHASE	PHASE	TASK	ACTIVE TASK	JAN	FEB	MAR	APR	MAY	JUN	JUL	AUG	SEP	OCT	NOV	DEC
SOIL PREPARATION	SOIL PREPARATION	A													
		B													
	SOW	C	X											52	9
	SEEDS/PLANT SEEDLINGS	D	X											26	81
PRUNING	DRY PRUNING	E	X										32	13	
		F	X										128	39	
	GREEN PRUNING	G	X		63	96	96	80	66						
		H	X		27	64	64	80	99						
HARVESTING	HARVESTING	I	X							106	106	80			
		L	X							106	106	80			
TREATMENTS	SOIL FERTILIZING	M													
		N													
	CROP TREATMENT	O													
		P													
TOTAL HOURS WORKED/MONTH by the homogeneous group					90	160	160	160	164	211	213	160	160	130	90

FIGURE 6.9 Calculation of hours worked per month by each member of the homogeneous group.

PHASE	TASK		JAN	FEB	MAR	APR	MAY	JUN	JUL	AUG	SEP	OCT	NOV	DEC	% vs. total hours worked/year	Nr. hours worked/ye ar	% vs. constant 1,760 hours worked/year
SOW SEEDS/PLANT SEEDLINGS	C	X											52	9	3.6%	61	3.5%
	D	X											26	81	6.3%	107	6.1%
DRY PRUNING	E	X										32	13		2.7%	45	2.6%
	F	X										128	39		9.8%	167	9.5%
GREEN PRUNING	G	X		63	96	96	80	66							23.6%	401	22.8%
	H	X		27	64	64	80	99							19.6%	334	19.0%
HARVESTING	I	X							106	106	80				17.2%	292	16.6%
	L	X							106	106	80				17.2%	292	16.6%
														TOTAL	**100%**	**1698**	**96%**

FIGURE 6.10 Calculation of total hours worked and prevalence (versus both total hours worked and constant of 1760 h/year) per task and per year for each worker.

PHASE	TASK	Active task	HOURS WORKED PER YEAR and TASK, by THE HOMOGENEOUS GROUP	HOURS WORKED IN REPETITIVE TASKS PER YEAR and TASK, by THE HOMOGENEOUS GROUP	% vs TOT HOURS WORKED YEAR (X)	% vs constant of 1,760 hours worked/year	Intrinsic OCRA checklist score (duration 8 hours with 1 meal break and 2 10-min. breaks) (Irit)
SOIL PREPARATION	A						
	B						
SOW	C	X	61	46	3.6%	3.5%	15.96
SEEDS/PLANT SEEDLINGS	D	X	107	80	6.3%	6.1%	21.28
DRY PRUNING	E	X	45	34	2.7%	2.6%	11.31
	F	X	167	125	9.8%	9.5%	27.93
GREEN PRUNING	G	X	401	301	23.6%	22.8%	9.31
	H	X	334	250	19.6%	19.0%	27.93
HARVESTING	I	X	292	219	17.2%	16.6%	18.62
	L	X	292	219	17.2%	16.6%	9.31
SOIL FERTILIZING	M						
	N						
			1698	1273	100%	96%	

FIGURE 6.11 Preliminary data required to convert the year into a fictitious working day representative of the entire year.

- Proportional duration of each task versus the total number of hours worked each month and over the entire year
- Proportional duration of each task versus the corresponding constant hours worked each month (160 h) and over the entire year (1760 h)

The calculations applied to the month employ the same criteria as for the year (Figure 6.12).

In order to complete the estimation of a fictitious working day representative of the year and each month of the year, certain procedures are required that involve recalculating the IRi scores (i.e., estimated for tasks lasting 8 h with one meal break and two 10-min breaks) to generate the new adjusted intrinsic risk index scores or IRic, reflecting the actual organizational conditions on the farm. The following parameters must be used:

- *Recovery multiplier*: This value is derived from the organizational data describing the hourly distribution of tasks on a modal day (Table 6.9); in the example, the score is 1.12.
- *Duration multiplier*: This value is derived from the duration of the working hours indicated in the fictitious working day representative of the whole year (Figure 6.13) or of each month of the year (Figure 6.14); it is calculated using the procedures shown in the relevant figures.

6.3.7.1 First Calculation Model: Time-Weighted Average (IR MultiMP)

The classic method for calculating the risk index with a daily repetitive task rotation features the following formula (as shown in section 6.2):

PHASE	TASK	Active ytsk	HOURS WORKED PER MONTH and TASK, by THE HOMOGENEOUS GROUP	NET DURATION OF REPEPTITIVE TASKS	% vs. TOT.HOURS WORKED MONTH **(X)**	% vs. constant of 160 hours worked/month	**Intrinsic OCRA checklist score** (duration 8 hours with 1 meal break and 2 10-min. breaks) **(Irit)**
GREEN PRUNING February	**G**	X	63	47	70%	39,3%	**9.31**
	H	X	27	20	30%	16,8%	**27.93**
		total	90	67	100%		
DRY PRUNING October	E	X	32	24	20%	20%	**11.31**
	F	X	128	96	80%	80%	**27.93**
		total	160	120	100%		

FIGURE 6.12 Preliminary data required to convert the month into a fictitious working day representative of each month of the year (To save space, the results for only 2 months are depicted).

Task	Total hours worked/year	Total hours worked/year in repetitive tasks	Duration of ***FICTITIOUS WORKING DAY*** in minutes, based on a constant of 220 working days/year	DURATION MULTIPLIER, calculated for the net duration of REPETITIVE WORK on the fictitious representative day of the year **(Dm_{tot})**	Recovery Multiplier obtained from Table 6.9
C	61	46	CALCULATION METHOD = 1,698 (total net repetitive working hours/year) / 220 (constant days worked/year) * 60		
D	107	80			
E	45	34			
F	167	125			
G	401	301			
H	334	250			
I	292	219			
L	292	219			
	1,698	**1,263**	**347,3**	**0,925**	**1.12**

FIGURE 6.13 Estimation of duration multipliers derived from the fictitious average working day representative of the whole year (See duration multipliers in Figure 6.1)

CONSTRUCTION OF FICTITIOUS DAY: February								
PHASE	TASK	Active ytsk	HOURS WORKED PER MONTH and TASK, by THE HOMOGENEOUS GROUP	% vs. TOT.HOURS WORKED / YEAR **(X)**	TOT. MINUTES OF REPETITIVE WORK IN FICTITIOUS DAY, PER REPRESENTATIVE MONTH **(Y)**		TOTAL DURATION MULTIPLIER for fictitious day	Recovery Multiplier obtained from Table 6.9
GREEEN PRUNING	**G**	X	47	70%	**203** (90 hours / month constant of 20 worked days* 60)		**0.75**	**1.12**
	H	X	20	30%				
		Total	90	100%				
CONSTRUCTION OF FICTITIOUS DAY: October								
DRY PRUNING	E	X	24	20%	**360** (160 hours / month constant of 20 worked days* 60)		**0.925**	**1.12**
	F	X	96	80%				
		total	160	100%				

FIGURE 6.14 Estimation of duration multipliers derived from the fictitious average working day representative of each month of the year (see duration multipliers in Figure 6.1) (To save space, the results for only 2 months are depicted.)

$$\text{IR MultiMP} = \left[\left(\text{IRir}_A \times \text{FT}_A\right) + \left(\text{IRir}_B \times \text{FT}_B\right) + \cdots + \left(\text{IRir}_N \times \text{FT}_N\right)\right] * \text{Dm}_{\text{tot}}$$

where:

$\text{IRir}_{A,B,..}$, N are the *intrinsic scores with recovery* obtained from the intrinsic OCRA checklists for the various workstations that the worker uses, calculated using the recovery multiplier corresponding to the actual distribution and duration of recovery times in the modal day.

FT_A, FT_B,.., FT_N represent the fractions of duration of the various repetitive tasks versus the total duration of repetitive work.

Dm_{tot} = *total duration multiplier*, relative to the net duration of all repetitive tasks in the shift (Chapter 4.1 and Table 6.3).

The general method for calculating the weighted average is to estimate risk exposure first for each month and then for the entire year as follows:

- Calculate the weighted average for each month.
- Add the values obtained for each month, then divide by 11 months (constant months worked per year). For the months in which the workers do not work (except for the holiday month constant), assign a representative OCRA checklist value of 2 (no risk).

Figure 6.15 shows all the data required to calculate the *weighted average applied to the fictitious day representative of each month of the year*, using the following steps:

- List the monthly tasks as a percentage of total hours worked/month: FT_j.
- List the IRic_J for the actual duration of the overall repetitive work (duration multiplier in Figure 6.14) and recovery multiplier in Table 6.8. The IRic_J scores are obtained (in the example shown here) using the following formula:

  IRic = IRi/1.33 (constant intrinsic recovery multiplier in scenario with one meal break and two 10-min pauses)/1 (constant intrinsic duration multiplier for 8 h = 1) * 1.12 (intrinsic recovery multiplier for one meal break and three 10-min pauses) * (duration multiplier for the fictitious day representative of each month; in this case, two different scores: February=0.85 and October=1)
- Multiply each OCRA checklist IR (IRic_J) by its proportional duration and subsequent summation of results:

$$\text{IR MultiMP} = \left(\sum \text{IRic}_J * \text{FT}_j\right) \tag{6.4}$$

PHASE	TASK	Active task	INTRINSEC OCRA checklist score IRi	INTRINSEC OCRA checklist score ADJUSTED for the total duration of repetitive work (fictitious day) and real recovery time (modal day) IRic	TOTAL WORKED HOURS/month (ACTIVE TASKS)	% on TOTAL WORKED HOURS/ month FT_j	(∑ IRic*FTj)
DRY PRUNING October	E		*11.31*	9.32	24	20%	9.32*20%
	F	X	*27.93*	21.76	96	80%	21,76*80%
				Weighted average October			19.27
GREEN PRUNING February	G	X	9.31	5.88	47	70%	5.88*70%
	H	X	27.93	17.64	20	30%	17.64*30%
			Weighted average February				9.41
Time-weighted average (∑) for each month of the year							

FIGURE 6.15 Data and formula required to calculate the time-weighted average (The monthly proportion of each task with respect to the total hours worked and the intrinsic OCRA checklist scores [IRic] adjusted for the recovery multiplier (Table 6.8) and the duration multiplier (Figure 6.14) on the fictitious day representative of each month. To save space, the results for only 2 months are depicted.)

TABLE 6.9
Final Calculation of the Annual Weighted Average Based on the Values Obtained for the Single Months, Versus 11 Months/Year Constant (If the workers had not worked all 11 months, an OCRA checklist score of 2 would have been assigned.)

	Jan	Feb	Mar	Apr	May	Jun	Jul	Aug	Sep	Oct	Nov	Dec
Right	0.0	9.4	13.1	13.1	14.5	16.4	11.8	11.8	10.9	19.3	14.5	13.1
				Time-Weighted Average							13.4	

Table 6.9 shows a summary of the weighted averages calculated for each month of the year and the final *representative value for the year*, calculated as illustrated here.

6.3.7.2 Second Calculation Model: Multitask Complex (IR MultiComp)

As illustrated here, if the daily repetitive task rotation has a duration of more than 90 min, the approach is based on the *task generating the highest overload as minimum.*

The procedure is based on the following formula, as explained in Section 6.2:

$$\text{IR MultiCom} = \text{IRic}_{1(\text{Dm1})} + (\text{IRic}_1 \times \text{K})$$

As mentioned earlier, if this mathematical model is applied to annual exposure, it is necessary to create a fictitious day representative of the whole year, which is somewhat complex, although essential to estimating the final risk.

In order to calculate the IR of each individual task based on its actual duration and not just with respect to the total duration of the repetitive tasks in the shift, it is necessary to calculate the fictitious duration in minutes, as if the year were comparable to a day.

This is done via the following steps (Figure 6.16):

- List the hours worked on each task during the year.
- Calculate their percentage versus the total hours worked over the year (FT_J).
- Find the overall duration of repetitive work (Y) in the average fictitious day representative of the whole year (Figure 6.12).
- Multiply (FT_J) by (Y).

The fictitious day for calculating risk using the multitask complex formula is thus reconstructed in detail; based on the fictitious duration of each task in minutes, as calculated, it is possible to obtain the following:

- Corresponding *Partial Duration Multiplier* (Dm_J) for each task (Figure 6.16)
- Total duration multiplier (Dm_{tot}) corresponding to the total duration in minutes of the fictitious day representative of the whole year (Figure 6.12)

PHASE	TASK	Ac tive task	REPETITIVE HOURSWORKED /YEAR	% vs. TOTAL HOURS WORKED / YEAR (FTJ)	TOTAL MINUTES OF REPETITIVE TASK IN FICTITIOUS DAY REPRESENTATIVE OF THE WHOLE YEAR (Y) (Table 6.16)	FICTITIOUS MINUTES (FM) AS IF THE YEAR WERE REDUCED TO A DAY (net duration of fictitious day representative of the whole year-Table 6.16) (fm$_i$= FTj* Y)	INTRINSEC OCRA checklist score ADJUSTED for the total duration of repetitive work (fictitious day) and real recovery time (modal day) IRic
SOW	C	X	46	3.6%	347	12	12,32
SEEDS/PLANT SEEDLINGS	D	X	80	6.3%		22	16,43
DRY PRUNING	E	X	34	2.7%		9	9,24
	F	X	125	9.8%		34	21,56
GREEN PRUNING	G	X	301	23.6%		82	7,19
	H	X	250	19.6%		68	21,56
HARVESTING	I	X	219	17.2%		60	14,37
	L	X	219	17.2%		60	7,19
		total	1,263	100%		347	
Summary: fictitious working day for the year						DURATION MULTIPLIER total fictitious day (Table 6.16)	0,925
						Recovery Multiplier derived from Table 6.9 (modal day of the year)	1,12

FIGURE 6.16 Estimation of the fictitious duration in minutes for each task present in the year, as if the year was reduced to a day (starting from the duration of the repetitive tasks present in the fictitious day representative of the whole year—Figure 6.12.) (As before, the recovery multiplier is the estimate deriving from the description of the representative day for the whole year shown in Table 6.8.)

- Recovery multiplier obtained using the estimate derived from the description of the modal day for the year shown in Table 6.8

Figure 6.17 shows all the data required to calculate the annual final risk index, applying the specific mathematical model.

For each active task, the following information is now available:

- Percentage versus total number of hours worked per year (FT_J)
- *Partial Duration Multiplier* (Dm_J) for each task
- Intrinsic OCRA checklist score (per side) recalculated using the partial duration of each task $IRic_{j(Dmj)}$
- Intrinsic checklist score by total duration $IRic_{j(Dm\ max)}$
- Maximum risk value calculated with respect to the partial duration $IRic_{1(Dm1)}$
- Maximum risk value calculated with respect to the total duration $IRic_{1(Dm\ max)}$
- $\Delta IRic_1$ = difference between the maximum risk value calculated with respect to total duration $IRic_{1(Dm\ max)}$ and maximum risk value calculated with respect to partial duration $IRic_{1(Dm1)}$
- K is the sum of:

Risk categories	FICTITIOUS MINUTES (FM) AS IF THE YEAR WERE REDUCED TO A DAY	% vs. total hours worked hours/year **(FTj)**	Duration multiplier for each task (fictitious day) **(Dm_J)**	IR OCRA checklist INTRINSEC scores for PARTIAL DURATION $IRic_{j(Dmj)}$	IR OCRA checklist INTRINSEC scores for TOTAL DURATION $IRic_{j(Dm\ max)}$
GREEN	141,7	41%	0,650	**5,1**	**7,2**
YELLOW	9,2	3%	0,100	**1,0**	**9,2**
LOW RED	12,5	4%	0,100	**1,3**	**12,3**
MEDIUM RED	184,0	53%	0,750	**15,1**	**18,6**
VIOLET (OR HIGH RED)	0,0	0%			
VERY HIGH RED	0,0	0%			
tot	347	100%		**17,6**	
15.5 + (18.6-15.5)* K= 17,6 K=((7.2°41%)+(9.2*3%)+(12.3*4%)+(18.6*53%))/18.6					

FIGURE 6.17 Calculation of the final index using the ir multicomp model (Right limb): we created six categories of risk from absent (green) to purple (very high risk). The minutes spent in each of the six categories were grouped. Within each category the average value of weighted risk was calculated for the category, considering the risk value of the tasks included in this category and their duration. From these values we build the fictitious working day, as representative of the year.

$$\left(\mathrm{IRic}_{1(\mathrm{Dm\ max})} * \mathrm{FT}_1\right) + \left(\mathrm{IRic}_{2\ (\mathrm{Dm\ max})} * \mathrm{FT}_2\right) + \cdots + \left(\mathrm{IRic}_{\mathrm{N}(\mathrm{Dm\ max})} * \ \mathrm{FT}_{\mathrm{N}}\right) \mathrm{IRic}_{1(\mathrm{Dm\ max})} \tag{6.5}$$

The final score thus obtained (IR MultiComp = 17.6 for the right limb) is not the same as the value obtained using the time-weighted average formula (IR MultiMP = 13.4).

Often in annual cycles, workers are exposed to many tasks. The formula of Multitask Complex is based on research of the worst task, for its duration (and the final value of the risk cannot be less than it). In the presence of many tasks, the% of the worst task duration would be reduced too much.

To correct this problem, the following criteria is used:

To create six risk categories from absent (green) to purple (very high risk).
The minutes spent in each of the six risk categories were grouped.
Within each category the average value of weighted risk was calculated for the category, considering the value risk of the tasks included in this category and their duration.

From these values we build the fictitious working day, as representative of the year.

Figure 6.18 compares the month-by-month results (for the right limb) using the two formulas, that is, weighted average versus multitask complex, as obtained automatically using the Excel template.

In order to better interpret the contrasting results, the two formulas have been used to calculate the IR for each month of the year, thus displaying the IR trends month on month.

With the risk index trends shown on a monthly basis, it is easier to comprehend the differences between the two formulas. Since the work is distributed differently over the various months of the year, the weighted average formula tends to flatten the peaks, determining lower risk scores than the multitask complex formula, which is never lower than the highest peak calculated for its actual duration.

As shown in Figure 6.18, the software calculates the exposure risk index IR using the two alternative formulas for each month of the year, generating not just the annual IR but also the monthly trend; this information is extremely useful for interpreting the final overall result.

Figure 6.19 provides an example of the multitask complex index calculation (IR MultiComp) for work-related risk exposure lasting just 2 months a year; the methods for estimating the fictitious day representative of the whole month are shown and the final IR risk index scores for both of the months worked (with reference, for the sake of simplicity, only to the right limb).

The calculations applied to the month use the same criteria as for the year. Since the modal working hours are the same as those used in the previous example, the recovery multiplier for the modal day shown in Table 6.8 is also used here.

6.3.8 Models for Analyzing Exposure in the Case of Several Crops per Year

If the workers on the farm are assigned to growing several crops (such as grapes and olives, or vegetables and fruit), the organizational data will need to be entered

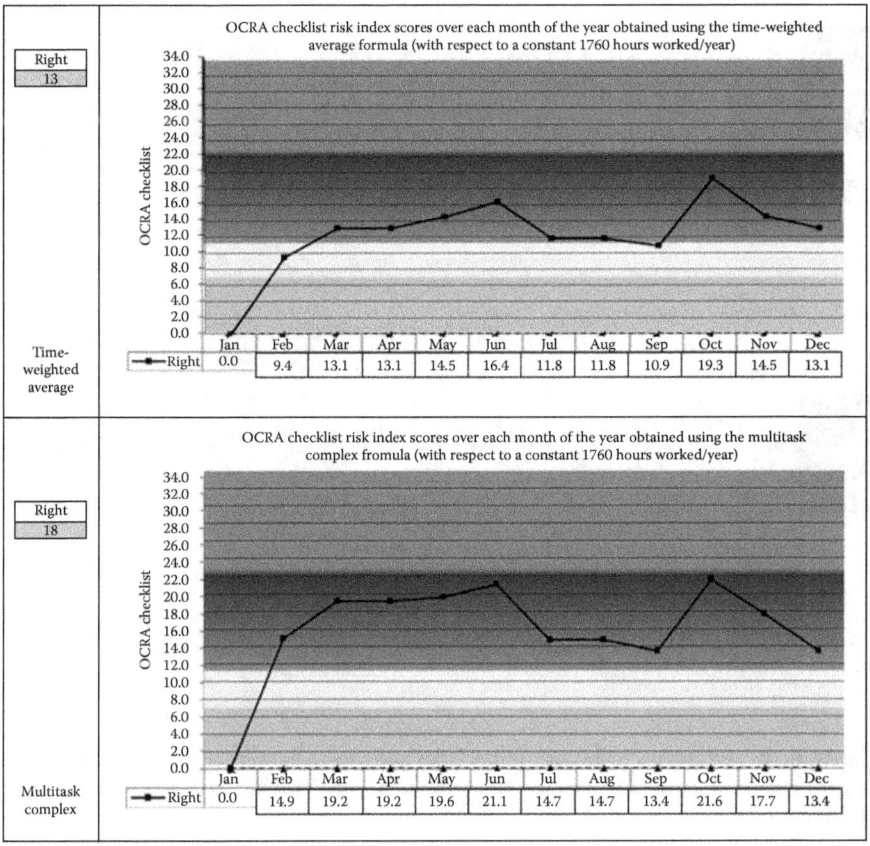

FIGURE 6.18 Risk index scores plotted by month over the whole year using both recommended formulas.

into the BASIC software (i.e., macrophases, phases, and tasks) together with the intrinsic OCRA checklist results (IRi) for all the different crops grown over the year; homogeneous groups will also have to be identified as described, as well as their monthly tasks over the year, even if the tasks involve different crops. Serious issues may emerge when several crops are combined and the number of tasks increases; the Excel software may not be able to deal with an excessively large amount of data. In such cases, it is necessary to use software tools that can handle such calculations.

6.3.9 Conclusions

When using the OCRA index and checklist to assess exposure to several repetitive manual tasks featuring potential biomechanical overload of the upper limbs, the traditional approach has been to refer to calculation models based on the *time-weighted average* concept. However, this approach is unsuitable to certain applications if not actually misleading, such as when there is high continuous exposure for approximately half the shift, followed by slight exposure for the rest of the shift. In such

CONSTRUCTION OF FICTITIOUS DAY: February								
PHASE	TASK	Active ytsks	HOURS WORKED PER MONTH and TASK, by THE HOMOGENEOUS GROUP	% vs. TOT HOURS WORKED/ YEAR (X)	TOT. MINUTES OF REPETITIVE WORK IN FICTITIOUS DAY, PER REPRESENTATIVE MONTH (Y)	FICTITIOUS MINUTES (FM) AS IF THE MONTH WERE REDUCED TO A DAY (X * Y) FOR EACH TASK ACTIVE	PARTIAL MULTIPLIER DURATION for each task	FINAL RISK INDEX
GREEEN PRUNING	G	X	47	70%	202 (90 hours / month constant of 20 worked days* 60)	141.75	0.65	
	H	X	20	30%		60.75	0.5	
		total	67	100%		202.5 (FM)		
Summary of fictitious working day, representative of the whole month, and **IR MultiCom** score for the right upper limb (FEBRUARY)						TOTAL DURATION MULTIPLIER for the day fictitious	0.75	14.9
						Recovery Multiplier obtained from Table 6.9	1.12	

TASKS								
PHASE	TASK	Active task	HOURS WORKED PER MONTH and TASK, by THE HOMOGENEOUS GROUP	% vs.TOT HOURS WORKED/ YEAR (X)	TOT. MINUTES REPETITIVE WORK IN FICTITIOUS DAY, PER REPRESENTATIVE MONTH (Y)	FICTITIOUS MINUTES (FM) AS IF THE MONTH WERE REDUCED TO A DAY (X * Y) FOR EACH TASK ACTIVE	PARTIAL MULTIPLIER DURATION for each task	FINAL RISK INDEX
DRY PRUNING	E	X	24	20%	360 (160 hours / month constant of 20 worked days* 60)	72.00	0.5	
	F	X	96	80%		288.,00	0.85	
		total	120	100%		360 (FM)		
Summary of fictitious working day, representative of the whole month, and **IR MultiCom** score for right upper limb (OCTOBER)						TOTAL DURATION MULTIPLIER for the fictitious day	0.925	21.55
						Recovery Multiplier obtained from Table 6.9	1.12	

FIGURE 6.19 Examples of IR risk index calculated for 2 months (February and October) using the multitask complex formula: How to obtain the fictitious day representative of the whole month.

cases, the weighted average does not reflect the peak continuous exposure of the half shift.

Based on multiple task load-lifting analyses already reported in the literature (Waters et al., 2007) and tested in practice, calculation models of both the OCRA index and OCRA checklist have been tested for analyzing multiple repetitive tasks based on the *task generating the highest overload as minimum* concept.

Based on experience acquired with different applications, the hypothesis—still to be confirmed by further research—has even been put forward that the calculation model based on the time-weighted average is still valid even if the task rotations are daily and generally fairly frequent (at least every 90 min) or when long tasks are broken down into subtasks. Conversely, if exposure to repetitive tasks entails much less frequent daily rotations (over 90 min) or nondaily rotations (i.e., weekly, monthly, or annually), it might be more effective to use the new calculation methods presented here.

The mathematical models suggested here could also be applied to scenarios where the task rotation is annual (or even monthly), after converting the data for the year (or month) into a fictitious working day.

Examples have been provided of methods for estimating the final risk index for annual multitask exposure in agriculture.

For the time being, the authors prefer to observe the outcome of the annual risk assessment, applying both mathematical models, since there is still little epidemiological data on the prevalence of upper limb work-related musculoskeletal disorders (UL-WMSDs) linked to risk evaluation studies.

Based on preliminary clinical findings, the multitask complex checklist formula appears to be more predictive of adverse health effects, at least for exposure covering a goodly portion of the year.

Figure 6.20 shows a scenario wherein a homogeneous group works on a crop for half of the year, but with very few hours and at very low risk for 3 out of the 6 months. The discrepancy between the scores obtained using the weighted average versus the multitask complex is quite remarkable. The first flattens the peaks on the months with low or no exposure, while the second is based on the estimate of the lowest risk task, considering its intrinsic duration. Here, the interpretation of the results for the attribution of risk is problematic. It goes without saying that health surveillance findings proving the exposure risk index scores would be most welcome, but in these cases, with workers exposed to risk for only half of the year, it might be difficult to attribute a disease or disorder to occupational factors when the worker's activities in the other half of the year are unknown.

In agriculture, scenarios such as the one depicted in Figure 6.20 are anything but rare.

Work may be organized in many different ways with respect to annual exposure; here are a few examples:

- Worker exposed to risk for only 1 or 2 months during harvesting of a specific crop (e.g., fruit in summer); worker employed by only one farm.
- Workers exposed to risk during the summer months during harvesting of several crops (e.g., different fruits) for one employer but on several farms.
- Workers exposed to risk practically all year round, working on more than one crop (e.g., grapes and fruit), on several farms (pruning).

The aim of the approach described in this chapter is not to determine the causes and effects of many common musculoskeletal diseases and disorders but rather to offer assessment options for tackling and discussing the problem at hand.

The use of the OCRA checklist for analyzing the risk of biomechanical overload of the upper limbs in farm work may still seem very complex or at least time-consuming. Arguably, many farming activities need monitoring all year round in order to analyze and evaluate each individual task, and often the number of tasks is extremely high.

However, it has been observed that when the same techniques are used for the same crops, the tasks are always the same, and thus they have the same intrinsic OCRA checklist scores. Hence it is possible to speak of *finite variable* farm work, that is, featuring uniform variables, which can be preset following an initial evaluation. Going forward, efforts will be made to develop basic software with predefined datasets tailored to different crops; to assess risk for a homogeneous group, the appraiser will only have to enter the most important farm-specific organizational

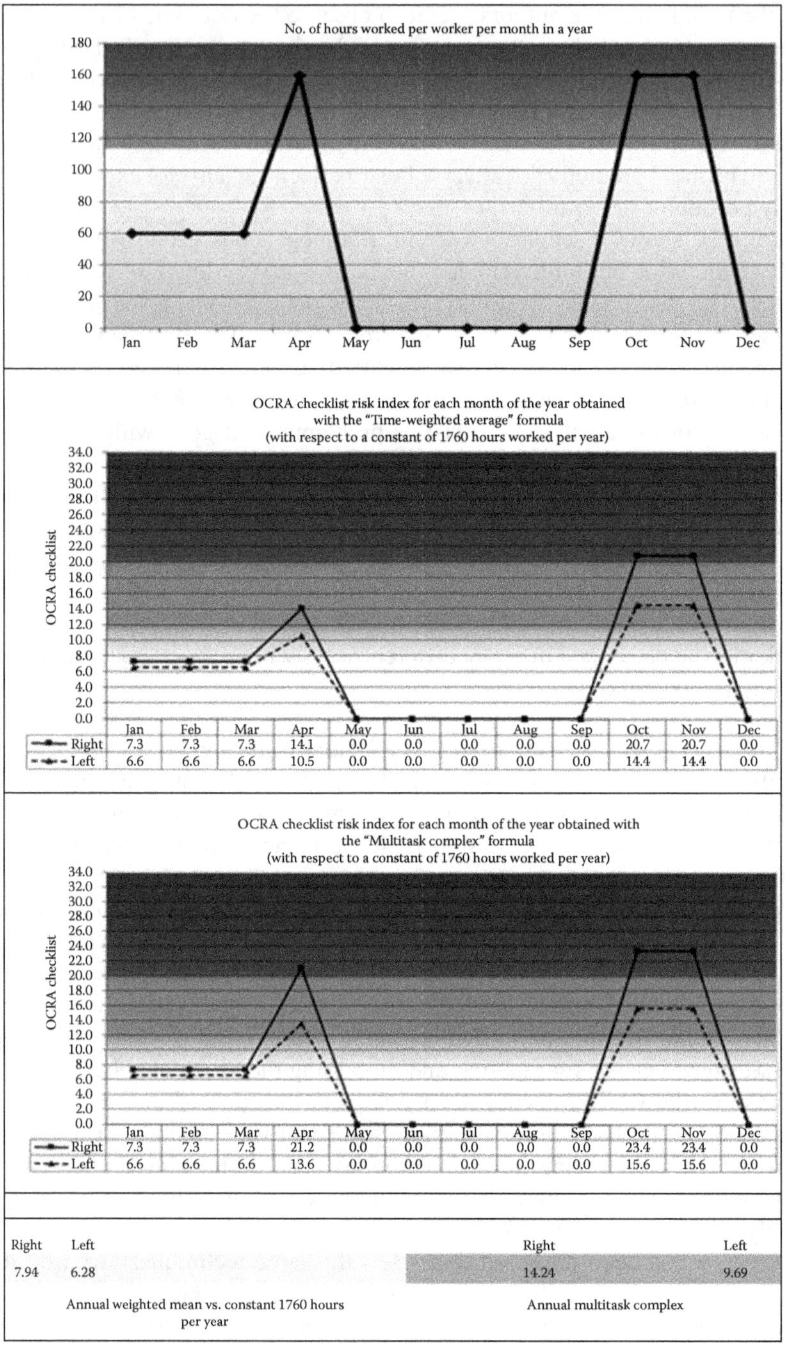

	Jan	Feb	Mar	Apr	May	Jun	Jul	Aug	Sep	Oct	Nov	Dec
Right	7.3	7.3	7.3	14.1	0.0	0.0	0.0	0.0	0.0	20.7	20.7	0.0
Left	6.6	6.6	6.6	10.5	0.0	0.0	0.0	0.0	0.0	14.4	14.4	0.0

	Jan	Feb	Mar	Apr	May	Jun	Jul	Aug	Sep	Oct	Nov	Dec
Right	7.3	7.3	7.3	21.2	0.0	0.0	0.0	0.0	0.0	23.4	23.4	0.0
Left	6.6	6.6	6.6	13.6	0.0	0.0	0.0	0.0	0.0	15.6	15.6	0.0

Right	Left	Right	Left
7.94	6.28	14.24	9.69
Annual weighted mean vs. constant 1760 hours per year		Annual multitask complex	

FIGURE 6.20 Risk index scores plotted by month over the whole year using both recommended formulas. Comparison between results for right and left limb. Farming activities involving only 3 months' work with very low exposure and 3 months with high exposure.

data. The data collection and management criteria have been validated and the structure has been defined; all that is needed now is a multitude of data and the necessary IT support.

6.4 WEEKLY AND MONTHLY MULTIPLE TASK ANALYSIS IN THE SERVICE SECTOR (SUPERMARKETS, CLEANING): TASK DISTRIBUTION AND DURATION OVER THE WEEK AND/OR MONTH. INTRINSIC RISK INDEXES AND FINAL EXPOSURE INDEX CALCULATION

6.4.1 Preliminary Organizational Analysis to Determine Whether the Task Cycle Is Weekly or Monthly

As stated in Sections 6.2 and 6.3, the process begins with the identification of the cycle within which an identical set of tasks recurs.

Work is typically organized on a weekly (or sometimes monthly) basis in the cleaning sector, or in supermarkets when the duration of daily shifts is modified; this obviously does not apply to annual "big cleans." Hospitals often organize tasks on a monthly basis, since shifts and tasks are distributed differently over three shifts both in in-patient wards and in kitchens. However, in many businesses, shifts and/or tasks are organized in weekly or monthly rather than daily scenarios.

In this section, the focus will be on two types of organizational analysis: weekly and monthly (some of which has already been described for the whole year, but here additional details will be provided as the analysis translates into a weekly study repeated for 4 weeks).

6.4.2 Analysis of Work Organization to Identify Tasks in the Period under Examination

It is anything but simple to identify tasks that may be very numerous and performed by different workers or groups of workers. At this initial stage, it is necessary to adopt the stepwise approach illustrated in Section 6.3 for the annual analysis, which involves:

- Identifying the specific operating area
- Breaking down the work into macrophases and phases; identifying all the relevant tasks
- Listing all the tasks required to complete the work in the period under examination (week or month), regardless of who performs them
- The assignment of tasks to workers (or homogeneous groups of workers performing the same tasks) and evaluation of exposure risk for each homogeneous group

It should be noted that if different "techniques" are used to perform the same operation, each one should be considered as a separate task and therefore must be

TABLE 6.10
Identification of Macrophases, Phases, and Tasks Carried Out Every Week by a Cleaning Service, Regardless of who Performs Them (Nonexhaustive Sample)

Macrophase	Phase	Task Description
Routine room cleaning	Floors	Pass squeegee over floor
		Mop floor
		Wash floor with cleaning cloth
	Furniture	Dust using dry cloth
		Wipe with damp cloth
	Windows and doors	Dust windows and doors
		Wash windows using telescopic pole
Collective area cleaning	Stairs	Pass squeegee over floor
		Mop stairs
		Wash floor with cleaning cloth
Routine cleaning of shared bathrooms	Hall, bathroom	Wash sinks
		Mop floor
	Cleaning services	Scrub toilet
		Replace toilet paper
		Scrub showers

listed (e.g., cleaning floors using a cloth, a mop, or an electrical appliance should be treated as different tasks).

It is extremely useful to film the most representative parts of the work and store the footage in the software.

Table 6.10 shows the breakdown of a hypothetical but not improbable classification of tasks into macrophases, phases, and tasks for a cleaning company (to save space, only very few phases and tasks are listed).

6.4.3 Analysis of Each Task Using the OCRA Checklist to Calculate the Intrinsic Score

Calculating the IR for a given task means evaluating each task as if it were the only one performed during the whole shift. A *constant shift* is characterized by:

- 430/460 net minutes of repetitive work (modal value 440; $Dm_{tot} = 1$)
- One 30-min meal break
- Two 10-min pauses

The IR calculation using the OCRA checklist applies to all of the tasks that characterize the job and to both the right and the left limb.

The final indexes must be entered into the appropriate sections of the software as well as all the scores relative to the various risk factors determined with the OCRA method (Table 6.11):

TABLE 6.11
Calculation of Intrinsic Risk Indexes for All the Tasks Making Up the Job, Regardless of Who Performs Them (calculate one index per limb)

Task	Recovery Multiplier	Recovery Score	Frequency	Force	Side	Shoulder	Elbow	Wrist	Hand	Stereotype	Total Posture	Additional	OCRA Checklist Intrinsic Indexes
A	1.33	4	6	0	R	12	0	1	8	3	15		27.93
B	1.33	4	3	0	R	2	0	1	4	3	7		13.30
C	1.33	4	2	1	R	1	2	3	6	3	9		15.96
F	1.33	4	9	0	R	12	0	0	8	0	12		27.93
G	1.33	4	5	0	R	0	0	0	2	0	2		9.31
H	1.33	4	9	0	R	12	0	0	8	0	12		27.93
I	1.33	4	6	0	R	8	0	0	8	0	8		18.62
L	1.33	4	3	0	R	0	0	0	4	0	4		9.31
M	1.33	4	3	0	R	0	0	0	2	0	2		6.65
N	1.33	4	4	0	R	0	0	0	3	0	3		9.31

- *No. of hours without adequate recovery* (always equal to 4, this being a predefined constant with respect to the intrinsic indexes)
- *Recovery multiplier* (always equal to 1.33, this being another predefined constant with respect to the intrinsic indexes)
- *Frequency score*
- *Force score*
- *Final posture score* (sum of shoulder, elbow, wrist, hand, and stereotypy scores). The stereotypy scores for each task must be defined using the OCRA method criteria. However, if the task rotation rate is very high, the stereotypy score can be excluded from the calculation of the final indexes; the software, in fact, allows that information to be supplied in order to retain or exclude it from the final risk indexes
- *Additional risk factor score*

With the organizational data indicated previously for agriculture (i.e., macrophases, phases, and tasks relating to the specific crop), and the OCRA checklist intrinsic risk scores for all the tasks identified (and for both the left and the right limb), the software is now ready to calculate the final indexes: The name will be: ERGOepmCHECKOCRAmultiMONTH-EN *"BASIC" FOR OPERATING SECTOR*. This preliminary data will form the basis for evaluating exposure risk for the homogeneous group(s).

The software includes a question about whether the rotations between various tasks are performed very quickly by the homogeneous group (such as room cleaning). If the answer is "yes," the software eliminates the calculation of stereotypy from all tasks.

6.4.4 Exposure Time Constants

Before tackling the organizational analysis of the homogeneous group, described here are the exposure constants that will be referred to for calculating the exposure time prevalences for the homogeneous group with respect to the various tasks. In this type of nondaily analysis, it is not uncommon to come across task distribution patterns that are irregular and erratic. It has been found to be useful to adopt several exposure constants representing the typical exposure level for the industry, as shown previously in Table 6.6, Section 6.3.

- Eight hours per day
- Five days a week
- Four weeks a year
- Eleven months a year

6.4.5 Identification of a Homogeneous Group

The next step is to assign the tasks to the workers, that is, to identify homogeneous groups.

For each specific area or operating sector, tasks will be assigned to different groups of workers.

When tasks of the same nature and duration are assigned to the same group of workers, we may speak of a *homogeneous group in terms of risk exposure*.

A homogeneous group may sometimes be made up of just one person if no other workers perform the same tasks qualitatively and quantitatively.

A group of workers may also appear to be performing the same tasks but over a longer or shorter period of time; in such cases, another homogeneous group must be created.

Using the software ERGOepmCHECKOCRAmultiMONTH-EN: *"BASIC" FOR OPERATING SECTOR*, open a file for each homogeneous group in the specific operating area; each homogeneous group will thus have a *BASIC FOR OPERATING SECTOR* file for assessing its specific exposure risk.

The Excel software proposes forms for an initial description of the personal details of the group (Table 6.12).

Now begins the actual risk assessment for each homogeneous group. This first section, focusing on the organizational analysis, needs to be broken down into its constituent parts.

TABLE 6.12
Definition of Homogeneous Group of Workers in an Operating Sector and Main Personal Details of Workers Included in Analysis

Sheet 1: Employer and Homogeneous Group

Name of company	XXXXXXXX
Address	YYYYYYY
Name of employee	ZZZZZZZZ
Total number of employees	20
Name of homogeneous group (job assigned to individual person or homogeneous group)	Daily cleaning
No. of workers performing the same tasks (if analysis involves a homogeneous group)	2
Short job description	All cleaning operations, but not periodical "big cleans"

Sheet 2: Names of Workers in Homogeneous Group

Personal Data		Date of Birth	Gender	Age	Date When Joined the Group	Date When Left the Group	Employed by the Company since
Surname name	XX						
Surname name	YY						
Surname name	ZZ						
Surname name	HH						
Surname name	ETC.						

6.4.5.1 Phase A: Shift Description

The organizational analysis of the work carried out by the homogeneous group begins by describing how the shifts are distributed. Table 6.13 shows an example of the qualitative description of how the shifts are distributed over a representative week, and sometimes month, when shifts are rotated in different ways over different weeks, as typically happens in hospitals.

6.4.5.2 Phase B: Description of Work Days in a Typical Week or Month and Estimation of Total Hours Worked per Day/Week or Month/Year

The next step is to carry out an analytical description of pauses and nonrepetitive work for each day of a "typical" working week representative of the week or of a "typical" working month (4 weeks) (Table 6.13).

As for a normal OCRA checklist, the analysis will highlight the shift duration, number and duration of pauses, and duration of nonrepetitive tasks in order to obtain the net duration of repetitive work and the first automatically generated score for the lack of recovery periods (recovery multiplier and duration multiplier).

Figure 6.21 shows an example of a graph depicting the organization of work days for a homogeneous group over a month: Reference is made to the same data shown in Table 6.14.

TABLE 6.13
Example of Shift Distribution over One Week of a Representative Month, if Shifts Change Each Week of the Month

Week 1	Mon	Tue	Wed	Thu	Fri	Sat	Sun
1 Shift	X	X					
2 Shift				X	X	X	
3 Shift							
Week 2	**Mon**	**Tue**	**Wed**	**Thu**	**Fri**	**Sat**	**Sun**
1 Shift		X	X	X			
2 Shift						X	X
3 Shift							
Week 3	**Mon**	**Tue**	**Wed**	**Thu**	**Fri**	**Sat**	**Sun**
1 Shift			X	X			
2 Shift		X			X	X	
3 Shift							
Week 4	**Mon**	**Tue**	**Wed**	**Thu**	**Fri**	**Sat**	**Sun**
1 Shift			X	X			
2 Shift	X	X			X		
3 Shift							

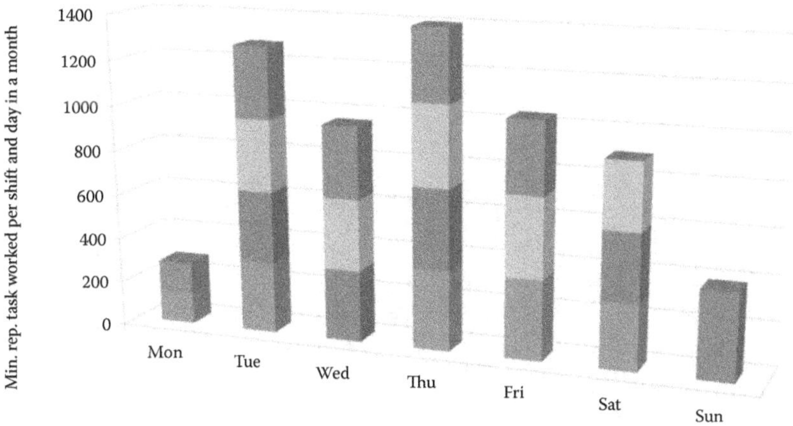

	Mon	Tue	Wed	Thu	Fri	Sat	Sun
4° week	140	320	320	320	320	0	0
3° week	0	320	319	360	360	300	0
2° week	0	320	320	360	0	300	390
1° week	140	320	0	360	360	300	0

FIGURE 6.21 Example of a bar graph depicting the net duration of a repetitive task in minutes for each day of the week in a representative month.

6.4.5.3 Phase C: Assignment of Tasks to a Homogeneous Group and Calculation of Proportional Duration in Each Individual Month

This step involves assigning the tasks performed by the homogeneous group (or individual exposed worker) both qualitatively and quantitatively.

While the assignment is relatively easy to define qualitatively, it may be more complex from the quantitative standpoint, especially if the tasks are very numerous (for instance, cleaning supermarkets or commercial kitchens, etc.).

The following tasks must be entered into the software for the *qualitative definition*:

- Tasks actually performed by the homogeneous group during the entire period—not yet broken down into weeks or months (placing an *X* for each task)
- Tasks actually performed daily by the homogeneous group

To obtain a *quantitative description* of the tasks, enter the percentage duration per week of each task into the software. The sum of the percentages for each day of the week must always add up to 100% (Table 6.15).

Extreme accuracy is not required for the proportional assignment of tasks; the employer, or even the members of the homogeneous group, should be able to provide this information.

Once the quantitative data has been provided, the software automatically estimates the number of minutes worked on each task every day of the week by applying the proportional task descriptions shown in Table 6.15 to the total number of hours worked per day/worker (Figure 6.21). An example is shown in Table 6.16.

TABLE 6.14

Example of a Description of Working Hours in a Week or, When Necessary, Representative Month, if Working Hours Are Different in Each Week of the Month

	Mon	Tue	Wed	Thu	Fri	Sat	Sun	Mon	Tue	Wed	Thu	Fri	Sat	Sun	Mon	Tue	Wed	Thu	Fri	Sat	Sun	Mon	Tue	Wed	Thu	Fri	Sat	Sun
Shift Duration (min)	180	420		420	420	360			420	420	420		360	480		420	420	420	420	360		180	420	420	420	420		
No. of Official Breaks		2		2	2				2	2	2					2	2	2	2				2	2	2	2		
No. of Actual Breaks		2		2	2	1			2	2	2		1	2		2	2	2	2	1			2	2	2	2		
Overall Duration of Actual Breaks Excluding Meal		30		30	30	20			30	30	30		20	20		30	30	30	30	20			30	30	30	30		
Actual Duration of Meal Break if Included in Shift		30		30	30				30	30	30			30		30	30	30	30				30	30	30	30		
Meal Break Not Included in Shift																												
Duration of Nonrepetitive Tasks																												
Net Duration Repetitive Tasks in a Shift	180	360		360	360	340			360	360	360		340	430		360	360	360	360	340		180	360	360	360	360		
Recovery Multipliers		1.20		1.20	1.20	1.33			1.20	1.20	1.20		1.33	1.33		1.2	1.2	1.2	1.2	1.33			1.20	1.20	1.20	1.20		

TABLE 6.15
Example of the Proportional Quantitative Assignment of Tasks over a Representative Week for a Homogeneous Group, with Respect to the Shift (If the cycle is monthly, the assignment must be repeated for all 4 weeks of the representative month)

		Monday	Tuesday	Wednesday	Thursday	Friday	Saturday	Sunday
Task	Weekly Task	1 Shift (%)	1 Shift (%)		2 Shift (%)	2 Shift (%)	2 Shift (%)	
AAA	X	3	3		3	3	20	
BBB	X	3	3		3	3	20	
CCC	X	3	3		3	3	20	
DDD	X	1	1		1	1	20	
EEE	X	1	1		1	1	20	
FFF	X	8	8		8	8		
GGG	X	50	50		50	50		
HHH	X	5	5		5	5		
III	X	3	3		3	3		
LLL	X	25	25		25	25		

The software will now calculate the critical figure enabling the final risk to be evaluated; the total number of hours worked by each homogeneous group on each task in the representative week (Table 6.16) or, if the analysis is monthly, for each week of the month and for the entire month (Table 6.17). This will produce their proportion with respect to the total number of hours worked and also with respect to the constant 2200 min per week (40 h/week) or 160 h/month (Tables 6.16 and 6.17).

In the example shown in Table 6.17, the total proportion of hours worked in the month versus the constant is 71.4%, that is, less than 100%; thus, all the percentage durations of each task are less than the intrinsic durations calculated versus the total hours actually worked over the month.

This means that the time worked is less than the constants.

This brings us to the main findings that can be used to convert a week or month into a fictitious day.

Once all the preliminary organizational information has been acquired, along with the data concerning risk, it is possible to apply the formulas for calculating the final risk indexes.

6.4.6 Application of Mathematical Models

6.4.6.1 Preliminary Preparation of "Fictitious Working Day" Representative of the Year and Each Month of the Same Year

As for risk assessments in the case of tasks that rotate daily (Section 6.2) or annually (Section 6.3), two models are provided here too for calculating the final exposure risk

TABLE 6.16
Example of Calculation of Hours Worked per Representative Week (Data Taken from Table 6.29): Preliminary Data for Obtaining the Fictitious Day Representative of the Week

COMPITO	Weekly Task	Monday	Tuesday	Wednesday	Thursday	Friday	Saturday	Sunday	No. of Minutes Worked in a Week per Task	% vs. Total Minutes/ Week	% vs. Weekly Constant (440 MIN * 5 days = 2200 min)
AAA	X	4.50	9.00		9.00	9.00	68.00		100	6.22%	4.52%
BBB	X	4.50	9.00		9.00	9.00	68.00		100	6.22%	4.52%
CCC	X	4.50	9.00		9.00	9.00	68.00		100	6.22%	4.52%
DDD	X	2.25	4.50		4.50	4.50	68.00		84	5.23%	3.81%
EEE	X	2.25	4.50		4.50	4.50	68.00		84	5.23%	3.81%
FFF	X	13.50	27.00		27.00	27.00	0.00		95	5.91%	4.30%
GGG	X	90.00	180.00		180.00	180.00	0.00		630	39.38%	28.64%
HHH	X	9.00	18.00		18.00	18.00	0.00		63	3.94%	2.86%
III	X	4.50	9.00		9.00	9.00	0.00		32	1.97%	1.43%
LLL	X	45.00	90.00		90.00	90.00	0.00		315	19.69%	14.32%
		180	360	0	360	360	340	0	1,600	100%	73%

TABLE 6.17

Reconstruction of the Month. (Calculation of total hours worked and prevalences [versus both total hours worked and a constant 160 h/month] per task and for each of the 4 weeks of the representative month and for the whole month. The data is thus acquired to transform the month into a fictitious day)

Active Tasks in the Month		Duration in Minutes of Each Task in Week 1	Duration in Minutes of Each Task in Week 2	Duration in Minutes of Each Task in Week 3	Duration in Minutes of Each Task in Week 4	Total Duration of Tasks in the Month in Minutes	Total Hours Worked in the Month	% vs. Total Hours Worked in the Month	% vs. Constant Month of 160 h
AAA	X	100	95	104	41	339	6	4.9%	3.5%
BBB	X	100	95	104	41	339	6	4.9%	3.5%
CCC	X	100	95	104	41	339	6	4.9%	3.5%
DDD	X	84	82	86	20	272	5	4.0%	2.8%
EEE	X	84	82	86	20	272	5	4.0%	2.8%
FFF	X	95	81	108	122	405	7	5.9%	4.2%
GGG	X	630	540	720	810	2700	45	39.4%	28.1%
HHH	X	63	54	72	81	270	5	3.9%	2.8%
III	X	32	242	36	41	350	6	5.1%	3.6%
LLL	X	315	485	360	405	1565	26	22.8%	16.3%
							114	100.0%	71.4%

index, one based on the time-weighted average and the other, the multitask complex, based not on average estimates but on the worst task score (calculated versus its actual duration), as the minimum exposure value, to be increased with respect to the value of the other tasks, again with respect to their duration.

Both mathematical models have been used and adjusted for calculating weekly or monthly multitask exposure risk, as illustrated.

In order to apply them to weekly and/or monthly exposure (a repetition of 4 weeks), as mentioned earlier, the period under examination is transformed into a single fictitious work day. The data for performing this conversion are summarized (Tables 6.16 and 6.17):

- Active tasks
- Duration of tasks in hours
- Percentage duration of each tasks versus the total hours worked in the week or month
- Percentage duration of each task versus the constant workload in the week (40, equal to 2200 min) or in the month (160)

In order to complete the estimation of a fictitious daily or monthly working day, certain procedures are required that involve recalculating the intrinsic risk index scores (IRi, i.e., estimated for tasks lasting 8 h with one meal break and two 10 min pauses) to generate the new adjusted intrinsic risk index values, or IRic, reflecting the actual organizational conditions in the workplace, using the formula provided in Section 6.2:

$$\text{IR MultiMP} = \left(\sum \text{IRic}_J * \text{FT}_j\right)$$

Therefore, the following parameters must be estimated:

- *Recovery Multiplier*: Take the value obtained from the organizational data describing the hourly distribution of tasks on a *representative day of the week or month* (Table 6.14); among those indicated for each day of the week (or month), choose the *modal* value that is 1.2 in the example
- *Duration Multiplier*: Take the value corresponding to the duration of the hours worked on the *average fictitious day* of the week and/or month, obtained as follows:
  - *For the week, divide the total number of minutes actually worked in the week (Table 6.18) by 5 working days/week (in the example, the fictitious day lasts 320 min and the corresponding duration multiplier is 0.925)*
  - *For the month (Table 6.19), divide the total number of hours worked in the month by the constant 20 working days/month. Multiplying the result by 60 produces the number of minutes of the fictitious working day representative of the month (in the example, the fictitious day lasts 342.5 min and the corresponding duration multiplier is 0.925).*

TABLE 6.18
Example of the Estimation of the Duration Multiplier Obtained from the Average Fictitious Day Representative of the Week (Table 6.30)

TASK	Active Task	Monday	Tuesday	Wednesday	Thursday	Friday	Saturday	Sunday	Total Minutes Worked per Task in a Week	Duration of *Fictitious Working Day* in Minutes, Representative of a Week	Duration Multiplier for Fictitious Day
AAA	X	4.50	9.00		9.00	9.00	68.00		100	Method of	
BBB	X	4.50	9.00		9.00	9.00	68.00		100	Calculation = 1600	
CCC	X	4.50	9.00		9.00	9.00	68.00		100	(total duration in	
DDD	X	2,25	4.50		4.50	4.50	68.00		84	minutes of	
EEE	X	2,25	4.50		4.50	4.50	68.00		84	repetitive tasks in a	
FFF	X	13.50	27.00		27.00	27.00	0.00		95	week)/5	
GGG	X	90.00	180.00		180.00	180.00	0.00		630	(constant days	
HHH	X	9.00	18.00		18.00	18.00	0.00		63	worked per week)	
III	X	4.50	9.00		9.00	9.00	0.00		32		
LLL	X	45.00	90.00		90.00	90.00	0.00		315		
TOT		180	360	0	360	360	340	0	1600	320	0.925

TABLE 6.19
Example of the Estimation of the Duration Multiplier Obtained from the Average Fictitious Day Representative of the Month (Table 6.31)

TASK	Active Tasks	Duration in Minutes of Each Task in Week 1	Duration in Minutes of Each Task in Week 2	Duration in Minutes of Each Task in Week 3	Duration in Minutes of Each Task in Week 4	Total Duration of Tasks in a Month in minutes	Total Hours Worked in a Month	Duration of Fictitious Working Day in Minutes, Representative of the Month	Duration Multiplier For Fictitious Day
AAA	X	100	95	104	41	339	6	Method of	
BBB	X	100	95	104	41	339	6	Calculation = 114.2	
CCC	X	100	95	104	41	339	6	(total hours of net	
DDD	X	84	82	86	20	272	5	duration of	
EEE	X	84	82	86	20	272	5	repetitive tasks in a	
FFF	X	95	81	108	122	405	7	month)/20	
GGG	X	630	540	720	810	2700	45	(constant of worked	
HHH	X	63	54	72	81	270	5	days in a month)*60	
III	X	32	242	36	41	350	6		
LLL	X	315	485	360	405	1565	26		
TOT		1600	1850	1780	1620	6850.0	114.2	342.5	0.925

6.4.6.2 First Calculation Model: Time-Weighted Average (IR MultiMP)

Figure 6.22 presents an example of how to calculate the weighted mean first for 1 week and then for the month. All the data is included for calculating the *time-weighted average*, according to the following formula, which is conceptually the same process as for the whole year:

- Calculate the *daily weighted average* for each day (as shown in Section 6.2), that is, take the product of each OCRA checklist score indicated here for its proportional duration, and add up all the results: ($\Sigma \mathbf{IRic}_j * \mathbf{FT}_j$).
- To obtain the *weekly weighted average*, add up the risk index scores (obtained with the OCRA checklist using the *weighted average* model) for each day of the week, and divide these by 5 (i.e., constant days worked per week). If the homogeneous group works less than 5 days a week, a risk score of 2 is assigned to nonwork days (i.e., no risk). This adjustment produces a more accurate final weekly exposure score, especially when the days worked do not correspond to the constant (e.g., when there are only 2 work days a week).
- To obtain the *monthly weighted average*, add up the risk index scores (obtained with the OCRA checklist using the *weighted average* model) for each individual work day in the 4 weeks, and divide the sum by 20 (i.e., constant days worked per month). Here too, assign a risk score of 2 to non-work days, when less than 20.

6.4.6.3 Second Calculation Model: Multitask Complex (IR MultiCom)

As illustrated here for the classic risk index calculation procedure, if the daily repetitive task rotation has a duration of more than 90 min, the approach is based on the *task generating the highest overload as minimum* concept. With this approach, the result will be at the least equivalent to the OCRA index for the most overloading task, in terms of its actual duration, and at the most equal to the OCRA indicator for the same task applied, however (only theoretically), to the overall duration of all the repetitive tasks examined and present in the shift.

The procedure is based on the following formula, as explained in Section 6.2:

$$\text{IR MultiComp} = \text{IRic}_{1(\text{Dm1})} + (\text{IRic}_1 \times \text{K})$$

As mentioned earlier, if this mathematical model is applied to weekly and/or monthly exposure, it is necessary to create *a fictitious day representative of the week and/or month*, which is somewhat complex, although essential to estimating the final risk.

Moreover, to apply this mathematical model, it is necessary to calculate the IR scores of each individual task also with reference to their actual duration (and not just with respect to the total duration of repetitive tasks in the shift). Hence, the *fictitious minutes of duration* must be obtained as if the week or month were comparable to a day.

	OCRA CHECKLIST RISK INDEX OBTAINED USING THE **TIME-WEIGHTED AVERAGE FORMULA** (RIGHT UPPER LIMB)				
DAY	**Week 1**	**Week 2**	**Week 3**	**Week 4**	**MONTH**
MON	10.8	0,0	0,0	10,8	
TUE	16.5	16,5	16,5	16,5	
WEN	0,0	16,5	16,5	16,5	
THU	16,5	16,5	16,5	16,5	
FRI	16,5	0,0	16,5	16,5	
SAT	9.2	9,2	9,2	0,0	
SUN	0.0	31.3	0,0	0,0	
Risk value for each week of the month	**13.9**	**18**	**15.1**	**15.4**	**15.6**

FIGURE 6.22 Example of time-weighted average for each week of the month and each month (To obtain the weekly weighted average, add up the risk index scores for each day of the week and divide these by 5 [i.e., constant days worked in the week]. If the workers work less than 5 days a week, assign a risk score of 2 to these days. To calculate the monthly weighted average, add up all the scores obtained for each individual day worked in the 4 weeks and divide the total by 20 [i.e., constant work days per month])

Figure 6.23 shows this broken down into two parts: Part A depicts the calculations relative to one sample week (it is not possible to show all 4 weeks in full) and Part B to the month. The fictitious minutes describing these representative days are obtained via the following steps:

- List of hours worked on each task for a week (for a weekly cycle)
- List of hours worked on each task for each week of the month and for the whole month, if the cycle is monthly
- Calculation of their proportion versus the *total number of hours worked* in the week, for each week of the month or for the whole month (FT_J)
- Duration of total repetitive work (Y) present in the average fictitious day representative of the week and/or month
- The product of (FT_J) by (Y)

ACTIVE TASKS	TOTAL MINUTES WORKED IN WEEK 1 BY HOMOGENEOUS GROUP	% VS. TOTAL HOURS WORKED IN WEEK 1 **(FTj)**	MINUTES OF TOTAL REPETITIVE WORK IN FICTITIOUS DAY REPRESENTATIVE OF WEEK 1 **(Y) (Table 6.9)**	FICTITIOUS MINUTES PER TASK IN FICTITIOUS DAY REPRESENTATIVE OF WEEK 1 **(fm_j= FTj * Y)**	DURATION MULTIPLIER FOR EACH TASK IN FICTITIOUS DAY Dm_J
AAA	100	6.%		**20**	0.200
BBB	100	6.%		**20**	0.200
CCC	100	6.%		**20**	0.200
DDD	84	5.%	1600/5 =320	**17**	0.200
EEE	84	5.%		**17**	0.200
FFF	95	6%		**19**	0.200
GGG	630	39.%		**126**	0.650
HHH	63	4%		**13**	0.100
III	32	2%	DURATION MULTIPLIER= 0.925	**6**	0.050
LLL	315	20%	RECOVERY MULTIPLIER= 1.21	**63**	0.05
	1600	100%		**320**	

ACTIVE TASKS	TOTAL HOURS WORKED IN A MONTH BY HOMOGENEOUS GROUP	% VS. TOTAL HOURS WORKED IN MONTH **(FTj)**	MINUTES OF TOTAL REPETITIVE WORK IN FICTITIOUS DAY REPRESENTATIVE OF MONTH **(Y) (Table 6.30)**	FICTITIOUS MINUTES PER TASK IN FICTITIOUS DAY REPRESENTATIVE OF THE MONTH **(fm_j= FTj * Y)**	DURATION MULTIPLIER FOR EACH TASK IN FICTITIOUS DAY Dm_J
AAA	6	5%		17	0.200
BBB	6	5%		17	0.200
CCC	6	5%		17	0.200
DDD	5	4%	114.2/20*60= 342	14	0.100
EEE	5	4%		14	0.100
FFF	7	6%		20	0.200
GGG	45	39 %		135	0.650
HHH	5	34%		14	0.100
III	6	5%	DURATION MULTIPLIER= 0.925	18	0.200
LLL	26	23%	RECOVERY MULTIPLIER= 1.22	78	0.500
	114.2	100%		**342.5**	

FIGURE 6.23 Estimation of fictitious duration in minutes of each task in the week and month (As if the week and month were reduced to a fictitious day). (The recovery multiplier is the modal multiplier between the days of the week or the days of the month [Table 6.14])

The fictitious day is thus reconstructed in detail for the multitask complex, whether it refers to each week or to the month; based on the fictitious duration of each task in minutes, as calculated here, it is possible to obtain the following:

- *Partial Duration Multiplier* (Dm_J) corresponding to each task
- *Total duration multiplier* (Dm_{tot}) corresponding to the total duration in minutes of the representative fictitious day
- *Recovery multiplier* using the modal one estimated in Table 6.14

Figure 6.24 shows all the data required to calculate the final risk index for the month, applying the specific mathematical model presented previously (the calculation for each week is identical).

$$\text{IR MultiComp} = \text{IRic}_{1(\text{Dm1})} + \left(\text{IRic}_1 \times \text{K}\right)$$

For each active task the following information is now available:

- Percentage versus total number of hours worked per week/month (FT_J)
- *Partial Duration Multiplier* (Dm_J) for each task
- The intrinsic OCRA checklist value (per side) recalculated using the *partial duration* of each task $\text{IRic}_{j(\text{Dmj})}$
- Intrinsic checklist value by *total duration* $\text{IRic}_{j(\text{Dm max})}$
- Maximum risk value calculated with respect to the partial duration $\text{IRic}_{1(\text{Dm1})}$
- Maximum risk value calculated with respect to the total duration $\text{IRic}_{1\ (\text{Dm max})}$
- ΔIRic1 = difference between the maximum risk score calculated with respect to total duration $\text{IRic}_{1\ (\text{Dm max})}$ and maximum risk score calculated with respect to partial duration $\text{IRic}_{1(\text{Dm1})}$
- Calculation of K according to the following formula:

$$\left(\text{IRic}_{1(\text{Dm max})} * \text{FT}_1\right) + \left(\text{IRic}_{2\ (\text{Dm max})} * \ \text{FT}_2\right) + \ldots + \left(\text{IRic}_{\text{N}(\text{Dm max})} * \text{FT}_\text{N}\right) \left(\text{IRic}_{1(\text{Dm max})}\right)$$

Since the example refers to a monthly cycle (with each of the 4 weeks featuring different work organization and risk exposure), the risk indexes are calculated for each week of the month and then for the whole month, using the two formulas (time-weighted average and multitask complex); the results include a view of trends for the indexes both day by day and week by week.

Often in weekly or monthly cycles, workers can be exposed to many tasks.

The formula of Multitask Complex is based on research of the worst task, for its duration (and the final value of the risk cannot be less than it). In the presence of many tasks, the% of the worst task duration would be reduced too much.

To correct this problem, the following criteria is used (see above Figure 6.17, for annual cycles):

To create six risk categories from absent (green) to purple (very high risk).

Risk categories	FICTITIOUS MINUTES (FM) AS IF THE YEAR WERE REDUCED TO A DAY	% vs. total hours worked hours/year (FTj)	Duration multiplier for each task (fictitious day) (Dm$_J$)	IR OCRA checklist INTRINSEC scores for PARTIAL DURATION $IRic_{j(Dmj)}$	IR OCRA checklist INTRINSEC scores for TOTAL DURATION $IRic_{j(Dm\ max)}$
GREEN	34	10%	0.350	2.5	6.5
YELLOW	31	9%	0.350	3.4	9.0
LOW RED	169	49%	0.650	8,6	12.2
MEDIUM RED	14	4%	0.100	1.6	15,2
VIOLET (OR HIGH RED)	18	5%	0.200	4.9	22.5
VERY HIGH RED	76	23%	0.500	16.4	30.4
tot	347	100%		23.9	
16.4 + (30.4-16.4)* K= 23.9 K=((6.5*10%)+(9*9%)+(12.2*49%)+(15.2*4 %)+ (22.2*5%) +(30.4*23%))/30.4					

FIGURE 6.24 Calculation of the final index using the IR MultiCom model (right limb): We created six categories of risk from absent (green) to purple (very high risk). The minutes spent in each of the six categories were grouped. Within each category the average value of weighted risk was calculated for the category, considering the risk value of the tasks included in this category and their duration. From these values we build the fictitious working day, as representative of the month.

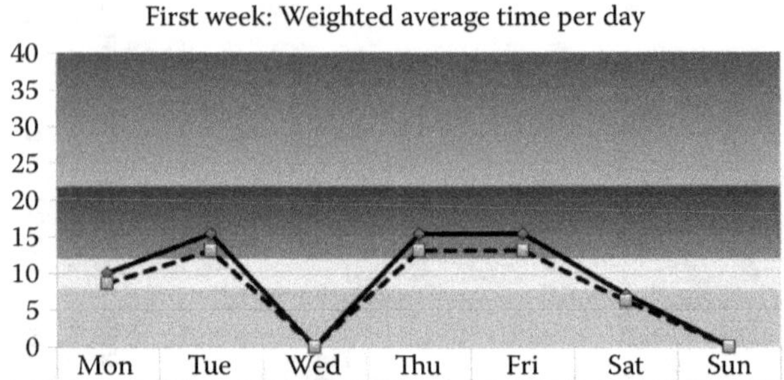

	Mon	Tue	Wed	Thu	Fri	Sat	Sun
R	10.8	16.5	0.0	16.5	16.5	9.2	0.0
L	9.3	14.2	0.0	14.2	14.2	8.2	0.0

R 13.9	L 12
Time-weighted average	

FIGURE 6.25 Example of trends for the risk indexes assessed each day of week 1. The risk indexes for the week are calculated for both limbs.

The minutes spent in each of the six risk categories were grouped.

Within each category the average value of weighted risk was calculated for the category, considering the value risk of the tasks included in this category and their duration.

From these values we build the fictitious working day, as representative of the week/month cycles.

Figure 6.25 shows the trends for the risk indexes assessed each day of week 1. The risk indexes are calculated for both limbs, using the weighted average time formula both recommended formulas; the results are different for the right and the left limb.

Figure 6.26 shows the trends for the risk indexes assessed each week of the month. The risk indexes for the whole month are calculated for both limbs, using both recommended formulas.

Figure 6.27 shows the trends for the risk indexes assessed each day in the 4 weeks of the month using the weighted average formula for right and left upper limbs.

With the risk index trends shown per day, per week, and for the entire month, it is easy to comprehend the differences between the two formulas. Since the work is distributed differently, the time-weighted average formula tends to flatten the peaks, determining lower risk scores than the checklist multitask complex formula, which is never lower than the highest peak, calculated for its actual duration.

6.4.7 Conclusions

Without reiterating the rationale underlying the application of two risk calculation models to these studies, as discussed in Section 6.3 on annual task rotations, emphasis is placed here on the difference between the two models applied to the weekly/monthly or annual cycle. While the formulas are identical, more precise organizational data is available for the weekly cycle analysis. In fact, it is possible to analyze each work day in the week of a weekly cycle. Therefore, more accurate scores and their corresponding

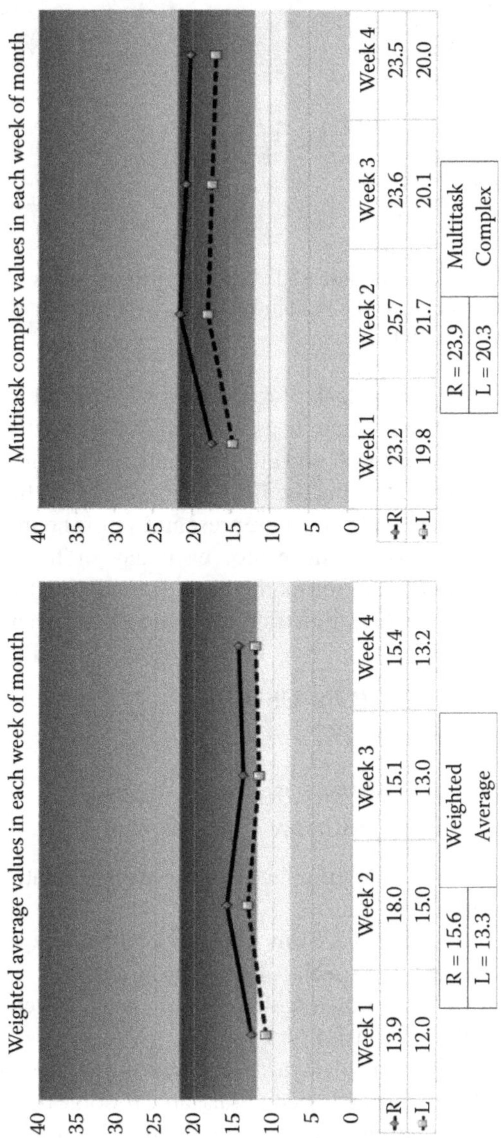

FIGURE 6.26 Trends for risk indexes assessed each week of the month. The risk indexes for the whole month are calculated for both limbs, using both recommended formulas.

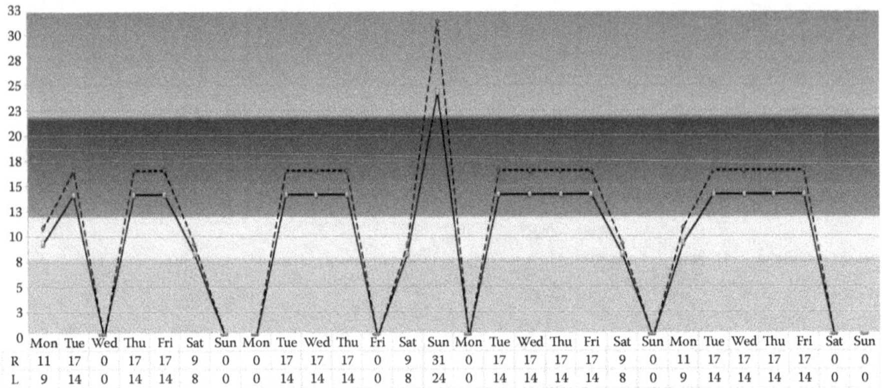

FIGURE 6.27 Trends for risk indexes assessed each day of the 4 weeks of the month. The weighted average formula is applied for the final index for each day of the month. Comparison between results for right and left limb.

multipliers are generated, such as the lack of recovery time. First the week is reduced into a fictitious working day, then the month, as was before done for the year.

For a monthly cycle, the approach toward calculating the corresponding risk indexes uses the same method for finding the IR of each week, then the whole month, again reconstructing a fictitious working day representative of the month.

The available data (precise risk indexes for each day of the week, for each week, and then for the whole month), forms the basis for building the useful graphs described earlier (Figure 6.27), which illustrate exposure trends even day by day.

6.5 STUDY OF TASK ROTATION USING THE OCRA MINICHECKLIST

6.5.1 OCRA Minichecklist Model for Jobs Featuring Several Repetitive Tasks (Multitask Work)

It is not unusual to use the OCRA minichecklist to assess situations in which a worker performs several different tasks.

Since it is difficult to perform such an evaluation manually with a view to conducting a risk assessment, a special Excel program has been developed for this purpose, which can be downloaded free from www.epmresearch.org: ERGOepmMINIcheckOCRAmultitask-EN.

Example 6.5.1 illustrates how to calculate risk when there is exposure to multiple tasks.

The OCRA minichecklist has already been illustrated in some detail in Chapter 5, and the criteria and mathematical models for calculating exposure to multiple tasks were explained earlier in this chapter (Sections 6.2, 6.3, 6.4); the same criteria and models have also been adopted for the OCRA multitask minichecklist. Therefore, the relevant analytical descriptions will not be repeated.

The Excel software provides a means of assessing a worker or homogeneous group of workers.

Overall shift duration (minutes)	420
Total duration of nonrepetitive tasks (e.g., cleaning, fetching supplies, etc.) in minutes	10
No. of actual breaks (recovery periods) during the shift, with a duration of at least 8 min (except meal break) that can be considered as recovery periods	2
Overall duration of all actual breaks (excluding meal break) in minutes	20
Actual duration of meal break if included in shift duration (in minutes)	30
No. of other breaks (i.e., meal break out of working time; travel time to/from different company locations). Mark one number only when these breaks last at least 30 min	

Description of rotation between tasks

Daily	Weekly	Monthly	Yearly	Net duration of repetitive work estimated in minutes
X				360

FIGURE 6.28 Example 6.5.1: Description of shift and pauses in order to study the net duration of repetitive work and recovery times.

The first step entails describing a working day (one shift) in order to identify the net duration of repetitive tasks, that is, the total duration of all the individual repetitive tasks performed by the worker and the distribution of meal breaks and pauses.

Figure 6.28 shows an example of a 7-h shift with a *net duration of repetitive work* of 360 min.

On the same page, the question is asked whether the task rotation begins and ends within one shift or whether the turnover cycle is weekly, monthly, or yearly.

The criteria and techniques for conducting an in-depth study of the turnover were illustrated in detail in the previous sections even when the cycle is longer than a day. With the OCRA multitask minichecklist, it is possible to carry out an albeit approximate evaluation, even when the task distribution changes over the different days of the week (weekly turnover), the month (monthly turnover), or the year (annual turnover).

Figure 6.29 shows the scheme used for indicating tasks and defining the proportion of time spent performing each one within the specific period, be it a day, a month, or a year.

Name of repetitive tasks performed by homogeneous group	There are no actual cycles but the same actions are always repeated	There are actual cycles	Cycle time in seconds (time per piece)	% duration, vs. total period considered
Task A	**Yes**			**15%**
Task B	**Yes**			**5%**
Task C	**Yes**			**50%**
Task D	**Yes**			**30%**
				100%

FIGURE 6.29 Example of a list of tasks and the proportion of time spent performing each one within the specific period, be it a day, a month or a year.

TASK A: Details concerning risk factors	Right side
There are no actual cycles: Indicate the duration of a representative period (in seconds)	60
Technical actions	**Few, low frequency**
Hand in pinch grip	Less than 1/3 of the time
Arm more or less at shoulder height	NO
Repetition of same actions/movements: Stereotypy	NO
Task requires moderate force (Borg score 3–4)	Less than 1/3 of the time
Pace determined by the machine	NO
Task B: Details concerning risk factors	**Right side**
There are no actual cycles: Indicate the duration of a representative period (in seconds)	60
Technical actions	Neither few actions, nor high frequency
Hand in pinch grip	Half of the time
Arm more or less at shoulder height	1/3 of the time
Repetition of same actions/movements: Stereotypy	NO
Task requires moderate force (Borg score 3–4)	NO
Pace determined by the machine	NO
Task C: Details concerning risk factors	**Right side**
There are no actual cycles: Indicate the duration of a representative period (in seconds)	60
Technical actions	**High frequency**
Hand in pinch grip	2/3 of the time
Arm more or less at shoulder height	1/3 of the time
Repetition of same actions/movements: Stereotypy	A good part of the time, but not all the time
Task requires moderate force (Borg score 3–4)	NO
Pace determined by the machine	NO
Task D: Details concernina risk factors	**Right side**
There are no actual: Indicate the duration of a representative period (in seconds)	60
Technical actions	**High frequency**
Hand in pinch grip	All of the time
Arm more or less at shoulder height	1/2 of the time
Repetition of same actions/movements: Stereotypy	All the time
Task requires moderate force (Borg score 3–4)	1/3 of the time
Pace determined by the machine	NO

FIGURE 6.30 Example 6.5.1: Description of the characteristics of each risk factor related to repetitive movements of the upper limbs for each task.

% of use	Net duration of repetitive task in shift	Duration multiplier	**Name of repetitive task**	**Recovery score**	**Frequency**	**Force**	**Side**	**Shoulder**	**Elbow**	**Wrist**	**Hand**	**Stereotype**	**Total posture**	**Additional**	**OCRA checklist**
15%	360	0.925	**Task A**	3	**5**	0	R	0	0	0	0	0	0	0	**5.6**
5%	360	0.925	**Task B**	3	**1**	0	R	**6**	0	0	3	0	6	0	**7.8**
50%	360	0.925	**Task C**	3	**9**	0	R	**6**	0	0	**4**	1.5	5.5	0	**18.1**
30%	360	0.925	**Task D**	3	**9**	2	R	**8**	0	0	**8**	3	11	0	**20.5**

FIGURE 6.31 Example 6.5.1: Intrinsic risk indexes for each task.

Result of the evaluation of exposure to repetitive work using the OCRA minicheckllist and multitask calculation miodel	
Time-weighted average	17.04
multitask complex	19.62

FIGURE 6.32 Example 6.5.1: Two different risk assessment results.

Figure 6.30 illustrates the characteristics of each risk factor related to repetitive movements of the upper limbs for each task, to be entered into the Excel spreadsheet used together with the minichecklist to calculate the intrinsic risk index of each task.

Therefore, a separate OCRA minichecklist has to be devoted to each task.

After filling in the first page (work organization, see previous figures), the organizational structure of the shift appears automatically on each subsequent page, devoted to the risk factors characterizing each task; the organizational structure will be identical to the one on the first page (Figure 6.28).

The intrinsic value, that is, risk score, is estimated as if the task were the only one performed for the entire duration of the shift.

Once all the tasks are entered on the page named "exposure calculation," the scores for the individual risk factors and the intrinsic risk score will appear for each task; these values are calculated based on the actual *net duration of the repetitive work* and the actual distribution of *recovery times* (Figure 6.31), along with the two final scores estimated using the weighted average formula and the multitask complex model (Figure 6.32).

7 Risk Mapping for Repetitive Movements of the Upper Limbs

Criteria, Techniques, and Objectives

7.1 INTRODUCTION

In the previous chapters, the assessment of risk caused by biomechanical overload of the upper limbs has been tackled using the most appropriate checklist model to suit the level of accuracy required by the specific investigation. Obviously, all the models feature a similar structure and the end result will be the Occupational Repetitive Actions (OCRA) checklist indicator, with scores assigned to the various risk factors present in the mathematical model.

These results represent an estimate of the individual risk associated with the activities performed at each workstation, but in order to correctly interpret the overall results (of the various settings analyzed), it has been necessary to develop a risk mapping tool to build a more global and thus more readily manageable view of the resulting trends.

The risk assessments for each workstation, obtained using different versions of the OCRA checklist, are entered into specially created Excel spreadsheets, tailored to the specific operation (production lines, homogeneous areas, departments, or the overall company), and broken down by gender. The criteria and procedures required for efficient and accurate risk mapping, which is also achieved through the use of the aforementioned spreadsheets, will be illustrated in this chapter. The risk mapping file ERGOepmOCRAchecklistMAP-EN can also be downloaded free from the website *www.epmresearch.org*.

7.2 DESCRIPTION OF THE WORK AND WORKFORCE AND IDENTIFICATION OF WORKSTATIONS WHERE REPETITIVE TASKS ARE PERFORMED

Before starting risk mapping, it is essential to briefly describe the way the work and workforce are organized and then to identify the workstations where repetitive tasks are performed that need to be entered (Chapter 2).

TABLE 7.1
General Description of the Company: Example Describing Operative Areas, Main Processes, and Work Shift Distribution

Brief Description of Production Setup, Operating Areas, and Shift Distribution

Company Name and Main Business

Company name	BETA							
Main business	Cutlery production							
	Departments and Manufacturing Areas							
Name	**Manufacturing Area**							
A 1	Laminating							
A 2	Lathe machining							
A 3	Assembly							
	Work Organization							
	Shift 1		**Shift 2**		**Shift 3**		**Daily Shift**	
Start	**Start**	**End**	**Start**	**End**	**Start**	**End**	**Start**	**End**
A 1	6 a.m.	12 p.m.	12 p.m.	6 p.m.	6 p.m.	12 a.m.		
A 2	6 a.m.	12 p.m.	12 p.m.	6 p.m.	6 p.m.	12 a.m.		
A 3							8 a.m.	5 p.m.

The initial descriptive part entails identifying departments or manufacturing areas where repetitive tasks are carried out, using the criteria set forth in Chapter 4. A number of preconfigured charts are provided to facilitate collecting this information. Table 7.1 shows an example of the information concerning the names of departments or production areas, type of production, and distribution of shifts in each department/area. Table 7.2 shows the total number of workers broken down by gender (and by department/area) and by type of workstation:

- Workers performing repetitive tasks on assembly lines
- Workers performing repetitive tasks on a bench (not line work)
- Workers performing nonrepetitive tasks (e.g., foremen, maintenance staff, or forklift truck drivers)

In order to more effectively interpret the results of an occupational health and safety program (if one is in place), it may also be useful to indicate the average age and average length of employment of each of the aforementioned categories of workers. It is also necessary to indicate the presence of any part-time workers.

7.3 REPETITIVE TASK MAPPING SOFTWARE

The risk mapping software program ERGOepmOCRAchecklistMAP-EN was created in cooperation with several companies, in particular Whirlpool and Electrolux,

TABLE 7.2
Example Describing Workforce, Broken Down by Department and/or Manufacturing Area, Age, Gender, Type of Workstation

		Worker Details											
		Male				Female				Full Time		Part Time	
Department/ Area	Total No.	No.	%	Average Age	Length of Employment (Years)	No.	%	Average Age	Length of Employment (Years)	No.	%	No.	%
A 1	50	50	100	30	8	0	0			50	100		0
A 2	50	30	60	38	10	20	40	30	8	50	100		0
A 3	100		0			100	100	40	15	80	80	20	20

	Task					
	Assembly Line		Individual Bench			
	With Repetitive Movements		With Repetitive Movements		Nonrepetitive Tasks	
Department/Area	No.	%	No.	%	No.	%
A 1	30	60	10	20	10	20
A 2			45	90	5	10
A 3			98	98	2	2
Total 200	30	15	153	76	17	8

whose occupational health and safety technicians implemented it in order to collect, interpret, and manage the results obtained with the OCRA checklist. Since this is a spreadsheet application, it is best suited to mapping risk in smaller operations. Highly complex software programs are available that offer more extensive data processing and management capabilities to meet all risk mapping requirements (e.g., Petra software by Lema Informatica).

Examples will be shown of the mapping procedures using the aforementioned program, which will more clearly illustrate how the criteria are applied.

The program is composed of the following sections:

- An initial section composed of several similar pages that serve to file the OCRA checklist results. The risk assessment data is entered here, broken down by up to a maximum of four departments or areas. It is advisable to use a separate page for each individual department or single large operating area.
- A second section summarizing the results of the mapping exercise, depicted both numerically and via graphs. The results are grouped by proportional risk levels and presented with respect to the entire company or to an individual department, area, or line and broken down by gender.

Each page containing the OCRA checklist results by department (page name: *map1*, *map2*, etc.) includes six *identical units*. The OCRA checklist scores for the task performed at each workstation must be entered on a separate horizontal line for each workstation. Each unit must refer to a line or identical operating area (for instance, the same types of lathe stands, preassembly benches, etc.). At the end, there is a table summarizing the outcomes of the risk assessment (for a specific line or homogeneous operating area) grouped by risk levels (no risk, slight risk, risk present, high risk).

The map only describes workstations where repetitive tasks are performed, regardless of whether there is any exposure risk.

The work performed at a workstation must be assessed without initially taking into account job rotations that the worker might be assigned to at other workstations; therefore, the repetitive task is analyzed as if the worker performs the same job for the entire shift. The initial exercise essentially involves *mapping workstation risk* rather than *mapping worker risk*. Obviously, unless the final outcome is task rotation, the two maps will coincide.

Each unit is first defined and named as a specific line or homogeneous operating area, and then the net duration of the repetitive work performed at each workstation is expressed in minutes in the shift under examination.

Reducing the net duration of the repetitive tasks in the shift will produce a proportional reduction in the final OCRA checklist risk index, which is adjusted by the corresponding *duration multiplier* (Chapter 4). It may be necessary to personalize the net duration of the repetitive task performed at each workstation, even though they belong to the same line or homogeneous operating area, if the duration differs significantly between the various workstations.

TABLE 7.3
Example 7.1: Risk Mapping Unit Reflecting the Results of an Assessment of LINE 1, Including Workstation Names and Organizational Details

Net Duration of Repetitive Task					
Minutes	**FullTime**				
440	Duration Multiplier = 1				
Total Cycle Time Duration (Task Time)	**No. of Actions**	**No. of Actions/min**	**% of Use**	**No.**	**Workstation Name**
		LINE 1			
30	10	20	30	P1	Workstation 1
30	15	30	100	P2	Workstation 2
60	40	40	100	P3	Workstation 3
60	30	30	100	P4	Workstation 4
40	35	52.5	100	P5	Workstation 5
60	30	30	100	P6	Workstation 6
60	20	20	100	P7	Workstation 7
60	20	20	100	P8	Workstation 8
30	30	60	100	P9	Workstation 9
60	25	25	100	P10	Workstation 10

Table 7.3 shows the first section of data required to complete the mapping exercise:

- Name of the workstation and task performed at the workstation
- Duration of the cycle
- Number of technical actions observed and counted in the cycle
- Frequency of actions per minute (calculated automatically by the software)

Although not all of this information is essential for computing risk indexes, the map can also be used as a strategic location for filing analytical data, especially with a view to future interventions to redesign the way that the work is organized.

The last column in Table 7.3 introduces the option of describing and proportionally weighting exposure risk for a workstation that is used only partially in the shift.

For instance, a certain process may be carried out at a certain workstation for only 30% of the shift.

By entering this usage as a percentage into the appropriate space, the final exposure index will be automatically decreased (in this case, this worker will be considered one-third of an exposed worker in the final calculation).

After completing the first section, the next step involves entering the data obtained by applying the OCRA checklist. Not only the final scores but also the partial values obtained for each risk factor must always be entered into the mapping software.

Table 7.4 shows an example of how all the assessment data obtained with the OCRA checklists is filed.

As reported in the chapter on compiling the OCRA checklist, only the dominant limb (most commonly the right one) is generally evaluated in the first risk mapping

TABLE 7.4

Form for Entering OCRA Checklist Results per Unit (Line or Homogeneous Operating Area) (Example 7.1)

Workstation Name	Recovery Multiplier	Hours without Recovery	Frequency	Force	Side	Shoulder	Wrist	Elbow	Hand	Stereotypy	Total Posture	Additional	Checklist Risk Index
						LINE 1							
Workstation 1	1.33	4	0	0	Right	1	0	0	2	0	2	1	3.99
Workstation 2	1.33	4	1	0	Right	1	0	0	6	1.5	7.5	0	11.31
Workstation 3	1.33	4	3	0.5	Right	1	2	2	4	0	4	0	9.98
Workstation 4	1.33	4	2	1	Right	1	2	2	3	0	3	0	7.98
Workstation 5	1.33	4	5	1	Right	2	2	3	2.5	0	3	0	11.97
Workstation 6	1.33	4	2	1.5	Bilateral	2	1	2	2	0	2	0	7.32
Workstation 7	1.33	4	0	0	Bilateral	0	0	0	0	0	0	0	0.00
Workstation 8	1.33	4	0	0	Bilateral	0	0	0	0	0	0	0	0.00
Workstation 9	1.33	5	8	1	Right	4	6	3	8	0	8	1	23.94
Workstation 10	1.33	0	1	0	Right	0	0	0	3	0	3	0	5.32

exercise. If the task is found to be symmetrical, then it will be specified as bilateral in the appropriate space (under the heading “side”: BIL).

In certain jobs, the right and left limb perform quite different actions, but both are prone to overload (e.g., cutting meat). In such cases, two different maps must be generated, one for each limb. Attention should be paid to avoid calculating the average between two limbs, or entering the data for both the right and left limb into the same mapping page.

The latter part of the unit includes all the organizational details that serve to weight the OCRA checklist values obtained so far.

Table 7.5 shows an example of how these details are compiled in order to achieve the weighting. The following information must be supplied for each workstation in the analysis so as to obtain the correct exposure assessment results, broken down into proportional risk levels:

- How many shifts use the workstation (in the example, certain workstations are used for two shifts, others for three, and some for just one shift).
- How many active workstations, identical to the one being analyzed, there are in a shift.
- How many workers in all operate at the workstation, or at a group of identical workstations, broken down by gender.

In Example 7.1 and Table 7.5, a total of 88 workstations have been identified as actually utilized. Of these, the breakdown by gender indicates that there are 24 female workers and 64 male workers. Comparing the overall number of workers actually using the workstations included in the analysis with the number obtained

TABLE 7.5
Example 7.1: Quantitative Description of Each Workstation by Number of Shifts, Number of Identical Active Workstations, and Number of Workers (Total and Broken Down by Gender)

OCRA Checklist Risk Index Scores	No. of Shifts per Day	No. of Identical Workstations	No. of Workstations with Repetitive Tasks		
			Total Workers	Female	Male
3.99	1	1	1.00	0.00	1.00
11.31	3	1	3.00	0.00	3.00
9.98	3	3	9.00	0.00	9.00
7.98	3	3	9.00	0.00	9.00
11.97	3	3	9.00	0.00	9.00
7.32	3	3	9.00	5.00	4.00
0.00	3	3	9.00	3.00	6.00
0.00	3	1	3.00	0.00	3.00
23.94	2	8	16.00	16.00	0.00
5.32	2	10	20.00	0.00	20.00
			88.00	24.00	64.00

from the map, it can be seen that the latter number appears to be lower; this is because the map does not count "on-call" workers or other staff stepping in to replace regular employees during breaks or periods off work. The last section of the risk assessment data filing unit suggests weighted average scores (for that particular unit) for each risk factor and the final OCRA checklist score, for all of the exposed workers, broken down by gender (Table 7.6). The example shows that the time-weighted average OCRA checklist score is 9.8 but with significant differences between males and females; the time-weighted average for females (OCRA checklist = 17.5) is higher than for males (OCRA checklist = 6.9).

After entering all the risk assessment data into the unit, the program processes the information automatically, dividing the score into exposure levels. Table 7.7 shows the results of the risk assessment, grouping the proportions into four exposure levels for both the group as a whole and broken down by gender. It is vital to evaluate exposure separately for men and women; to ensure an accurate interpretation of the occupational health surveillance results, it can be useful to refer to their respective exposure levels.

At the bottom of each mapping page is a summary of the risks present in the department or overall operating area that the page refers to. The summary is the same as the risk classification model for each unit, as presented in Table 7.7.

The mapping program includes a page entitled *summary*, which groups together and compares the four pages that sum up the various areas/departments. This is where the final risk distribution is calculated by exposure levels for all the repetitive tasks analyzed within the company.

The final summary is the same as the risk classification model presented within each section, as presented in Table 7.7.

The last page of the mapping program, named *graphs*, automatically generates pie charts depicting the data in the summary page; these refer to both the final results for each department or overall operating area and for the entire company.

Figure 7.1 shows the graph corresponding to Example 7.1 of Table 7.7, with the distribution broken down by risk level relative to the entire group, while Figure 7.2 shows the same data broken down by gender.

7.4 RISK MAPPING FOR WORKSTATIONS ARRANGED IN MANUFACTURING ISLANDS OR MULTITASK WORKSTATIONS

When production facilities feature *manufacturing islands* (i.e., when one or more workers complete the assembly or a part at several workstations) or when several tasks are performed at the same workstation (e.g., when pieces of the same part are worked at the same workstation and the tasks feature different degrees of risk exposure), generally speaking, the analysis is more complex and the cycle time may be longer, lasting for many minutes.

To analyze the risk of biomechanical overload in both such settings, it is essential to:

TABLE 7.6
Example 7.1: Time-Weighted Average Scores for Each Individual Risk Factor and Final OCRA Checklist Score, versus Exposed Workers (Total and Broken Down by Gender)

Workstation Name	Recovery Multiplier	Hours without Recovery	Frequency	Force	Side	Shoulder	Wrist	Elbow	Hand	Stereotypy	Tot. Posture	Additional	Checklist Risk Index
					LINE 1								
	1.33	4.0	2.9	0.6		1.4	1.8	1.5	3.5	0.1	3.6	0.2	9.8
Females (average)	1.33	4.0	5.8	1.0		3.1	4.2	2.4	5.8	0.0	5.8	0.7	17.5
Males (average)	1.33	4.0	1.9	0.4		0.8	0.9	1.1	2.7	0.1	2.9	0.0	6.9

TABLE 7.7
Classification of Risk Exposure in Four Levels for the Entire Group and Broken Down by Gender

	Total	%	No. Males	%	No. Females	%
High	16	18	0	0	16	67
Present	12	14	12	19	0	0
Borderline or very low	18	20	18	28	0	0
Absent or acceptable	42	48	34	53	8	33
Total	88		64		24	

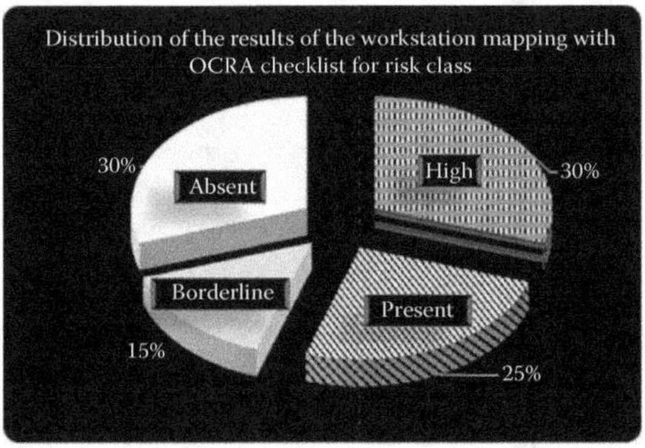

FIGURE 7.1 Example 7.1: graph classifying exposure at workstations, broken down into four risk levels for the entire company.

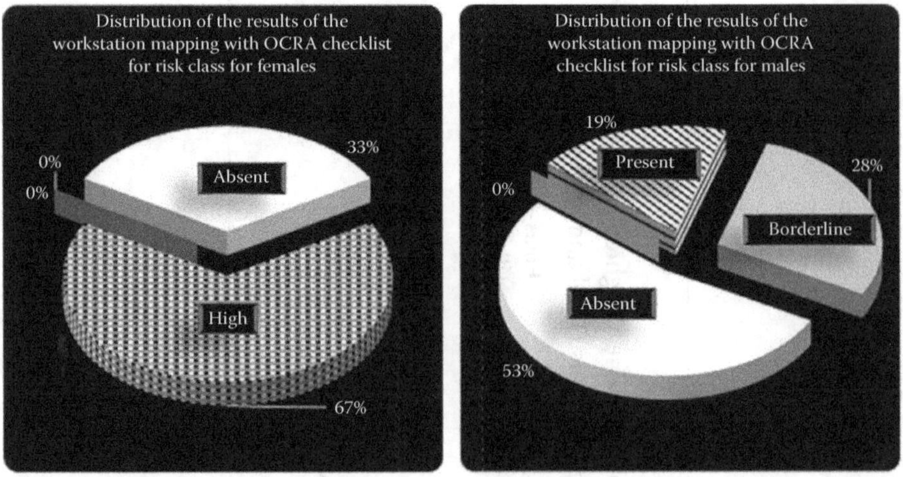

FIGURE 7.2 Example 7.1: graph classifying exposure at workstations, broken down into four risk levels and by gender.

- Identify the various tasks that constitute the manufacturing island or that are performed to produce variants of the same product at the same workstation.
- Analyze each task with its own OCRA checklist.
- Estimate the proportional duration of each task versus the net duration of the repetitive work in the shift.

Basically, this is a genuine multitask analysis, as described in Chapter 6.

It is possible to add this type of analysis to the risk mapping program, using a single unit to enter the assessment scores of each task constituting the island or multitask workstation.

The program automatically calculates the average weighted score for the island (or multitask workstation) based on the proportional duration of each task and the intrinsic risk scores for the individual tasks making up the whole job.

Tables 7.8 and 7.9 show an example (Example 7.2) of risk mapping for a manufacturing island.

Table 7.8 shows the purely organizational data; once the tasks have been identified, it is important to correctly fill in the data under the heading *percentage of utilization*. If the total net duration of repetitive work is 100%, the figure to be entered in the column will be the percentage of time devoted to each individual task. The sum must cover the repetitive task as a whole.

Table 7.9 shows the results of the risk assessment exercise; the OCRA checklist scores are entered here for each task. The utmost attention should be devoted to the column indicating the number of workstations (*no. workstations*); for example, if there are 10 workers per shift working at the assembly island, this same number is entered for all the tasks. The software automatically times how long each worker spends on each task. Then, as usual, the data is broken down by gender.

TABLE 7.8
Example 7.2: Description and Timing of Tasks Making Up an Island

Total Cycle Time Observed	No. of Technical Actions	Frequency (Actions/min)	% of Use in Shift	No.	Workstation Name
					Island 1
30	10	20	5	1	Task 1
30	15	30	20	2	Task 2
60	40	40	10	3	Task 3
60	30	30	5	4	Task 4
40	35	52.5	5	5	Task 5
60	30	30	5	6	Task 6
60	20	20	10	7	Task 7
60	20	20	5	8	Task 8
30	30	60	30	9	Task 9
60	25	25	5	10	Task 10
			100		

TABLE 7.9
Example 7.2: Risk Assessment for a Manufacturing Island. OCRA Checklist Scores for Each Task Identified (Intrinsic Values) in the Overall Job

Workstation Name	Hours without Recovery	Frequency	Force	Side	Shoulder	Wrist	Elbow	Hand	Stereotypy	Posture	Additional	Checklist Risk Index	No. of Shifts per Day	No. of Identical Workstations	Total Workers
								LINE 1							
Task 1	4	0	0	D	1	0	0	2	0	2	1	3.99	3	10	30
Task 2	4	1	0	D	1	0	0	6	1.5	7.5	0	11.31	3	10	30
Task 3	4	3	0.5	D	1	2	2	4	0	4	0	9.98	3	10	30
Task 4	4	2	1	D	1	2	2	3	0	3	0	7.98	3	10	30
Task 5	4	5	1	D	2	2	3	2.5	0	3	0	11.97	3	10	30
Task 6	4	2	1.5	B	2	1	2	2	0	2	0	7.32	3	10	30
Task 7	4	0	0	B	0	0	0	0	0	0	0	0.00	3	10	30
Task 8	4	0	0	B	0	0	0	0	0	0	0	0.00	3	10	30
Task 9	5	8	1	D	4	6	3	8	0	8	1	23.94	3	10	30
Task 10	0	1	0	D	0	0	0	3	0	3	0	5.32	3	10	30
Average scores	4.0	3.4	0.5		1.8	2.3	1.5	4.6	0.3	5.0	0.4	12.5			

The risk score for the manufacturing island is generated by the total weighted score appearing at the bottom of the final OCRA checklist risk score column for each task (in Table 7.9, the total average weighted score is 12.5).

It is essential to store the risk scores for each individual task (i.e., the intrinsic task risk) in order to effectively redesign the work. This type of analysis serves to identify which task(s) or factor(s) within the task are most responsible for overloading the upper limbs in that particular work area.

It is worth pointing out that this evaluation procedure can also be adopted when different models of the same product are assembled at the same workstation, that is, with differences between the tasks and therefore different exposure levels.

7.5 RISK EXPOSURE MAPPING FOR INDIVIDUAL WORKERS WHEN WORK FEATURES TASK ROTATION

The risk mapping program described so far defines the intrinsic risk in relation to work performed at a specific workstation, as if the workstation were utilized by an individual worker to carry out only one repetitive task for the entire shift.

The next step consists in analyzing the exposure level of each individual worker. Organizationally, there are two possible scenarios: The worker either stays at the same workstation all the time or moves around from one workstation to another. In the first scenario, the exposure risk due to biomechanical overload of the upper limbs will correspond to the risk index for the task(s) performed at the (only) workstation that the worker uses.

Conversely, if the worker rotates between several workstations (Figure 7.3), the two formulas illustrated in Chapter 6 must be used to estimate the worker's actual exposure risk to multiple tasks.

The Excel spreadsheet does that calculation. It uses both the weighted mean formula (when the task rotation takes place in less than 90 min) and the multitask complex formula (when the task rotation takes place over more than 90 min).

Once the tasks performed by the worker in the shift have been identified (e.g., task A, task B, task C), their proportional duration of each task should be established (e.g., task A takes up 20% of the time; task B, 70% of the time; task C, 10% of the time). Having previously calculated the risk scores for each task with the OCRA checklist (score A, score B, score C), it is now possible to obtain the two final exposure scores.

The workplace mapping spreadsheet includes a section for analyzing worker exposure. Naturally, this part is closely linked to the risk map presented for analyzing workstation risk. However, in this section, the focus of the analysis is not the workstation but the worker.

Tables 7.10 and 7.11 show examples of how this analysis is applied. Table 7.4 (Example 7.1) shows the risk assessment for an assembly line, describing the OCRA checklist scores for each workstation along the line, as if each workstation were used for the entire shift.

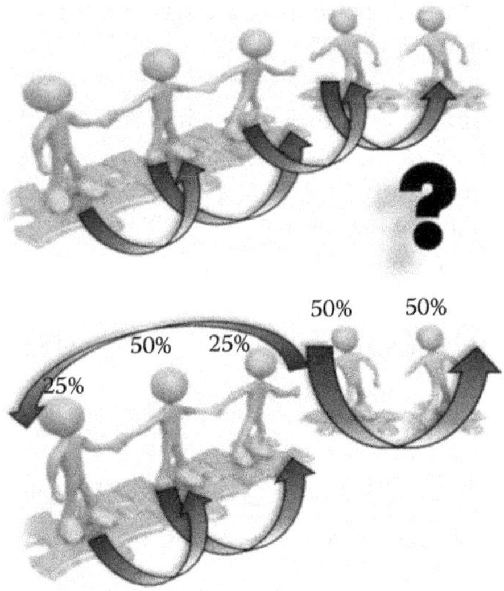

FIGURE 7.3 From the workstation to the worker; task rotation between different workstations (location and duration).

Table 7.10 focuses on the workers, and shows that there is no task rotation at the workstations along the line: The percentage of time spent at each workstation is therefore 100%. In this instance, the risk map for the task(s) performed at the workstations is identical to the risk map for the workers (see Figures 7.1 and 7.2).

A task rotation program will be all the more effective when risks are minimized by properly redistributing risk exposure (Figure 7.3).

An efficient software program can also be used to determine the *best possible rotation* via a series of simulations (Table 7.11 and Figure 7.4); as a result, no worker will be exposed to elevated risk levels.

7.6 CLOSING REMARKS CONCERNING THE INITIAL MAPPING OF RISK ASSOCIATED WITH REPETITIVE MOVEMENTS AND EXERTION OF THE UPPER LIMBS

We have described the various methods and solutions that are available for carrying out initial risk mapping.

Having reached the end of the chapter , we believe it may be useful to briefly summarize and list the relevant criteria, procedures, and objectives.

Risk mapping is thus essential first for estimating and then for managing risk.

Mapping is the first step toward introducing a risk management process that entails:

TABLE 7.10
Example 7.1: Exposure Level for Workers on the Line Illustrated in Table 7.4 without Task Rotation

Worker					Workplace A%				Workplace B%				
No.	Surname	Name	Age	Gender	Workstation No.	Workstation Name	OCRA Checklist Risk	% Time	Workstation No.	Workstation Name	OCRA Checklist Risk	% Time	Individual Risk Exposure
1	AA	AA	45	m	P1	Post. 1	4.0	100		–	0		4.0
2	BB	BB	15	f	P2	Post. 2	11.3	100		–	0		11.3
3	CC	CC	34	m	P3	Post. 3	10.0	100		–	0		10.0
4	DD	DD	66	m	P4	Post. 4	8.0	100		–	0		8.0
5	EE	EE	25	f	P5	Post. 5	12.0	100		–	0		12.0
6	FF	FF	40	m	P6	Post. 6	7.3	100		–	0		7.3
7	GG	GG	44	m	P7	Post. 7	0.0	100		–	0		0.0
8	HH	HH	24	m	P8	Post. 8	0.0	100		–	0		0.0
9	II	II	25	f	P9	Post. 9	23.9	100		–	0		23.9
10	LL	LL	26	m	P10	Post.10	5.3	100		–	0		5.3

TABLE 7.11
Optimization of Exposure Risk For Workers on a Line by Introducing Adequate Task Rotation (no more than 90 min)

Worker					Workplace A%				Workplace B%				
No.	Surname	Name	Age	Gender	Workstation No.	Workstation Name	OCRA Checklist Risk	% Time	Workstation No.	Workstation Name	OCRA Checklist Risk	% Time	Individual Risk Exposure
1	AA	AA	45	m	P1	Post. 1	4.0	50	P2	Post.2	12.5	50	7.6
2	BB	BB	15	f	P2	Post. 2	11.3	50	P1	Post.1	7	50	7.6
3	CC	CC	34	m	P3	Post. 3	10.0	100		–	0		10.0
4	DD	DD	66	m	P4	Post. 4	8.0	100		–	0		8.0
5	EE	EE	25	f	P5	Post. 5	12.0	50	P7	Post. 7	4	50	6.0
6	FF	FF	40	m	P6	Post. 6	7.3	100		–	0		7.3
7	GG	GG	44	m	P7	Post. 7	0.0	50	P5	Post. 5	13	50	6.0
8	HH	HH	24	m	P8	Post. 8	0.0	50	P9	Post. 9	23	50	12.0
9	II	II	25	f	P9	Post. 9	23.9	50	P8	Post. 8	4	50	12.0
10	LL	LL	26	m	P10	Post. 10	5.3	100		–	0		5.3

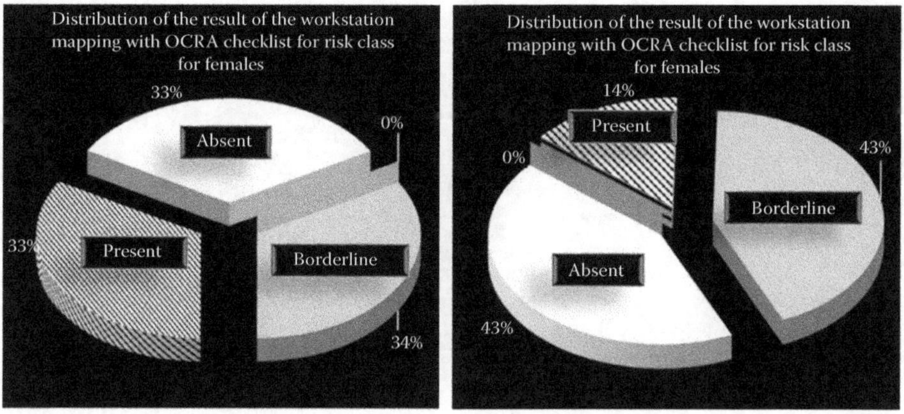

FIGURE 7.4 Pie charts depicting four levels of exposure risk, broken down by gender, after the introduction of optimized task rotation between workstations (compare with Figure 7.2).

- An initial risk assessment of jobs involving repetitive tasks and an initial forecast of the expectations of workers with work-related musculoskeletal diseases of the upper limbs (UL-WMSDs).
- An initial comparison between the risk level and the prevalence of individuals with UL-WMSDs.
- Defining a cause-and-effect relationship between UL-WMSDs and jobs performed in the workplace.
- Creating corrective actions to reduce risk and consequently health issues, with discussions and the setting of priorities.
- Possibly introducing task rotation also with a view to reducing overall exposure risk.
- Hopefully, bringing workers with UL-WMSDs back into the workforce in jobs with minimal or only slight exposure, depending on their condition, via risk mapping and discussions with department heads and occupational health professionals.

8 Analysis of Complex Tasks
Applications

8.1 INTRODUCTION

In Chapter 6, the focus was on analyzing exposure when a job involves multiple repetitive tasks. This chapter now goes on to provide examples of multitask exposure analysis in work situations that are as common as they are difficult to evaluate, because they derive from a complex organizational structure.

The following *complex jobs* will be analyzed here:

- Cleaning services
- Supermarkets
- Commercial kitchens
- Commercial laundries
- Building construction: Builders and plasterers
- Hospitals (jobs involving repetitive movements of the upper limbs)

This chapter is devoted to these special jobs, to provide anyone involved in areas that involve analyzing exposure due to biomechanical overload with information (and residual doubts) and application procedures that we hope will be useful.

All these jobs have the following in common:

- The tasks making them up are not structured as in a factory, with clear-cut cycles and a definite number of parts.
- The tasks are extremely numerous.
- There is considerable variation between weekly/monthly and daily tasks.
- Task duration may also vary considerably.
- The duration of the tasks and the percentage of time they take up within the shift are not always known.

As a result of all these factors, risk evaluations conducted according to current regulations tend to generate unreliable and incomplete assessments, with risk often either under- or overestimated.

This may also be blamed on an underlying lack of organizational analyses, which are as difficult as they are indispensable in the aforesaid sectors, and in the service industry in general.

In our experience, risk assessments have been disappointingly inaccurate and biased in the aforementioned sectors due to difficulties in identifying:

- All the various tasks (listing tasks).

- Who performs what tasks and when (identifying homogeneous groups).
- The duration and sequences of all the various tasks.

Oddly enough, this last problem has proved to be the hardest one to solve. Unless the experts are able to determine the duration of the many tasks performed in the shift, how will it be possible to determine exposure and subsequent risk levels?

At times it can almost feel like "mission impossible"!

In reality, the need to evaluate numerous tasks is not an insurmountable challenge, because the Occupational Repetitive Action (OCRA) method can provide all the answers; however, the analysis takes time, and this needs to be factored in.

From the outset, the analyst must realize that detailed preliminary organizational studies are essential for accurately identifying tasks. Of course, the analysis should also include identifying homogeneous groups and turnover.

The jobs carried out in the sectors under examination are undoubtedly difficult to assess, but they all share one advantage: The tasks involved hardly differ from company to company. If one were to analytically evaluate all the tasks making up the same job, and consider all the various techniques for performing the same tasks to be separate, it would never be necessary to reassess them (e.g., considering the six or more different techniques that exist for cleaning floors as six or more different tasks).

What we aim to do, in these sectors, is to gradually build up a kind of database containing "tasks with preassessed intrinsic risks" for use across the industry.

However, one of the main challenges has been establishing the duration of tasks.

To solve this basic problem, experts in the various sectors were asked to cite the factors they take into consideration when defining the *time lines embedded in a contract*, in the building, cleaning services, or commercial kitchen industries.

They mentioned the *average task duration per unit of measurement*, for example, to sweep a floor measuring 10 m^2 using a sweeper with a rubber blade, wash a window measuring 10 m^2, or paint a certain section of wall using a roller.

We, therefore, came up with the idea of estimating how long it took to perform the tasks, using actual data based on the work contracted out to the individual worker or homogeneous group of workers.

That having been explained, we shall now illustrate the structure and aims of the next lengthy and highly detailed sections.

In the sections on cleaning services, commercial kitchens, supermarkets, and hospitals—the ones we defined as "mission impossible"—we will not be providing risk exposure scores, but rather:

- A highly detailed organizational analysis of the work including the identification of macrophases, phases, and tasks. This structure will be considered representative of the sector. The annexes also include lists of tasks with descriptions of their number and content: they may seem countless but we believe these lists may help appraisers, especially in the early stages of task identification, before performing the risk assessment proper.

- Techniques for identifying homogeneous groups.
- Other organizational studies needed to identify the activity performed by homogeneous groups.
- Units of measurement and explanations of how to use them to define task duration.

It goes without saying that with such a vast volume of data to handle, the final risk scores cannot be calculated manually. Another aim of ours is in fact to provide simple tools (albeit at times a little less than simple, given the massive amounts of data) to allow operators less experienced in conducting exposure risk assessments to estimate risk by simply entering organizational data and the figures associated with the contract (which the company can generally provide). For certain sectors, this is not a utopia: for cleaning services and commercial kitchens, for example, it is already possible with the ad hoc software program Petrasoftware (Lema Informatica, 2015), while for others…the way has been paved.

With regard to the other two sectors that this chapter deals with, commercial laundries and painters and plasterers, we will not just be explaining how to conduct an organizational analysis, but also illustrating the results of the assessments.

The coming sections of this chapter are not likely to be a light read: We will be presenting detailed organizational analyses that some will arguably find complex and heavy going. Alas, it cannot be helped if the reader wishes to first analyze and understand in order to manage risk in these sectors.

8.2 CLEANING SERVICES

8.2.1 Introductory Remarks on the Main Features of the Cleaning Services Sector (ISPESL, 2004)

Over the past few years, cleaning services have become a thriving sector both in Italy and abroad. The reasons lie in the increasing tendency of the public to outsource this work and in better value for money, leading many more clients, including small businesses, to outsource work to cleaning contractors rather than nonprofessionals. Over the past 10 years, the service sector in general, including cleaning services, has changed to better meet market needs, often tailoring offerings to a specific sector, ranging from private and commercial cleaning to remediation and maintenance services in general, thus tailoring the work to the client's demands.

The sector is composed primarily of very small businesses with elevated employee turnover. This makes it difficult to produce standard risk profiles and standard working practices. With the exception of mid-sized businesses and larger corporations, there is often a poor understanding of specific risks, especially biological and chemical risks. Education and training are carried out almost exclusively by large- and medium-sized enterprises or consortiums.

In Italy in 2001, there were 36,726 cleaning companies registered with the local chambers of commerce, including 20 types of businesses each with a different legal status, such as

- One man companies: 62.71%
- Limited liability cooperatives: 12%
- General partnerships: 8.32%
- Limited liability companies: 7.51%
- Limited partnerships: 5.86%

According to the European Federation of Cleaning Industries, in Italy there are 8,000 cleaning contractors set up as businesses, employing approximately 400,000 workers (approximately 2% of the national workforce).

Based on official chamber of commerce data, around 80% of Italian cleaning contractors are small- and medium-sized enterprises with an ill-defined organization and use of human resources. According to ISPESL (Italian National Institute for Occupational Safety and Prevention), the average number of employees per cleaning firm is 2.4, a little over 2 per company. This means that the sector is highly fragmented, with a very large number of firms employing a very small number of workers.

ISPESL also provides some interesting information about workplace accidents and exposure risk in the sector.

These are the main causes of accidents, in decreasing order of importance:

- Falling or slipping (48.6%)
- Dislocations, strains, and sprains ("putting your foot down wrong") (30.3%)
- Improper movements in general ("knocking against something") (11.2%)

Based on an analysis of workplace accident reports, broken down by age and gender, it appears that women have more accidents than men.

Work-related risks can be listed as follows, in order of priority:

- Slipping and falling from a height (scaffolding or ladders)
- Use of chemicals and contact with biological agents
- Use of electrical devices and machinery
- Load handling and repetitive movements

Cleaning services are provided in a wide range of locations, the most common being the following (Figure 8.1):

- Offices (including home offices)
- Health-care facilities
- Manufacturing plants

In manufacturing alone, cleaning services perform numerous tasks; according to ISPESL, the most common are the following:

- Dusting and waste collection
- Sweeping floors, manually or with vacuum cleaners
- Floor washing manually

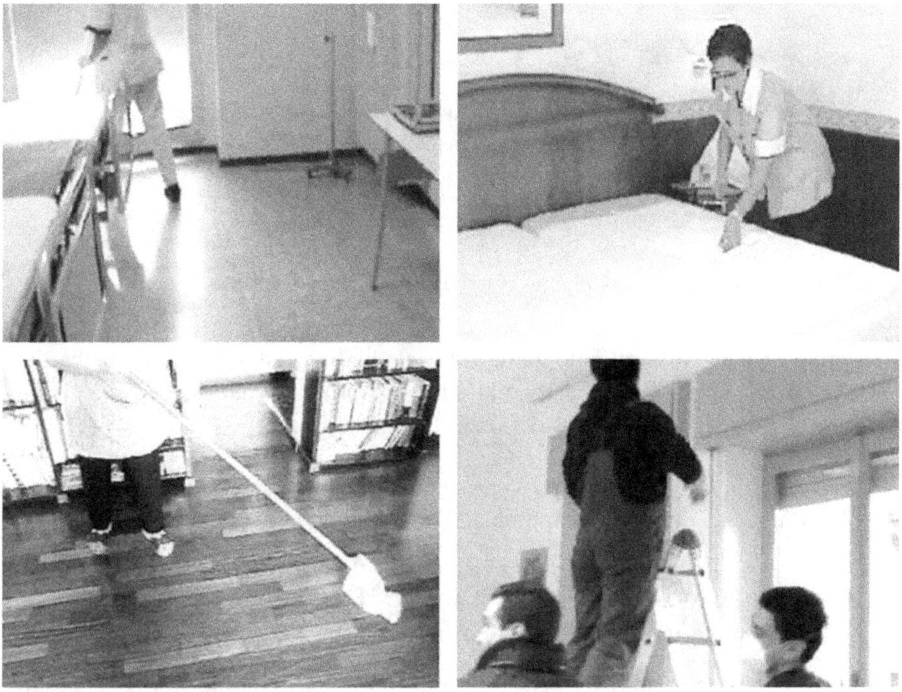

FIGURE 8.1 Common examples of cleaning tasks.

- Inspection of dusting and sweeping
- Floor washing using electric scrubbers/dryers
- Floor waxing and dewaxing
- Cleaning vertical surfaces
- Cleaning and disinfecting toilets and bathrooms

8.2.2 Introduction to Emerging Issues in the Assessment of Risk Caused by Biomechanical Overload of the Upper Limbs

In the cleaning services sector, there are quite a number of emerging issues in the field of risk assessment. It is in fact particularly difficult to quantitatively and qualitatively identify what could be defined as a routine working day due to the following factors:

- The large number of tasks performed
- Extreme variability in the way they are performed
- Countless different workplaces (often with different workplaces in the same shift)
- Highly variable working hours (different part-time (PT) shifts on different days of the week)
- Difficulties in determining the net duration of repetitive tasks (constant changes in workplaces, schedules, rooms, etc.)

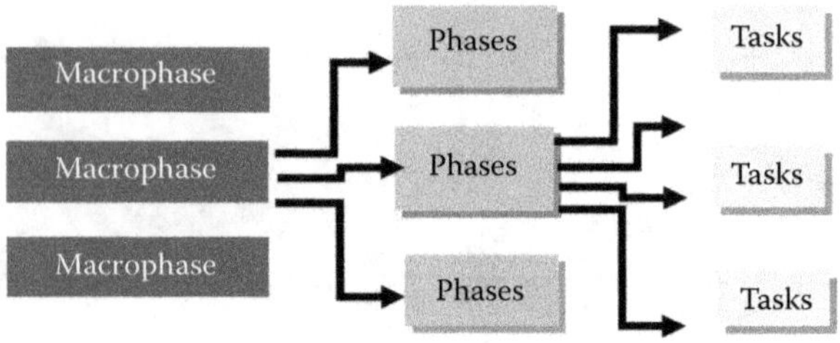

FIGURE 8.2 Breakdown of work into macrophases, phases, and tasks.

In order to achieve a reliable method for evaluating risk due to biomechanical overload of the upper limbs, a special software program had to be created that would

- Allow even nonprofessionals to carry out a reliable risk assessment.
- Lend itself to rapid risk mapping and risk map updates.

8.2.3 Classification of Operations Broken Down into MacroPhases, Phases, and Tasks

Given the extreme variability of cleaning jobs, in order to define worker exposure, it has been necessary to break down work into *macrophases*, *phases*, and *tasks*, as shown in Figure 8.2.

Table 8.1 shows the main macrophases into which cleaning work can be broken down, based on organizational studies carried out in offices, schools, hotels, and hospitals. The analysis should be completed by reviewing other situations such as factories or laboratories.

To explain how tasks are classified within macrophases and phases, we illustrate the procedure by way of an example, as shown in Tables 8.2 and 8.3. Once the macrophases have been identified, each phase is subsequently broken down into actual repetitive tasks.

Table 8.2 shows how the macrophase "routine and periodic bedroom cleaning" has been found to consist of five phases.

Table 8.3 shows the "bathroom cleaning phase", listing all the various tasks.

It should be noted that all possible methods and techniques for performing the tasks have been considered: for example, washing floors without using electrical devices includes four different methods, from using a mop or a cloth (attached or nonattached), to a completely manual method, in a squatting position (see Figure 8.3).

Routine bathroom cleaning actually consists of 15 separate tasks! (see Table 8.3 and Figure 8.4).

Table 8.4 provides another example of tasks classified by phase; this macrophase is the "periodic room cleaning" phase (see Figure 8.5).

So far, about a hundred tasks have been classified with respect to daily and periodic cleaning activities.

TABLE 8.1
Main Macrophases Identified in Organizational Analyses for Cleaning Duties in Offices, Schools, Hotels, and Hospitals

Routine and/or Thorough Cleaning
Offices (with or without toilet)
Bedrooms (with or without toilet)
Routine and/or Thorough Cleaning of Common Areas
Common areas (excluding bathrooms and stairs)
Stairs
Routine and/or Thorough Cleaning of Shared Space (Open-Space Offices, Meeting Rooms, Classrooms, etc.)
Open-space offices
Conference rooms
Meeting rooms
Classrooms with tables (primary, middle, and high school)
Classrooms with desks
Kindergarten classrooms, dining rooms, and other recreational (play) areas
Kindergarten nap rooms
Storage rooms
Waste transportation trolleys
Outside areas
Shared bathrooms
Daily and Periodic Window Cleaning in All Areas
Special Cleaning in Common Areas and Rooms: Venetian Blinds, Roller Blinds, Elevators
Handling/Transporting Cleaning Equipment
Handling/Transporting Luggage

TABLE 8.2
Five Phases Making Up the Macrophase "Routine and Periodic Bedroom Cleaning"

Macrophase: Routine and Periodic Cleaning of Bedrooms (with or without Toilet)	
Phase 1	Remove waste and dirty linens from bedroom and/or bathroom
Phase 2	Clean bathroom
Phase 3	Make/remake bed
Phase 4	Routine room cleaning
Phase 5	Check and refill minibar

8.2.4 Stepwise Presentation of the Model for Analyzing Biomechanical Overload Risk for the Upper Limbs

The procedures for collecting the data required to estimate exposure risk for cleaning staff will now be illustrated step by step, together with the criteria for calculating the final risk indexes.

TABLE 8.3
"Bathroom Cleaning" Phase: Listing all Tasks

Bathroom Cleaning: List of Tasks
Wash floor manually
Wash floor using brush and cloth (not attached)
Wash floor using mop (moderate force to squeeze)
Wash floor with cloth (attached to long handle)
Clean mirror with cloth
Clean sink
Clean shower cubicle
Clean bath tub
Clean toilet
Clean bidet
Clean bathroom walls manually with cloth
Replace bathroom amenities (shampoo etc.)
Replace towels
Replace toilet paper, paper hand towels
Replace soap dispenser
Scrub toilet bowl with cleaning product
Clean toilet brush and holder
Steam-clean and dry bathroom
Wash floor using mop (maximum force to squeeze)

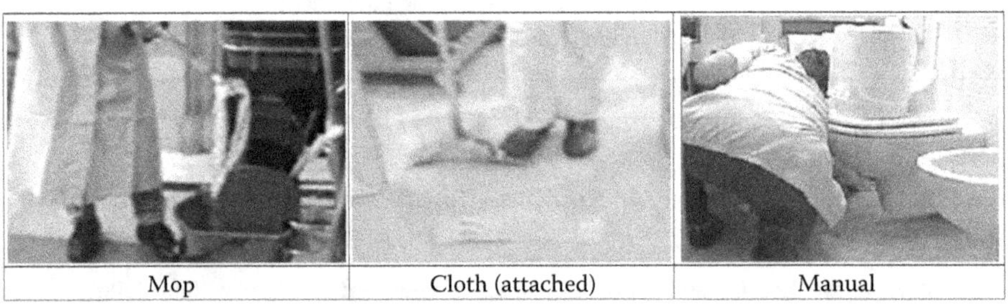

FIGURE 8.3 Several possible methods and techniques for cleaning bathroom floors.

Since the calculations are very complex, a software program has been developed to support the analysis, which will be described later.

8.2.4.1 Step 1: Identification of Homogeneous Group

The definition of a homogeneous group was provided in Chapter 6, Section 6.3.5.

When *tasks of the same nature and duration* are assigned to the same group of workers, we may speak of a *homogeneous group in terms of risk exposure*. A homogeneous group may sometimes be made up of just one person, if no other workers perform the same tasks qualitatively and quantitatively.

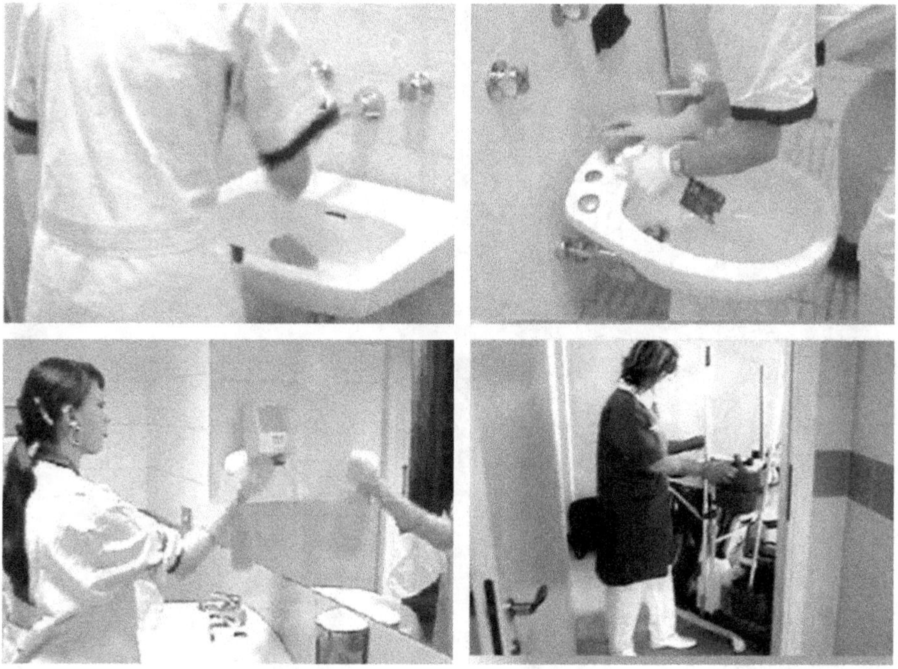

FIGURE 8.4 Several tasks performed in bathroom cleaning.

TABLE 8.4
Tasks Classified by Phase in "Periodic Room Cleaning" Macrophase

Macrophase: Periodic Room Cleaning		
Phase 1	Removal and replacement of furnishings	
Phase 2	Ceilings and walls	
	Task 1	Wash all ceilings
	Task 2	Remove cobwebs from walls and ceilings using extendable pole
	Task 3	Wash walls and windows using squeegee
	Task 4	Remove and replace ceiling lamp using ladder
	Task 5	Dust lighting fixture with feather duster using ladder
	Task 6	Clean ceilings using vacuum cleaner
	Task 7	Dust ceilings using feather duster
	Task 8	Vacuum ventilation/air conditioning vents
	Task 9	Move ladder
Phase 3	Clean furniture and other objects	
Phase 4	Clean floors	

FIGURE 8.5 Several tasks involved in periodic cleaning activities.

Among cleaners, homogeneous groups are generally very small, sometimes composed of just one or two workers.

8.2.4.2 Step 2: Location of Working Areas—Area Managers, Contracts, Workplaces for a Specific Homogeneous Group

The exposure analysis is further complicated by the fact that the same worker or homogeneous group of workers often covers multiple locations during the same shift.

To accurately locate where the worker is working, a few definitions are provided regarding the specific terminology used in the sector (at least among the larger cleaning contractors):

- *Area*: Geographical area assigned to and managed by a supervisor in charge of a group of workers
- *Contract*: An agreement with a business or organization to provide cleaning services
- *Premises*: Locations, covered by a contract, in which cleaning services are provided.

Figure 8.6 shows an example.

Table 8.5 shows an example of a cleaning contract with locations assigned to a homogeneous group of two workers. The workers cover three different premises

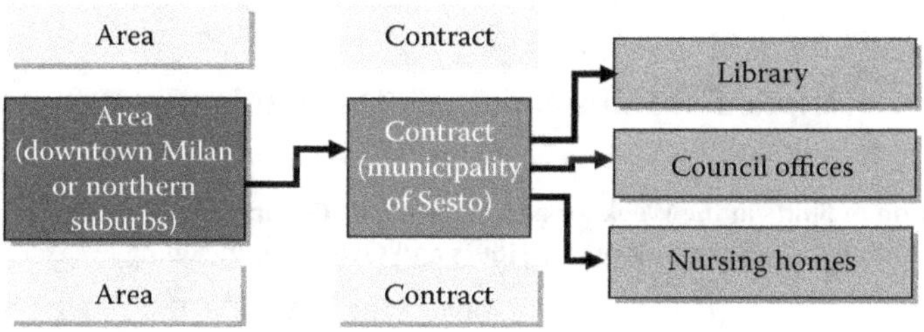

FIGURE 8.6 Description of workplace and workers with definition of area, contractor, and locations.

TABLE 8.5
Example of a Cleaning Contract with Locations Assigned to a Homogeneous Group of Two Workers over Different Days of the Week

Area manager		**PP**						
Name of worker or homogeneous group		Giovanna G. and Francesca R.						
Client		Municipality of SXXX						
No. of workers in the homogeneous group (persons performing the same work with the same work schedule)		2						
Brief description of the tasks performed by the employee or the homogeneous group		Routine office cleaning						
No. of locations covered		3						
	Client Name	**Municipality of SXXX for Locations 1-2-3**						
	Locations Covered by Each Contract	**Mon**	**Tue**	**Wed**	**Thu**	**Fri**	**Sat**	**Sun**
Location 1	Chapel x	X						
Location 2	Office building in the main square	X	X	X	X	X	X	
Location 3	School gym at XXX St.			X				
	No. of locations covered	2	1	2	1	1	0	0

under the same contract, 6 days a week. They work in different places on different days.

It is therefore necessary to analyze how the work is distributed over an entire week: Just 1 day would be insufficient for the purposes of the exposure analysis. The multitask exposure features a weekly cycle for routine (or daily) cleaning and a monthly/annual cycle for thorough or periodic cleaning.

The layout depicted in Table 8.5 is similar to that of the software. The example refers to a homogeneous group, and includes preliminary organizational data concerning the premises at which the work is performed, on different days of the week.

8.2.4.3 Step 3: Work Schedule for Each Day of the Week and Distribution of Breaks

Since the cycle is weekly, the procedures for describing the duration of shifts and the distribution of breaks for each day of the week are those illustrated in Chapter 6, Section 6.4.

TABLE 8.6
Example of Description of Shifts and Break Distribution in a Representative Week

Distribution of Shifts in the Week for a Homogeneous Group (Total Duration of Shift in Minutes, for Regular Working Hours and Overtime, Including Breaks)

	Mon	Tue	Wed	Thu	Fri	Sat	Sun
Shift duration (min)	300	180	300	180	180	300	
No. of breaks during the shift lasting at least 8 min (except meal breaks) that can be considered as valid recovery periods	1	0	1	0	0	1	
Overall duration of all valid breaks (excluding meal break) in minutes	15	0	15	0	0	15	
Duration of meal break if included in shift duration (min)							
No. of other breaks (i.e., meal break before or after working time; travel time to/from different company locations). Enter a number only if these breaks last at least 30 min							
Total duration of breaks (if included in shift duration)	15	0	15	0	0	15	0

An example is shown in Table 8.6. The work is performed at several locations; therefore, it is important to indicate whether there are breaks of at least 30 min between the various premises. Generally speaking, such breaks are not included in the shift and should be treated as unpaid *pauses/meal breaks.*

However, if travel between locations is included in the shift and takes longer than 8 min, such breaks should be regarded as valid recovery periods and reported as such in the forms provided.

Table 8.6 shows the presence of shifts of different durations during the week. The distribution of hours within shifts and over different days of the week tends to vary considerably in cleaning companies.

Figure 8.7 provides a graphic depiction of the duration of weekly shifts (total hours worked over the week), ranging from 2 to 40 h. The most common shift durations range from 15 to 21 h a week (accounting for 44% of the shifts in the large company taken here as an example). Therefore, to avoid overestimating risk, it is essential to estimate the number of hours worked per week in the exposure analysis. The data have been gleaned from one of the largest Italian cleaning companies.

8.2.4.4 Step 4: Calculation of Net Duration of Repetitive Work—Nonrepetitive Work and Unsaturation

It is never easy to calculate the actual time spent performing repetitive tasks, such as the cleaning jobs described here. Unlike factory work, these tasks are not performed in just one place or at a limited number of workstations, but involve constant travel to and from different locations.

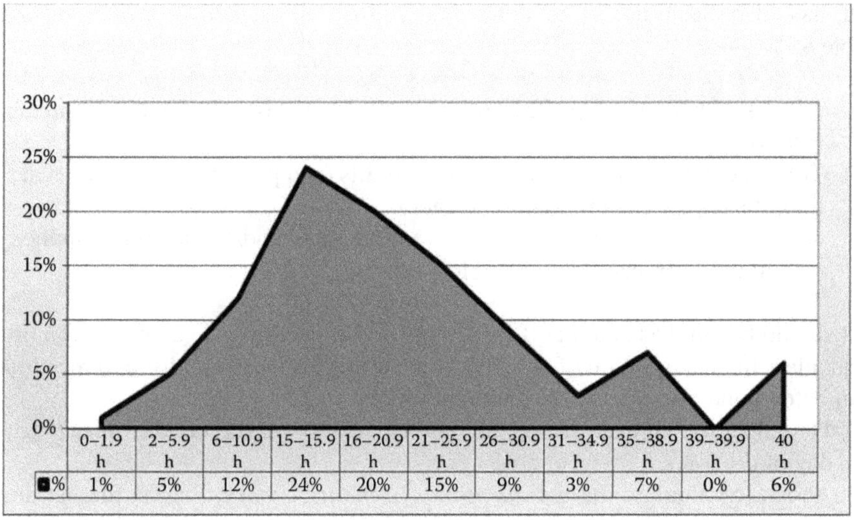

FIGURE 8.7 Example of hours worked per week and prevalence in a large Italian cleaning company with over 3000 employees.

When drafting tenders, technical experts know that they must factor in *unsaturation*, that is, *time wasted* either waiting or dealing with unexpected events. For example, one large cleaning business has estimated the following unsaturation factors (Table 8.7):

- Moving from room to room
- Every 100 m^2 when cleaning common areas (shared bathrooms, corridors, stairs, foyers, etc.) or very large rooms (conference halls, meeting rooms)
- Moving from floor to floor (waiting for elevators, etc.)

TABLE 8.7
Time Constants Defining Unsaturations and Routine Nonrepetitive Tasks

Time Constant Unsaturation	
Time Constants that Every Business Should Estimate on a One-Off Basis for its Employees	**Minutes**
Minutes of average unsaturation for moving from room to room (for rooms measuring 30 m^2)	0.5
Minutes of average unsaturation for working in common areas (measuring 100 m^2)	1
Minutes of average unsaturation for moving to different floors	3
Time Constant for Area of Nonrepetitive Tasks	
Time spent setting up cleaning carts (in minutes) at each individual location	2
Time spent changing into and out of overalls (PPE) in minutes, for each individual location	3
Average time spent retrieving keys and/or tools and/or cleaning supplies (in minutes), for each individual location	2

It has also been useful to determine *routine nonrepetitive work constants* (Table 8.7):

- Time spent setting up cleaning carts (in minutes), for each individual location
- Time spent changing into and out of overalls (and personal protective gear; in minutes), for each individual location
- Average time spent retrieving keys and/or tools and/or cleaning supplies (in minutes), for each individual location

It should be emphasized that if these time constants are to be employed, in order to calculate the net duration of cleaning time, every business will have to reestimate them beforehand *specifically* for their employees.

Table 8.8 shows an example of the final calculation of the *net cleaning time* for each day of the week.

In short, to calculate the *net duration of repetitive work*, that is, the net duration of cleaning tasks, the following difficult but necessary estimations must be carried out:

- *Duration of nonrepetitive tasks*, including such routine tasks as changing into and out of overalls, setting up carts, retrieving supplies and keys, and so on, as well as any other supervisory tasks (supervisors, school janitors, etc.)
- *Unsaturations* primarily due to moving from one room to another or from one floor to another
- *Breaks*, if present

8.2.4.5 Step 5: Description and Duration of Tasks

As illustrated in Section 6.4.5, the next step entails assigning tasks to the homogeneous group, and determining the duration of each individual task on each day of the week. Of course, the intrinsic OCRA checklist score will be calculated for each task.

Table 8.9 shows an example of tasks performed by a homogeneous group. For the sake of visual clarity, the example is relatively simple and includes a limited number of tasks.

The *duration of each task* needs to be defined or estimated. Some have described the process of gathering this essential information as "*mission impossible*"!

When asked about the duration of their various tasks, cleaners and their supervisors often point out why it is so difficult to come up with a reliable answer; for example:

- People work at different speeds and some cleaners are slack and anything but thorough in performing their work
- Although they may not perform strenuous "big cleans," they do carry out thorough cleanings once a week or more, or once a month

TABLE 8.8
Final Calculation of the Net Duration of Cleaning Work for Each Day of the Week

Analysis of Work Organization in the First Week of One Representative Month		Distribution of Shifts in the Week for a Homogeneous Group (Total Duration of Shift in Minutes, for Regular Working Hours and Overtime, Including Breaks)						
		Mon	Tue	Wed	Thu	Fri	Sat	Sun
Shift	Shift duration (in minutes)	300	180	300	180	180	300	0
Nonrepetitive tasks	Setting up cart	4	2	4	2	2	2	0
	Changing into and out of overalls	6	3	6	3	3	3	0
	Retrieving keys and/or tools and/or cleaning supplies	4	2	4	2	2	2	0
	Other time spent performing nonrepetitive tasks (supervision, etc.)							
	Total nonrepetitive tasks in the shift (min)	14.0	7.0	14.0	7.0	7.0	7.0	0.0
	% Nonrepetitive tasks vs. total shift duration	5%	4%	5%	4%	4%	2%	
Unsaturation	Minutes of average unsaturation in changing room	5	5	5	5	5	5	0
	Minutes of average unsaturation in changing areas of shared services	5.2	5.2	5.2	5.2	5.2	5.2	0.0
	Minutes of average unsaturation for moving to different floors	7.5	7.5	7.5	7.5	7.5	7.5	0
	Total unsaturation in the shift (in minutes)	17.9	17.9	17.9	17.9	17.9	17.9	0.0
	% Unsaturation vs. shift duration	6%	10%	6%	10%	10%	6%	
Breaks	Actual duration of breaks in the shift (min)	15	0	15	0	0	15	0
	Net duration of repetitive tasks, estimated	253	155	253	155	155	260	0

TABLE 8.9
Example of List of Tasks Performed By a Homogeneous Group before Quantitative Breakdown over the Various Days of the Week

Routine and Thorough/Periodic Room/Bedroom Cleaning			Mon	Tue	Wed	Thu	Fri
Remove Waste and Dirty Linen from Bedroom and/or Bathroom							
Empty rubbish bins	Daily	100%					
Routine Room/Bedroom Cleaning							
Open/close windows and curtains	Daily	100%					
Sweep floors using broom and long-handled dust pan	Once every 5 days	20%					
Wipe down with spray and damp cloth (furnishings, doors, windows, up to door height)	Once every 5 days	20%					
Wipe down with cloth dampened and squeezed out (furnishings, doors, windows, up to door height)	Daily	100%					
Tidy up room (move and replace objects)	Daily	100%					
Clean Furniture and Other Objects							
Wipe down radiators with damp cloth	Once every 5 days	20%					
Clean Floors							
Wash floor using mop (moderate force to squeeze)	Once every 5 days	40%					
Clean Common Areas (Indoor and Outdoor)							
Clean and Wash Floors							
Sweep floors using broom and long-handled dust pan	Once every 5 days	20%					
Mop floor (moderate force to squeeze)	Once every 5 days	20%					
Clean Sanitary Ware in Common Area Bathrooms							
Wipe down mirrors	Daily	100%					
Wash sink	Daily	100%					
Clean toilet	Daily	100%					
Clean bidet	Daily	100%					
Replace toilet paper, paper hand towels	Three times every 5 days	60%					
Refill liquid soap	Once every 5 days	20%					
Empty rubbish bin	Daily	100%					
Daily and Periodic Window Cleaning in All Areas							
Wash doors and windows inside and/or outside using short-handled squeegee	Once a month	3.3%					

- Task durations are guesstimated rather than calculated on the basis of the same precise timing standards used in the manufacturing industry (MTM, chronometric, etc.)

It is literally impossible to time each task in the shift objectively and thus carry out a risk assessment in the usual way (Chapter 6.4).

8.2.4.6 Step 6: Calculation of Intrinsic Task Duration, by Unit of Measurement, and Total Durations of All Tasks Present in the Shift

Lengthy discussions were conducted with the supervisors and area managers of the aforesaid large Italian cleaning operation, and they were asked several times how they estimated the duration of a contract and therefore its cost. They all gave the same answer: *it takes experience*! In an effort to understand what criteria their experience was based on, it emerged that it all revolved around *task durations defined by predefined units of measurement*. Here are some examples of commonly used units of measurement:

- *Classic unit of measurement for offices* (maximum two or three desks, two rubbish bins, six chairs): The representative area is 30 m^2. For each task performed in this space, an *average execution time per unit of measurement (room size) of 30* m_2 is identified: for example, in a typical room measuring 30 m^2, the time required to wash the floor using a squeegee, or the time required to mop the floor, or the time required to dust the desks, and so on.
- *Unit of measurement for open-space* offices (with more than three desks): The representative area is 40 m^2. For each task performed in this space, an average execution time per unit of measurement (room size) of 40 m^2 is identified.
- *Unit of measurement for common areas* (corridors, bathroom floors, tiled surfaces): The representative area is 20 m^2. For each task performed in this space, an average execution time per unit of measurement of 20 m^2 of surface area cleaned is identified.
- *Unit of measurement for sanitary ware* (toilets, sinks, showers) in common areas (objects to be cleaned): for instance, the average execution time for cleaning a sink or a toilet, and so on.

Table 8.10 lists a number of routine room cleaning tasks (floor cleaning phase) and the *average execution time for each task* with respect to the room unit (average room measuring 30 m^2).

Once the specific units of measurement for the various tasks were determined, representing the basis for estimating average execution times for the various tasks by their respective units of measurement (Table 8.11), the operators calculated the intrinsic duration, by unit of measurement, of each task. It should be stressed that the *average durations* were estimated using a stopwatch. The experts observed workers performing their tasks and reported that they were neither too fast nor too slow; in

TABLE 8.10
Examples of Intrinsic Durations (in minutes) of Floor Cleaning Tasks per Unit of Measurement (30 m² Room)

Clean Floors	Average Task Duration for Room Measuring 30 m² (min)
Vacuum foyer and corridors	2
Wash carpets	
Wash floor using brush and cloth (loose)	0.5
Wash floor using brush and cloth (loose)	
Scrub floor manually with squeegee	6.58
Scrub floor with long-handled squeegee	1
Wash floor with single disk floor cleaner	4
Water extraction after washing floors with single disk floor cleaner	3
Empty water tank (50 kg)	0.45
Wash floor with scrubber drier machine	3
Wash floor with high-pressure cleaner	
Wax floor with floor wax applicator	4
Polish floor with electric floor polisher	3
Pass industrial drum vacuum cleaner	3
Wash floor using mop (moderate force to squeeze)	2.5
Wash floor using mop (maximum force to squeeze)	2.5

TABLE 8.11
Some of the Most Common Units of Measurement Used to Determine Intrinsic Task Duration

Periodic room cleaning	30 m²	Duration of each task in each representative average room measuring 30 m²
Routine room cleaning	30 m²	Duration of each task in each representative average room measuring 30 m²
Cleaning of common areas (shared bathrooms, corridors, stairs, foyers, etc.)	20 m²	Duration of each task for 20 m² room (same for bathroom tiles)
Cleaning of open space offices, meeting rooms, classrooms	40 m²	Duration of each task for floors measuring 40 m² (same for tiles)
Bathrooms/toilets, radiators, light fittings	Each piece	Time for each piece cleaned/washed (toilet, sink, shower, etc.)

their opinion, the work was carried out properly, bearing in mind that the correct speed can also influence the quality of the work.

The experts reported that reanalyzing and defining the durations of the various tasks was not only valuable for achieving a more objective risk assessment, but also gave a more objective estimation of the duration of the tasks and therefore the cost of the contract.

TABLE 8.12
Calculation of Task Duration in the Shift Based on the Intrinsic Task Duration per Unit of Measurement

Tasks	A. Intrinsec Duration for 30 m² (min)	B. Square Metres Actually Cleaned in a Shift by a Worker	C. Net Duration of Repetitive Tasks in the Shift (min)
A	2	300	2*300/30 = 20
B	5	300	5*300/30 = 50
C	10	1000	10*1000/30 = 166
Total time in minutes performing repetitive tasks in a shift (i.e., sum of net duration of each repetitive task present in the shift)			236

Table 8.12 shows how to calculate the duration of a task in a shift based on the intrinsic duration of the task by unit of measurement, as described in this section.

The example shows three routine room cleaning tasks and their respective intrinsic durations (column A). Column B indicates the surface area in square meters to be cleaned.

Column C shows the formula for calculating the *average duration* of each repetitive task in the shift. By adding up the durations of each individual task, one calculates the total duration of the work in the shift.

As part of the application of the methodology, the cleaning company assessed the intrinsic duration of all the cleaning tasks in the analysis, versus the specific units of measurement indicated in Table 8.11. The job took quite some time but is performed *only once*.

In fact, the tasks have been identified in such detail and are technically so simple that these durations can be used as constants to work out the net duration of cleaning tasks even at other companies.

As to correctly identifying tasks, it should be stressed that every activity (or phase) can be carried out in different ways, which may generate different exposure risks for the upper limbs. Take the floor washing phase, for instance: Each method used to execute the phase gives rise to a different task. The following is a list of different ways of washing floors:

- With a brush and cloth (loose)
- With a cloth (loose)
- With a mop (that needs squeezing)
- With scrubber drier machine
- With high-pressure cleaner, and so on

Technically, the tasks are very simple, and therefore it is hardly difficult to analyze and determine the average intrinsic time needed to wash the relative unit of measurement, for instance a typical 30 m^2 room or 20 m^2 corridor (units for common areas) or for 40 m^2, the unit for open-space offices.

8.2.4.7 Step 7: Calculation of Total Task Duration When the Job Frequency Is Not Daily

Table 8.9 shows that there is one more complication to be taken into account when estimating the durations of the various tasks in the shift: that is, tasks performed on a nondaily basis. Better-organized cleaning companies divide workers into different teams: Some perform routine cleaning work and others do heavier periodic work. Even certain routine cleaning activities may not be performed daily, as shown in Table 8.13 (e.g., office floors are notoriously not washed every day, but once a week). Cleaning contracts include shifts in which workers perform daily tasks as well as periodic ones that are distributed throughout the week. It is not easy to calculate such partial durations in a *weekly cycle*, as there may be uncertainties as to when and how such periodic tasks are actually performed. An example is provided in Table 8.14. Task A (as calculated previously in Table 8.12) lasts 20 min (since the area to be cleaned measures 300 m^2): The work is performed 5 days a week, over a weekly shift; therefore it is considered as 20 min a day. Task B also involves cleaning the same area, but takes 50 min and is performed only once a week. The day the work is performed is not known, nor how it is distributed during the week; therefore, the task is "spread" over the whole week, which equates to 10 min of activity per day. Task C lasts 166 min and is performed once a month: Using the same procedure as before, theoretically it would equate to 5.5 min a day.

All this produces a *fictitious week* that includes the equivalent durations of all the daily and less frequently performed tasks (remembering that the durations are all calculated on the basis of the relevant contract).

Table 8.15 shows a more complete version of Table 8.9.

8.2.4.8 Step 8: Definition of Contract Assigned to a Homogeneous Group

As seen previously, the duration of all the tasks can be obtained and objectively defined based only on the type of contract assigned (i.e., number of square meters per room and/or number of square meters per common area and/or number of individual units, etc.).

Table 8.16 shows a description of a contract compiled for a homogeneous group working in three locations.

TABLE 8.13
Method for Calculating the Duration of Tasks Performed Periodically over a Weekly Cycle: Percentage Needing to Be Represented as Daily, "Spread" over the Entire Week

Once every 6 months	0.55%	Once a week for a working week of 6 days	16.67%
Once every 4 months	0.82%	Once a week for a working week of 5 days	20.00%
Once every 3 months	1.10%	Twice a week for a working week of 6 days	33.33%
Once every 2 months	1.64%	Twice a week for a working week of 5 days	40.00%
Once every month	3.29%	Three times a week for a working week of 6 days	50.00%
Twice a month	6.58%	Three times a week for a working week of 5 days	60.00%

TABLE 8.14
Method for Calculating the Duration of Nondaily Tasks

Task	A. Intrinsic Duration for 30 m² (min)	B. Square Meters Cleaned in a Shift	C. Duration of Repetitive Tasks in the Shift (min)	Task Frequency	Mon	Tue	Wed	Thu	Fri
A	2	300	2*300/30 = 20	Daily (100%)	20	20	20	20	20
B	5	300	5*300/30 = 50	Once a week (20%)	10	10	10	10	10
C	10	1000	10*1000/30 = 166	Once a month (3.29%)	5.5	5.5	5.5	5.5	5.5

TABLE 8.15
Example of List of Tasks Performed by a Homogeneous Group (see also Table 8.9) Complete with All Task Durations

Routine and Periodic Room/Bedroom Cleaning			Mon	Tue	Wed	Thu	Fri
Remove Waste and Dirty Linen from Bedroom and/or Bathroom							
Empty rubbish bins	Daily	100%	3.62	3.62	3.62	3.62	3.62
Routine Room/Bedroom Cleaning							
Open/close windows and curtains	Daily	100%	1.17	1.17	1.17	1.17	1.17
Sweep floors using broom and long-handled dust pan	Once every 5 days	20%	2.33	2.33	2.33	2.33	2.33
Dust using damp cloth and spray (furnishings, doors, windows, up to door height)	Once every 5 days	20%	3.50	3.50	3.50	3.50	3.50
Dust using cloth dipped in liquid detergent and squeezed (furnishings, doors, windows, up to door height)	Daily	100%	2.33	2.33	2.33	2.33	2.33
Tidy up room (move and replace objects)	Daily	100%	1.40	1.40	1.40	1.40	1.40
Clean Furniture and Other Objects							
Wipe down radiators with damp cloth	Once every 5 days	20%	0.70	0.70	0.70	0.70	0.70
Clean Floors							
Wash floor using mop (moderate force to squeeze)	Twice every 5 days	40%	7.00	7.00	7.00	7.00	7.00
Clean Common Areas (Indoor and Outdoor)							
Clean and Wash Floors							
Sweep floors using broom and long-handled dust pan	Once every 5 days	20%	0.67	0.67	0.67	0.67	0.67
Wash floor using mop (moderate force to squeeze)	Once every 5 days	20%	1.00	1.00	1.00	1.00	1.00
Clean Sanitary Ware in Common Area Bathrooms							
Wipe down mirrors	Daily	100%	0.33	0.33	0.33	0.33	0.33
Wash sink	Daily	100%	0.67	0.67	0.67	0.67	0.67
Clean toilet	Daily	100%	2.00	2.00	2.00	2.00	2.00
Clean bidet	Daily	100%	0.33	0.33	0.33	0.33	0.33
Replace toilet paper, paper hand towels	Three times every 5 days	60%	0.90	0.90	0.90	0.90	0.90
Refill liquid soap	Once every 5 days	20%	0.10	0.10	0.10	0.10	0.10
Empty rubbish bin	Daily	100%	0.75	0.75	0.75	0.75	0.75
Daily and Periodic Window Cleaning in All Areas							
Wash doors and windows inside and/or outside using short-handled squeegee	Once a month	3.3%	0.54	0.54	0.54	0.54	0.54

TABLE 8.16

Description of a Contract Compiled for a Homogeneous Group Working in Three Locations

Municipality of SXXX Locations Attended to by the Worker or Homogeneous Group: Cleaned Spaces in Square Meters	Total Cleaned in Contract Locations (m²)	Total Square Meters Assigned to a Worker or Homogeneous Group under this Contract	% vs. Total Square Meters	No. of Workers in Homogeneous Group	Rooms to Clean (m²)	Common Areas/ Bathrooms to Clean (m²)	Storage Areas to Clean (m²)	Open Space Areas to Clean (m²)	Large Meeting Rooms to Clean (m²)	Small Meeting Rooms (m²)
Parish hall x	1000	605	61%	2	100	100	20	100	1	14
Downtown offices	366	366	100%	2	100	100	20	30	2	14
School gym A	350	350	100%	2	100	100	20		0	20

Baths

No. Offices

Municipality of SXX Locations Attended to by the Worker or Homogeneous Group: Cleaned Spaces in Numbers of Units to Be Cleaned	Bathrooms Common Areas (m²)	No. Bathrooms	No. Mirrors	No. Sinks	No. Toilets	No. Bidets	No. Showers	No. Bath Tubs	No. Bath Tubs
Parish hall x	20	5	5	5	5				2
Downtown offices	20	5	5	5	5				2
School gym B	20	5	5	5	5				2

Municipality of SXX Locations Attended to by the Worker or Homogeneous Group: Cleaned Spaces in Numbers of Units to Be Cleaned or in Square Meters	Classrooms with Individual Desks (m²)	Classrooms with Tables (m²)	No. Rooms with Beds	Ave. No. Beds Occupied	No. Floors Cleaned	Outdoor Areas (m²)	Staircases (m²)	No. Windows with Venetian Blinds	No. Windows with Roller Blinds	Windows (m²)	No. Elevators and Goods Lifts	Kindergartens: Classrooms (m²)	Kindergartens: Sleep Areas (m²)
Parish hall x	100	140			1	10	0	0	6	0	0	0	0
Downtown offices	100				3	10	20	8	0	0	3	0	0
School gym C	100				1	10	0	8	0	0	0	0	0

The quantitative information to be collected may seem quite substantial, but it is worth emphasizing that all the data can be gleaned from the contract and the work schedule assigned to the cleaners.

The technical data appearing in the previous table quantify the actual areas to be cleaned and represent the basis for calculating exposure times.

8.2.4.9 Step 9: Checking the Consistency of the Organizational Data

Step 4 describes how to calculate the net duration of repetitive work (net time devoted to cleaning), by subtracting unsaturations, nonrepetitive work (both routine and non-routine), and breaks (both official and nonofficial) from the duration of the shift. This can be called the *estimated net duration of the repetitive work.*

Steps 6 and 7 describe how to calculate the duration of each task in the shift (whether daily or periodic), by multiplying the intrinsic task duration (determined on the basis of its specific unit of measurement) by the quantity of the contract actually assigned. By adding up the durations of the tasks for all the days of the week that the workers work, it is possible to obtain the average net time actually needed to perform the cleaning, as per the contract. This will be called the *net duration of repetitive work calculated on the basis of the quantitative contract data.*

Before assessing risk, it is essential to compare the two results to check the consistency of the organizational information collected so far. In fact, the *risk calculation will be based on the duration of the repetitive work.*

Based on experience with larger cleaning service companies, it seems that there are often major discrepancies between these two values.

Table 8.17 shows a few examples:

- *Example 1*: The duration calculated on the basis of the contract is longer than the estimated net duration. This means that in this case the area manager assigned either too many tasks or excessively large areas or units to be cleaned.
- *Example 2*: The duration calculated on the basis of the contract is 30% less than the estimated net duration. This means that in this case the area manager assigned either too few tasks or excessively small areas or units to be cleaned. The best use is not being made of the contract and there is too much downtime.
- *Example 3*: The duration calculated on the basis of the contract is the same as the estimated net duration. This means that in this case the area manager has assigned the right number of tasks with respect to the area or number of units to be cleaned. The contract is optimized.

These three examples show that the twofold procedure for estimating and comparing the duration of net cleaning tasks can be extremely useful for designing contracts correctly and subsequently for assessing risk. As has already been pointed out, there is not a great deal of understanding in terms of organizational timing analysis even among large-scale cleaning companies, and contracts are often awarded based on the personal experience of the various area managers rather than on a set of pre-established common criteria.

TABLE 8.17
Examples of Comparisons between Net Duration of Estimated and Calculated Repetitive Work (Cleaning)

Example 1: Organizational Study of First Week in a Representative Month of the Year	Shift Distribution over a Week for the Homogeneous Group (Total Shift Duration on Each Day of the Week, Considering Regular Working Hours and Overtime, Including Breaks)						
	Mon	Tue	Wed	Thu	Fri	Sat	Sun
Shift duration (min)	300	300	300	300	300	0	0
Estimated net duration of repetitive task	180	180	180	180	180	0	0
Net duration of repetitive tasks evaluated using average times calculated considering the actual square meter or units to be cleaned under the contract	334	334	334	334	334	0	0
Difference in %	+46%	+46%	+46%	+46%	+46%		
Example 2: Organizational Study of First Week in a Representative Month of the Year							
Shift duration (min)	150	150	150	150	150	150	0
Estimated net duration of repetitive task	122	122	122	122	122	122	0
Net duration of repetitive tasks evaluated using average times calculated considering the actual square meter or units to be cleaned under the contract	93	93	93	93	93	93	0
Difference in %	−30%	−30%	−30%	−30%	−30%	−30%	
Example 3: Organizational Study of First Week in a Representative Month of the Year							
Shift duration (min)	480	480	480	480	480	0	0
Estimated net duration of repetitive task	275	275	275	275	275	0	0
Net duration of repetitive tasks evaluated using average times calculated considering the actual square meter or units to be cleaned under the contract	272	272	272	272	272	0	0
Difference in %	1%	1%	1%	1%	1%		

8.2.4.10 Step 10: Calculation of Intrinsic Risk Indexes for Each Individual Task

As in the case of any multitask assessment, it is necessary to calculate the risk associated with each individual task using the OCRA checklist.

The intrinsic risk index must be estimated as if the task were performed for up to 440–460 min in the shift, with one meal break and two 10-min breaks (see Chapter, Section 6.4).

8.2.4.11 Step 11: Calculation of the Final Risk Index

As for calculating risk in a task rotation scenario, here there are also two models for calculating the final exposure risk index. The first is based on the time-weighted average model. The second, the multitask complex model (Chapter 6), is instead based on the worst task score, based on its actual duration, taken as the lowest exposure score, to be increased with respect to the scores for other tasks, also based on their duration.

Both of these models have been utilized to calculate weekly multitask exposure risks for cleaners. As extensively illustrated in Chapter 6, Section 6.4, to enable them to be applied to weekly exposure, it is necessary to convert the week into a single *fictitious workday equivalent.* The data required for this conversion is briefly listed as follows:

- Active tasks (i.e., those performed every day of the week, and those performed periodically).
- Task duration: This can be difficult to measure and is less than objective when reported through interviews. It is far better to calculate task duration based on the quantitative data gleaned from the contract, that is, the average execution times.
- The proportional duration of each task with respect to the total hours worked over the week.
- The proportional duration of each task with respect to the constant of hours worked over the week, that is, 40.

8.2.5 Presentation of a Few Evaluations for a Representative Italian Sample

A few sample surveys were undertaken in northern and southern Italy on cleaners performing daily tasks (plus a few weekly or monthly tasks, but none of the classic hefty annual "big cleans") to assess exposure risk due to repetitive movements, utilizing the OCRA multitask checklist and the models previously illustrated.

The surveys analyzed exposure risk for various types of weekly shifts, ranging from 10 to 40 h of exposure, among a representative sample of companies. Table 8.18 summarizes the main organizational data (working hours, net duration of repetitive work based on the type of contract, considering standard durations assigned to homogeneous groups, units, m^2 of space to be cleaned, etc.) and the results of the risk assessment using the OCRA checklist tailored to measuring weekly exposure.

Based on these data, it was immediately noticed that there was a discrepancy between the number of hours included in the various weekly shifts and the surface area (m^2) covered by the cleaning contract. In northern Italy, there is an almost linear relationship between the duration of the weekly shift and the space to be cleaned, while in the south the relationship is not linear, as shown in Figure 8.8.

More specifically, in the correlation between the space (in m^2) to be cleaned under the contract and the weekly shift duration, as illustrated in Figure 8.9, the relationship between the two functions is entirely random (R2 = .15).

Conversely, matching the space (in m^2) to be cleaned under the contract with the *net duration of repetitive work* (Figure 8.10), estimated based on the intrinsic units of task duration, the trend begins to appear more linear (R2 = .7). This would arguably confirm the accuracy of the estimated intrinsic units of task duration utilized.

TABLE 8.18
Main Organizational Data and Results of the Risk Assessment Using the OCRA Checklist for a Representative Sample of Italian Cleaning Service Employees

Location: Northern Italy	Location: Southern Italy	Hours to Be Worked in a Shift (Official Total Shift Duration)	Shift Duration in Minutes	No. of Days Worked over the Week	Net Duration of Repetitive Tasks (Estimated)	No. of Breaks	Square Meters to Clean Based on the Contract	OCRA Checklist Multitask Risk Index Right (Upper Limb)	OCRA Checklist Multitask Risk Index Left (Upper Limb)
	X	10	120	5	35	1	150	5.2	3.3
X		15	150	6	93	1	500	5.8	2.9
X		17.5	210	5	248	1	1351	9.0	5.2
	X	18	240	5	123	2	800	6.4	3.8
	X	18	180	6	83	1	367	5.4	2.7
X		20	240	5	142	3	880	5.8	2.9
X		22.5	270	5	382	0	2165	10.0	6.4
	X	25.5	270	5	304	0	1380	12.1	9.5
X		30	300	5	334	0	1806	11.4	6.6
	X	37.5	450	5	104	3	453	6.6	3.6
X		40	480	5	272	3	2904	10.0	7.1
	X	40	480	5	104	3	453	6.9	3.7

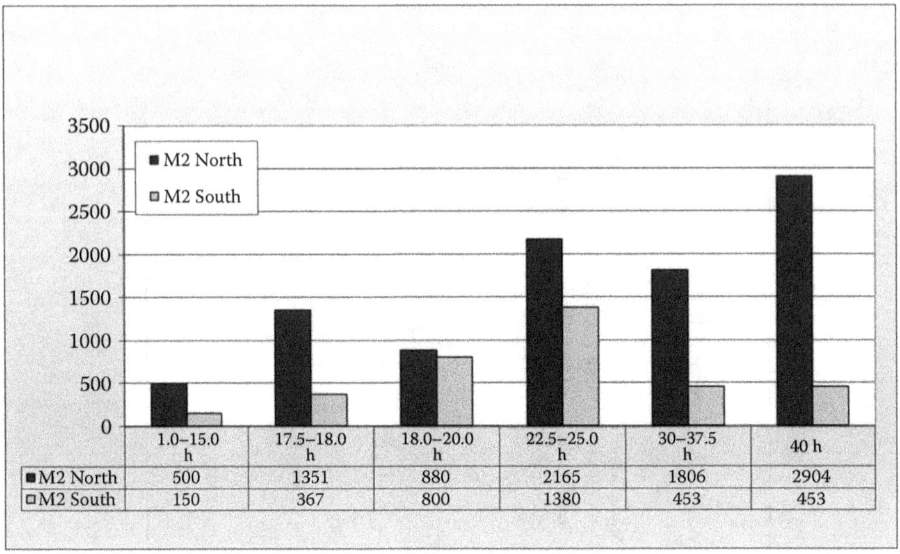

FIGURE 8.8 Relationship between square meter cleaned and shift duration among a representative sample of Italian cleaners.

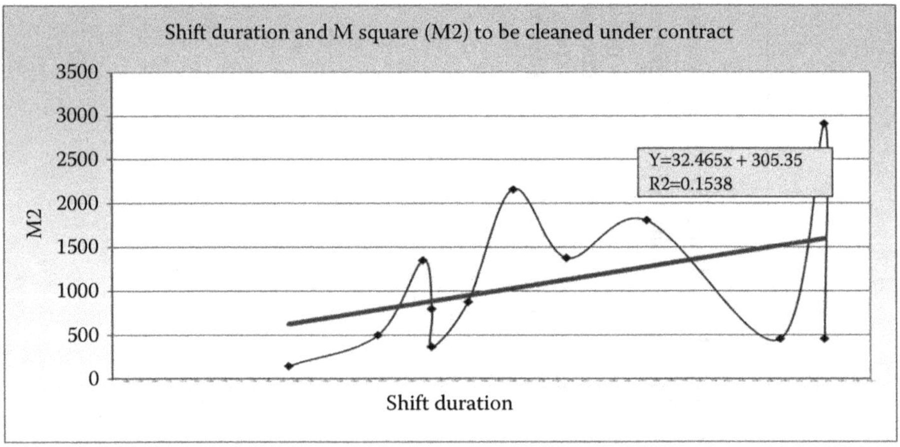

FIGURE 8.9 Statistical relationship between total shift duration and surface area (in square meters) to be cleaned under the contract.

Figure 8.11 clearly shows that it is not possible to relate official working hours to risk exposure. The graph clearly shows that by analyzing the trend for the shift duration and risk index functions, there is no correlation whatsoever (R2 = .1).

Figure 8.12 goes on to tell us that it is instead possible to relate the *net duration of repetitive work* (estimated on the basis of the contract and the preassigned time units) to risk exposure. The graph shows that by analyzing the trend for both of the aforesaid functions, not only is there a correlation but it is very significant (R2 = .86).

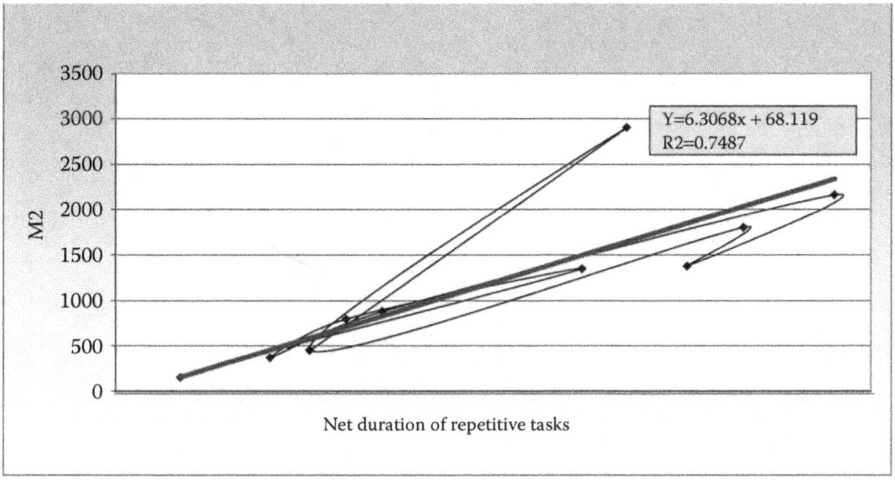

FIGURE 8.10 Statistical relationship between net duration of repetitive work and surface area (in square meters) to be cleaned under the contract.

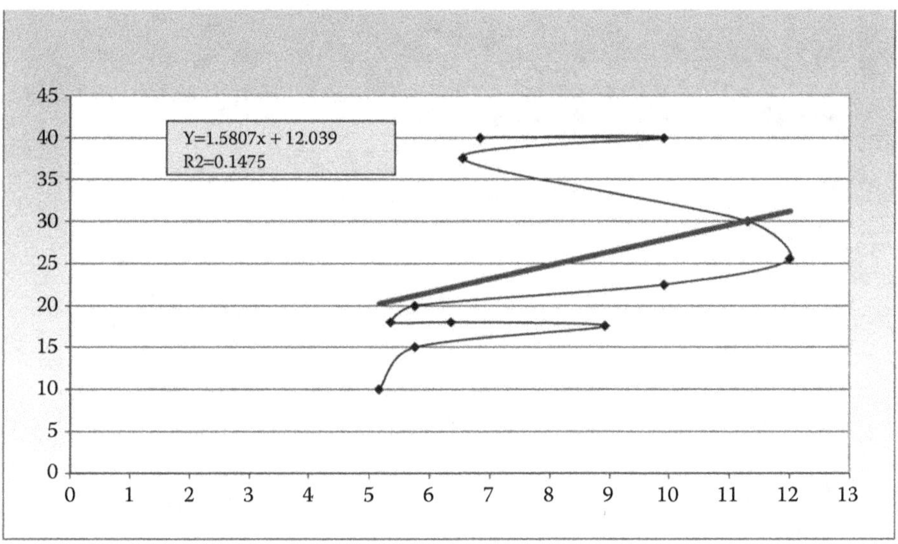

FIGURE 8.11 Statistical relationship between total shift duration and risk indexes.

Conversely, it is not advisable to directly relate the surface area to be cleaned under the contract (in m^2) to exposure risk. Figure 8.13 clearly shows that by analyzing the curve for these two functions, there is a correlation but it is not fully significant (R2 = .6).

This is due to differences in the types of tasks assigned under the contract, which may significantly affect the net duration of the cleaning tasks.

Going back to Table 8.18, it is worth noting and emphasizing that risk seems to appear in shifts of more than 22 h a week when, however, the *calculated net duration*

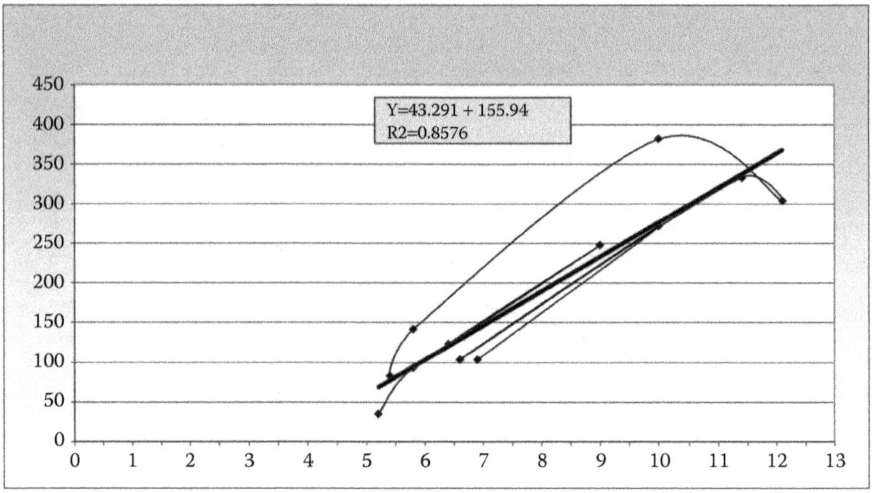

FIGURE 8.12 Statistical relationship between net duration of repetitive work and risk indexes.

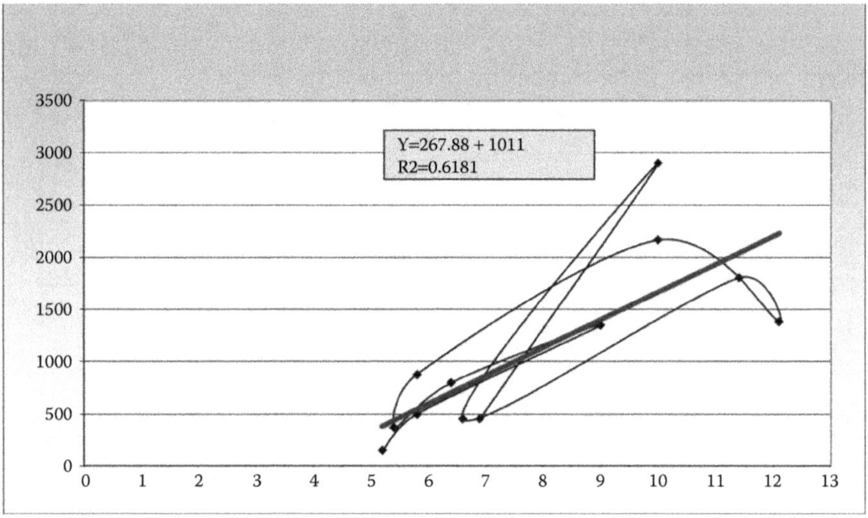

FIGURE 8.13 Statistical relationship between surface area to be cleaned (in square meters) and risk indexes.

of repetitive work assigned by the contract is higher than in the actual shift (in other words, the tasks should take longer than the average preassigned times; therefore, theoretically, there is no opportunity for workers to take breaks. Accordingly, based on the graph depicting the average distribution of shifts across northern and southern Italy (Figure 8.14), it appears that 64% of workers performing daily tasks work weekly shifts of up to 20 h, most of whom are classified green for risk, but some of whom are classified yellow (borderline).

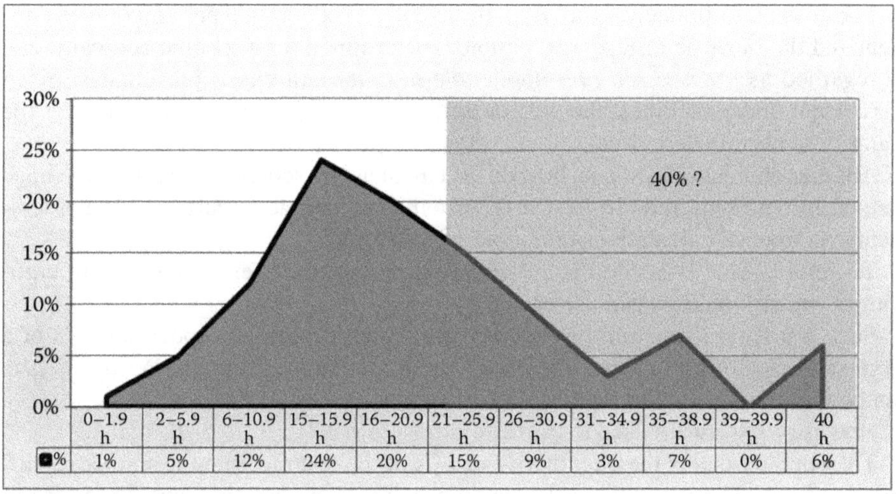

FIGURE 8.14 Prevalence of weekly duration of cleaning services (by duration category) in a large cleaning services company.

However, for shifts of more than 20 h a week, the exposure level cannot be estimated as it is too heavily affected by the net duration of repetitive tasks, which are highly variable and thus cannot supply a reliable direct relationship with exposure risk levels. Therefore, the exposure level must be estimated.

8.2.6 Summary and Operational Suggestions

Assessing the risk of biomechanical overload in cleaners may be a somewhat daunting prospect. There are countless tasks, some daily but many others less frequently; information about their actual duration in the shift may be difficult to acquire; the same jobs may be performed in several different ways; the work may be carried out in different locations with very different shifts (both qualitatively and quantitatively); and so on. All this makes it hard, if not downright impossible in some cases, to gather the necessary information. However, since the number of potentially work-related musculoskeletal disorders and diseases is anything but negligible, there is definitely a need to assess exposure risk levels as objectively as possible, so as to accurately determine medicolegal responsibilities.

In this chapter, we have illustrated how to achieve this objective: by calculating exposure time for each individual task based on the quantitative content of the contract rather than on the duration of the shift.

This has entailed certain other assessments, such as estimating the average execution time of each individual task per predefined unit of measurement. Moreover, the methodology calls for the OCRA checklist to be calculated for each task and for both limbs (i.e., an estimation of the intrinsic task scores).

The question is, must these lengthy evaluations be repeated for every cleaning company that needs analyzing?

The answer, fortunately, is no. Both the average task duration per unit of measurement and the intrinsic task scores, although determined for a specific company, can be regarded as *preassessed and applicable to other situations*; presumably, given our current understandings, the preassessed data we have gathered should cover the majority of cleaning services.

Another obvious question is, how do we calculate the scores, considering the huge amount of data required? In fact, it is virtually impossible to carry out the assessments via manual calculations or simple spreadsheets.

Therefore, what is needed is a software program that even nonexperts in ergonomics and risk indexes can use easily.

Such a software program has been created, and it meets the specific needs of a large-scale cleaning company. The details for accessing the program (*Petrasoftware*) can be found in the bibliography (Lema Informatica, 2015).

How is this software used to evaluate exposure risk?

The operator enters the quantitative details of the contract into the software: shift duration on each day of the week; any nonroutine, nonrepetitive jobs; the name of each task; and the frequency of each task.

A large library of possible cleaning tasks has been predetermined and completed by the intrinsic risk assessment of each task with the OCRA checklist.

Both final risk scores will appear automatically (i.e., the weighted mean and multitask complex), also as a *final report*, to be produced in the event of a medicolegal issue.

Interestingly, if the preliminary data are entered into the software as soon as the employee is hired, together with the various shifts assigned as part of the contract, the employer and occupational health advisors will have at the ready the exposure risk scores for the employee and thus also the relevant responsibilities.

What once was regarded as mission impossible—risk assessment in the cleaning services sector—has now become possible!

As has been seen, unless the proposed criteria are adopted, exposure risk may be underestimated, but is all too often overestimated, especially when the real duration of cleaning tasks is based not necessarily on the length of the shift but rather on the quantitative content of the contract.

So now what remains to be done? Clearly, the task durations per unit of measurement need to be discussed further and integrated with the findings of experts working in other cleaning services, and more task evaluations might also be added to those presented here. This would extend the database of preassessed tasks to include all the main variants of the current tasks, and provide a more comprehensive but simpler risk assessment process for this sector.

8.3 LARGE-SCALE RETAILERS AND REPETITIVE MANUAL TASKS AMONG SUPERMARKET WORKERS

8.3.1 Magnitude of the Issue: Epidemiological Data

Large-scale retailing is a modern type of trade where goods are sold through a chain of large grocery stores, supermarkets, hypermarkets, and so forth.

Outlets or points of sale are generally classified by channel according to the amount of space (in m^2) actually allocated to sales, without calculating common areas, parking lots, and so on, and on the assortment of goods on sale.

Large-scale retailing can be broken down into the following sales channels:

- *Hypermarket*: Retail premises measuring over 2500 m^2.
- *Supermarket*: Retail premises measuring between 400 and 2500 m^2.
- *Convenience store*: Retail premises measuring between 100 and 400 m^2.
- *Discount store*: This type of store does not stock branded products.
- *Cash and carry*: A store open only to wholesalers.
- *Traditional retail stores*: Stores that sell products made by many different brands, in premises measuring up to 100 m^2.
- *Self-service specialist drug stores*: Stores selling home and personal-care products.

The census released in 2001 by the Italian Statistics Bureau (ISTAT) reported that there were 32,513 employees in the large-scale retailing sector and 266,757 people employed in the retail trade: specifically, 36,367 in hypermarkets, 171,512 in supermarkets, 32,503 in superettes, 22.875 in food stores, and 3,500 in frozen food outlets.

In the literature, there are numerous reports stressing the close link between upper limb disorders, primarily carpal tunnel syndrome (CTS), and cervical pain and the job of supermarket cashiers. Conversely, there are few studies on spinal conditions among supermarket workers due to prolonged standing.

In particular, Bonfiglioli et al. (2005, 2007) observed increases in CTS among cashiers as opposed to teachers. In these studies, 40% of cashiers were found to suffer from bilateral (BIL) CTS while 18.4% of teachers had BIL CTS, 8.2% on the right side and 7.7% on the left (Figure 8.15).

The same studies also showed that the greatest increase in CTS was among full-time (FT) cashiers, versus both PT cashiers and teachers. The 2005 study detected CTS in 29% of FT cashiers and 13% of PT cashiers, but only in 9% of teachers (Figure 8.16).

Based on the 2007 study, Figure 8.16 indicates that 31% of FT cashiers and 19.3% of PT cashiers have CTS, versus 16.3% of teachers.

Another study (Panzone et al., 1996) narrows down the type and location of upper arm disorders among supermarket cashiers. The problems primarily affect the shoulders (right 19.2%, left 15.6%), followed by the wrists (right 11.7%, left 7.2%). The conditions include CTS (right 10.5%, left 12%), epicondylitis (right and left 11.7%), other elbow disorders (right 5.8%, left 15.6%), and De Quervain syndrome (right and left 5.8%) (Figure 8.17).

Margolis et al. (1987) compare musculoskeletal disorders among supermarket workers in different positions, including (Figure 8.18) cashiers, deli counter staff, stock clerks, fresh produce staff, and office workers.

The study draws a distinction between conditions affecting the neck, the upper limbs in general, and specifically the hands.

In 2005, a Japanese study found an association between musculoskeletal symptoms and workers assigned to cold stores (Inaba et al., 2005).

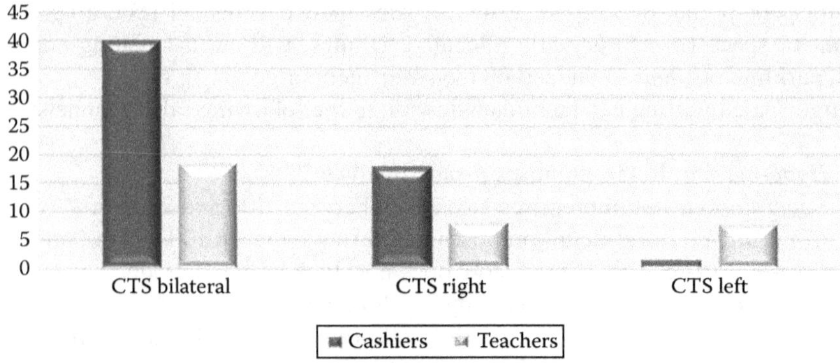

FIGURE 8.15 Comparison between the prevalence of CTS in supermarket cashiers and teachers (Bonfiglioli et al., 2005, 2007).

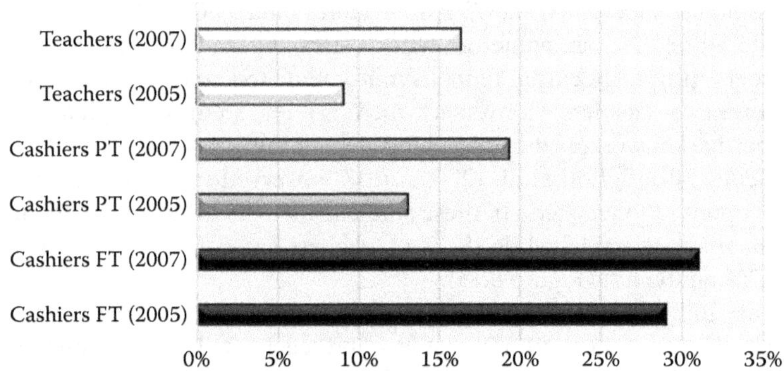

FIGURE 8.16 Percentage of conditions due to CTS in full-time and part-time cashiers compared with teachers (Bonfiglioli et al., 2005, 2007).

However, there are still very few studies providing a comprehensive comparison between risk assessments and the results of clinical assessments for all supermarket jobs with regard to musculoskeletal conditions due to manual load handling and/or repetitive movements among supermarket workers in general. So far it would seem that only two such studies have been published in the literature:

- The first is by Osorio et al. (1994), where the authors clearly document the presence of CTS among workers highly exposed to strenuous and repetitive movements, especially among cashiers, butchers, pastry cooks, and bakers.
- The second is by Draicchio et al. (2007), who used the Revised Niosh Lifting Equation (RNLE) and other methods to measure risk due to manual

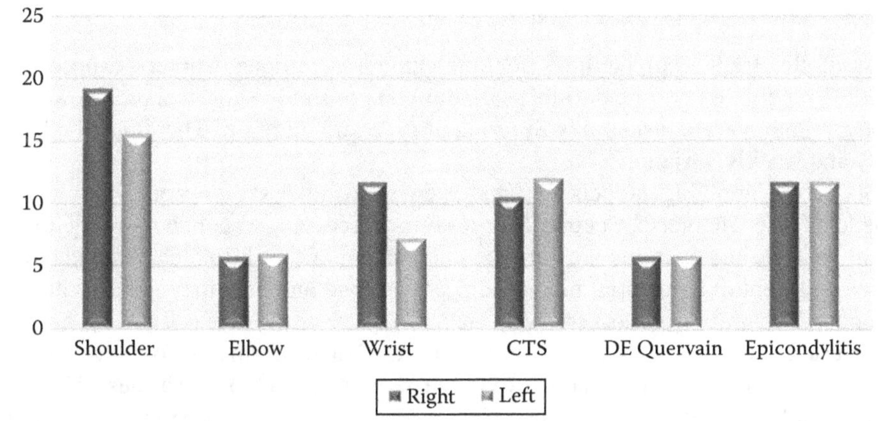

FIGURE 8.17 Prevalence of musculoskeletal disorders among supermarket cashiers (Panzone et al., 1996).

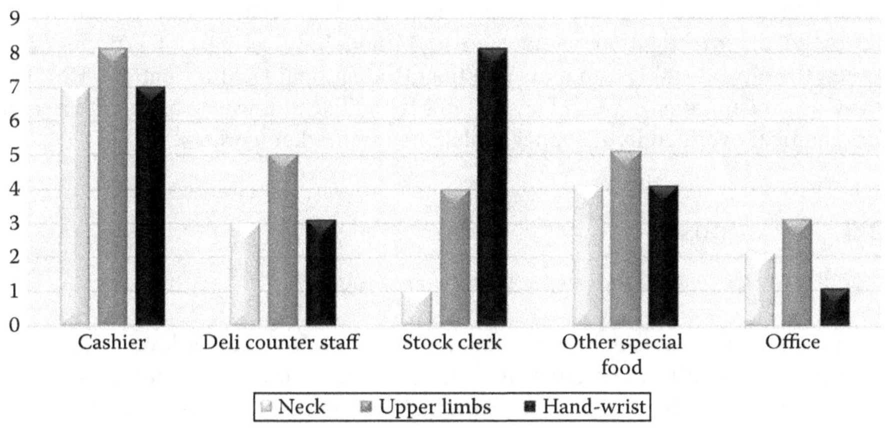

FIGURE 8.18 Percentage of musculoskeletal disorders among supermarket workers broken down by job description (Margolis et al., 1987).

materials handling in the fresh produce section of a medium-sized store (Draicchio et al., 2007).

To close this section on epidemiological data, the lack of data on supermarkets and the challenges involved in assessing risk confirm the need to undertake more systematic research into risk due to biomechanical overload and the specific damage that it causes.

We will next suggest an effective and specific methodology for analyzing risk based on more recent experiences.

8.3.2 Risk Analysis Procedures

Most of the traditional methods for analyzing risk among workers exposed to multiple repetitive tasks tend to focus on daily exposure; in supermarkets, however, exposure may vary considerably in terms of type and duration. There may be both daily and weekly variations.

So before applying models for analyzing exposure risk for weekly multitask schedules (see Chapter 6, Section 6.4), it is first necessary to conduct an organizational analysis and examine the duration and content of the shifts, that is, to identify which tasks entail a potential biomechanical overload and how they are assigned to homogeneous groups of workers.

As usual, the aim is to create an easy-to-use "tool" (in the form of software or spreadsheets) for automatically estimating exposure risk with respect to the various tasks.

This tool represents a "best practice" for use by numerous professionals, including accident prevention officers and occupational health experts, insurance companies, and so on, who are involved in preventing and managing risk due to biomechanical overload, as well as for medicolegal purposes, such as achieving an objective risk assessment to recognize occupational disorders.

The data presented here and the assessment tool derived from it have been developed based on investigations conducted in Milan, Italy, as part of a partnership between the city's local health units and the Occupational Health Centre (CEMOC) at the Clinica del Lavoro "Luigi Devoto" in Milan. The aim was to assess the risk of biomechanical overload in the upper limbs of supermarket workers.

8.3.3 Background

The following *11 operational areas* have been identified:

1. *Cashiers*: Operate at checkout counters fitted with a scanner level with the till. Occasionally, the scanner is vertical, or there may be a hand-held scanner, and the cashier's duties may also include bagging the client's purchases.
2. *Stock clerks*: Move merchandise of varying weight from the warehouse to the store and place it on shelves of varying height.
3. *Grocery clerks*: Stock the dairy section and move cold goods from the refrigerator to the counter.
4. *Butchers*: Cut and wrap meat portions, and restock the meat counter and refrigerator.
5. *Deli counter*: Cut, weigh, and wrap cheese portions; bone and wrap raw ham (prosciutto); move and position deli meats and cheeses; slice deli meats using manual or electric slicer.
6. *Fresh produce*: Wrap and price fruit and vegetable portions; place price tags and position fruit/vegetable trays on display shelves; position individual portions or small containers of fruit and vegetables on shelves; close plastic containers.
7. *Fish counter*: Clean fish, prepare and slice large fish portions, replenish fish counter display, weigh and price fish.

8. *Bakery*: Load frozen loaves into oven, package and sell small bread at bakery counter.
9. *Cleaners*: Clean shelves and rotisserie.
10. *Merchandise loading/unloading*: Unload trucks and transport merchandise using manual or electric pallet jack; handle empty pallets.
11. *Other merchandise handling*: Move other merchandise using assorted carts, trolleys, pallet jacks, and so on.

Before presenting the results of the risk assessment, we will comment on the outcomes of risk assessments carried out on several supermarkets that primarily used the OCRA index and the OCRA checklist (Table 8.19).

Table 8.19 lists the results of risk assessments carried out on several tasks at five supermarkets.

The data show highly variable results between the five supermarket chains, despite the work being largely identical. Moreover, many jobs are not assessed at all; therefore, the risk mapping is not exhaustive.

Essentially, Table 8.19 suggests that

- For cashiers, the risk assessment carried out using the OCRA index or OCRA checklist (the latter in just one supermarket) reported variable results (green, yellow, and red).
- For fresh produce staff, the risk level was medium (yellow).
- For butchers, except for two cases of slight risk, the risk level was medium red.
- The deli counter showed the most variable risk levels, ranging from slight to medium–high.

8.3.4 Identification of Tasks and Assessment of Intrinsic Risk: Results of the Analysis

CEMOC, the occupational health center at the Clinica del Lavoro in Milan, conducted a more systematic analysis. The study began with visits to several small, medium, and large supermarket chains in Milan and ended with interviews with management in order to accurately identify the relevant operational areas and tasks.

During the initial visit, each individual task carried out by supermarket staff was analyzed and filmed. The videos were later assessed using the OCRA checklist, for both the right and left arms, to calculate the *intrinsic task risk* (i.e., as if the worker performed the same task for the whole 8h shift, with one meal break and two breaks lasting at least 8–10 min each—see Chapter, Section 6.4). A total of 67 tasks were identified and analyzed.

The various operational areas are listed in the following, along with the risk factors and intrinsic OCRA checklist results.

8.3.4.1 Cashier

There are various types of checkout counters fitted with different types of scanners: flat (horizontal), vertical, or hand-held. Cashiers also have different job descriptions,

TABLE 8.19
Comparison between Risk Assessments Performed in Several Milanese Supermarkets by Managers Primarily Using the OCRA Index

		Supermarket				
	Job	E.	G.	S.	C.	L.
		OCRA	OCRA	OCRA	CHECK	OCRA
Cashier	Cashier only	2.6	1.14		18.7	
	Cashier and stock shelves	1.8				
Fresh produce	Unload pallets	2.7				
	Wrap and price	3.2				
Butcher	Handle individual portions (trays)		0.66			
	Wrap and replenish	4.5–7				
	Cut and debone		4.7			
	Prepare and wrap meat			1.8		
	Prepare meat			4.99	18	
Deli counter	Cut and portion soft and hard cheeses	14.3		6.75		
	Wrap, weigh, and price cheeses	4.75	2.12			
	Handle deli meats/cheeses			0.76	8.5	
	Portion cooked food	1.75				
	Wrap, weigh, and price cooked food	5.25				
	Debone thin end of prosciutto ham	13.9		3.8		
	Debone thick end of prosciutto ham	18.25				
Grocery	Sales assistant					2.2
Fish counter	Clean fish		0.16			
Bakery	Wrap baguettes		1.6			
	Bag loaves		2.8			
	Label bread		1.5			
	Wrap baguettes/bag and price bread		2			
	Prepare bread		2.2			
	Prepare pizza/focaccia		2.2			
Pastry counter	Prepare sponge cakes		2.2			
	Prepare jam tarts		2.5/2.9			
	50% sponge cakes and 50% jam tarts		2.3			
Hot food	Prepare chickens for rotisserie		1.4			

which may include bagging merchandise for customers. At the supermarkets in our study, the checkout counters had horizontal scanners and the cashiers did not bag merchandise (as is generally the case in Italy) (Figure 8.19).

To collect data on work cycles (i.e., average duration and operations performed), with 1 customer representing one work cycle, 25 customers were analyzed at four different checkouts to obtain a representative sample.

Four subtasks were identified for each cycle/customer, and the OCRA checklist was used to measure the risk and respective mean duration of every subtask for each representative average customer/cycle.

FIGURE 8.19 Cashier using horizontal scanner without bagging customer's shopping.

Based on the results of the analysis, the following information emerged: average duration of each customer/cycle, average number of items bought, average number of "heavy" items, average waiting time within the customer/cycle.

The average duration of a customer/cycle was approximately 2.5 min. In percentage terms, this is the duration of the relative subtasks: scanning merchandise (63%); payment by cash or credit/bank card (19%); scan customer's loyalty card (4%); wait for payment (14%).

These were the intrinsic OCRA checklist values corresponding to each subtask making up the average customer/cycle:

- 10.6—scan customer loyalty card
- 21.9—scan items
- 13.3—payment
- 0—wait for payment

Given the percentage duration of these subtasks, the time-weighted average representing the intrinsic risk for this job (featuring a net duration of 420–440 min, with two 10-min breaks and a meal break) is 16.1 (see Table 8.20).

Cashiers obviously spend some time waiting for customers (the main component of total unsaturation); daily average waiting times will vary throughout the day.

With regard to the different days of the week, the highest saturation levels are reported on Thursday, Friday, and Tuesday; the lowest is on Monday.

Besides this information, other research involving a large supermarket chain has confirmed the average customer/cycle scores: Here, the average unsaturation rate is estimated to be between 17% and 19%.

The distribution of the unsaturation rate (per customer/cycle) is random and nonscheduled and the customer waiting time is variable (but seldom over 8 consecutive min), therefore the unsaturation neither modifies nor reduces the score for the *lack of recovery time* risk factor. Consequently, there are no *recovery times within the cycle* in the "customer at checkout" cycle (though they are often mistakenly calculated!).

TABLE 8.20
Intrinsic Risk Score (OCRA Checklist) for Cashier's Job Obtained as a Time-Weighted Average of Four Subtasks, for Cash Registers with Flatbed Scanner and No Bagging of Customer's Shopping, for a Shift Duration of 440–460 min, with 2 × 10-min Breaks and One Meal Break

% of Use	Task on Line or Homogeneous Area	Recovery Multiplier	Recovery Score	Frequency	Force	Side	Shoulder	Elbow	Wrist	Hand	Stereotype	Total Posture	Additional	OCRA Checklist
4%	Customer card	1.33	4	3	0	BIL	3	3	0	5	0	5	0	10.64
63%	Flatbed scanner	1.33	4	8	0.5	DX	3	5	3	5	3	8	0	21.95
14%	Payment	1.33	4	5	0	DX	3	2	2	5	0	5	0	13.30
19%	Wait for payment	1.33	4	0	0	BIL	0	0	0	0	0	0	0	0.00
	Average		4.0	5.9	0.3		2.4	3.6	2.2	4.1	1.9	5.9	0.0	16.1

8.3.4.2 Stock Clerk

Stock clerks perform many tasks that involve moving merchandise using devices such as four-wheeled carts and manual or electric forklifts, and stocking/filling shelves manually.

The merchandise may weigh anywhere from a few grams, in the case of sale signs and labels, to 12 kg, in the case of detergents and crates of bottles. Heavier weights are quite rare. The items are moved to and from different heights.

For heavier items weighing more than 3 kg, it will also be necessary to assess the manual load-handling risk (using the RNLE suggested by ISO standard 12228-1).

This job encompasses the following tasks (Figure 8.20):

1. Move crates/boxes (6–12 kg) manually to shelves
2. Stock shelves with items weighing up to 2 kg
3. Fill and transport cardboard boxes weighing between 6 and 12 kg using four-wheeled carts
4. Open cardboard boxes/items
5. Position items weighing between 6 and 12 kg along aisles
6. Position special offer signs
7. Stock shelves with bottled water
8. Return all unwanted items left at checkout counter to appropriate shelf locations

The risk assessment for the various tasks performed by stock clerks leads to OCRA checklist scores ranging from 16 to 20 (red). Collecting unwanted items left at the checkout counter is associated with slight risk (11; yellow), while stocking shelves with bottled mineral water is high risk (27; purple).

Stock clerks use both upper limbs in their work; therefore, the values calculated for the left side are the same as for the right side.

8.3.4.3 Grocery Clerks

Tasks include stocking and replenishing refrigerated grocery counters with fresh food including cheese and dairy, salami, ham, and so on (Figure 8.21).

Generally speaking, grocery counter tasks are handled not by stock clerks but by deli staff. The usual stock clerk risks are present, in addition to the additional very high risk associated with taking items out of refrigerators—here, the OCRA checklist score may be as high as 25.5 (purple).

8.3.4.4 Butchers

Butchers perform numerous tasks that involve cutting entire sides of meat suspended on hooks, slicing meat, portioning chickens, and so on, as well as preparing trays. In most of the supermarkets in the study, the meat arrives partly precut; in this case, the butchers do not have to handle and cut heavy loads of meat such as whole sides.

The butchers are also in charge of replenishing and stocking the meat counter.

FIGURE 8.20 Various tasks performed by stock clerks.

FIGURE 8.21 Various tasks performed at the grocery counter.

The tasks performed by the butchers in our study include the following:

1. Halve chickens using a cleaver and place on trays
2. Remove chicken breasts and place on trays
3. Remove chicken thighs and place on trays
4. Prepare and clean pieces of meat
5. Cut large portions of meat into smaller ones and place on trays
6. Slice meat into steaks using electric slicer and place on trays
7. Cut pork chops using a knife and place on trays
8. Cut through backbones using a knife to make chops and slice meat using a knife
9. Cut through backbones using cleaver to make chops and slice meat using a knife
10. Pack trays of precut meat
11. Pack chicken breasts in cardboard boxes
12. Pack sausages on trays
13. Wrap, weigh, and price meat trays
14. Restock meat counter
15. Restock meat fridge
16. Cut half sides suspended from ceiling hook
17. Debone femurs

Figure 8.22 shows 17 tasks, analyzed using the OCRA checklist for the left and right upper limbs.

The task associated with the highest risk is cutting sides of meat suspended from hooks and deboning the femur, both of which require the use of considerable force, especially deboning "prosciutto" hams. Workers are also obliged to adopt awkward postures with the arm held above shoulder height (purple).

Other tasks that are high-risk (purple) due to the frequency with which they are performed include packing meat cuts on trays, packing chicken breasts in cardboard boxes, and wrapping full trays; the OCRA checklist score for these tasks is 24. Other tasks for supermarket butchers range from slight to medium, with OCRA checklist scores of between 12 and 21.

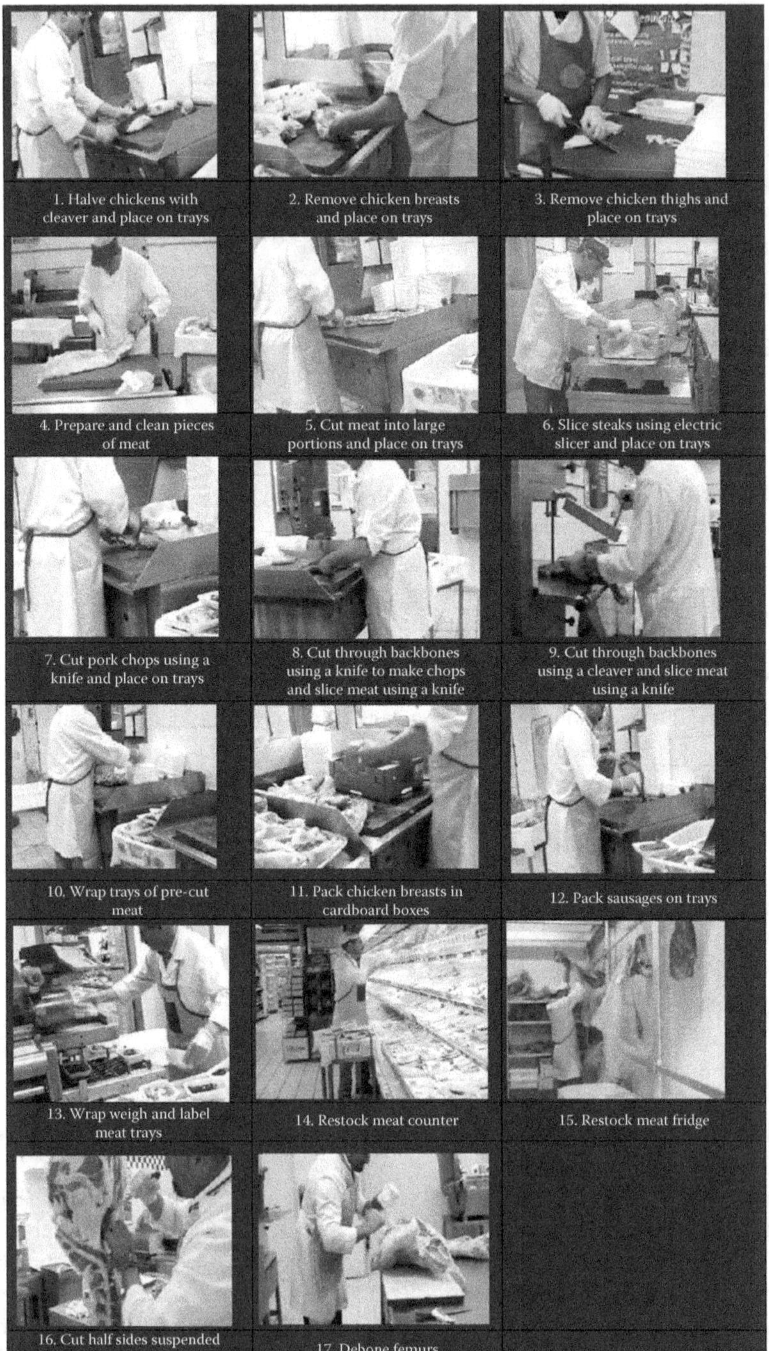

FIGURE 8.22 Various tasks performed at the meat counter analyzed using the OCRA checklist for the left and right upper limbs.

Comparing the checklist values for the left and right sides, it can be seen that the dominant hand experiences the highest biomechanical overload.

For standard meat cutting operations, the left side may be at higher risk than the right side depending on the position of the left hand. When cutting larger pieces of meat, while the right hand is holding the knife (grip), the left hand is holding the meat (pinch or palmar grasp).

Often, butchers wear a metal mesh glove or some kind of protective safety gloves. These gloves are indispensable for preventing accidents but make it harder to grip objects and may become slippery, requiring even more force.

8.3.4.5 Deli Counter

The deli counter performs numerous tasks that also differ in terms of risk: from deboning "prosciutto" hams to slicing and packaging deli meats and cheeses.

The main tasks carried out by deli counter staff are the following:

1. Cut hard cheeses into large pieces using a knife
2. Pack and wrap cheese portions (low machine)
3. Pack and wrap cheese portions (tall machine)
4. Open large packages of cheese
5. Weigh and price cheeses
6. Prepare prosciutto hams for slicing on electric slicer
7. Wrap prosciutto hams
8. Move and position deli meats and cheeses on shelves (weight: over 8–10 kg)
9. Slice deli meats on manual slicer and prepare (slice, package, weigh, label)
10. Slice deli meats on electric slicer and prepare (slice, package, weigh, label)
11. Cut small rounds of soft and semi-soft cheeses
12. Debone prosciutto hams
13. Move and position smaller deli meats and cheeses on shelves

Figure 8.23 shows the typical tasks performed by the deli counter staff.

Certain tasks, such as deboning prosciutto hams and cutting hard cheeses manually, require peak force and are extremely high-risk (OCRA checklist scores between 23 and over 30).

When shelves are taller than the deli staff, lifting and lowering deli meats weighing more than 8–10 kg may subject the shoulders to extreme biomechanical overload (OCRA checklist score: 43.5—purple).

The other deli counter operations are classified as red (medium risk, scores between 14 and 20), except for moving and positioning smaller deli meats and cheeses (weighing less than 3 kg) on shelves, which has an OCRA checklist score of 7 (green).

Deli counter tasks may be carried out at the front or in the back of the store. The risk score will be identical in either case. The only difference is the net duration of exposure, which is higher for back-of-store tasks since there is no unsaturation time, due to waiting for customers, for example.

FIGURE 8.23 Various tasks performed at the deli counter.

8.3.4.6 Fresh Produce

In the fresh produce section, workers perform a variety of tasks, such as (Figure 8.24):

1. Package lettuce or other fruit and vegetables
2. Pack and wrap watermelon
3. Price fruit and vegetables
4. Position price tags
5. Move and position hampers on display units
6. Position fruit and/or vegetables (individually or in small packs in display cases)
7. Position prebagged fruit and/or vegetables
8. Close plastic crates for fresh produce

Workers in the fresh produce section of supermarkets are at very high risk with respect to repetitive movements of the upper limbs when moving and positioning hampers full of fruit or vegetables on display units; the OCRA checklist score is 25.5 (purple). While some tasks have negligible or no risk (i.e., positioning individual

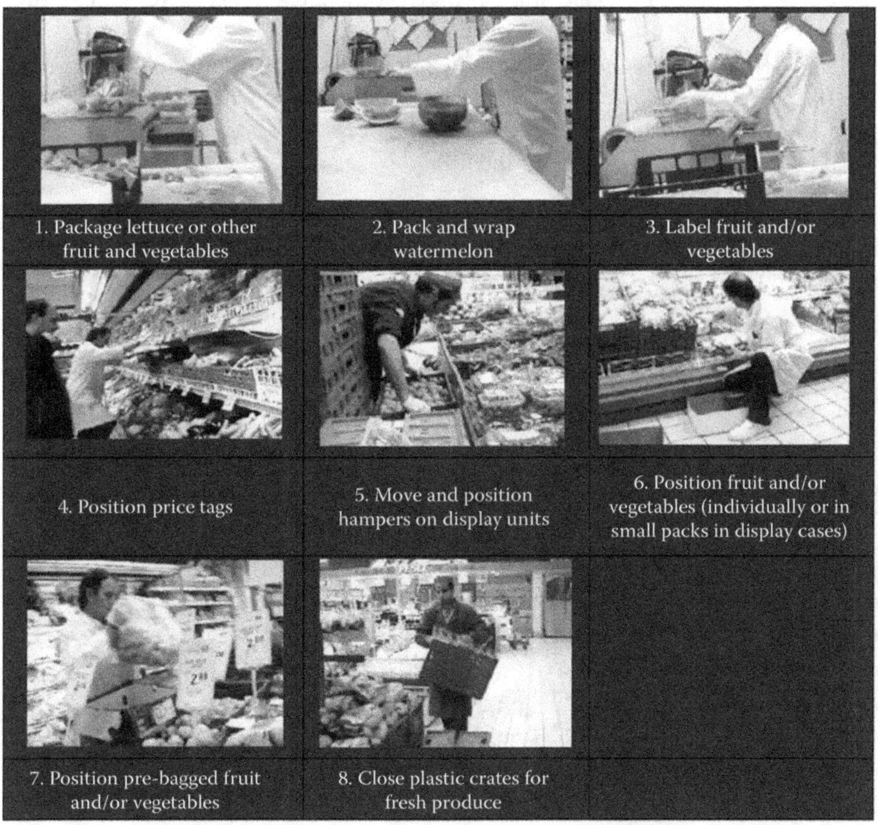

FIGURE 8.24 Various tasks performed in the fresh produce section.

pieces of fruit or vegetables or small groups on display units), most tasks are classified red (OCRA checklist scores: between 14.5 and 22).

8.3.4.7 Fish Counter

Workers at the fish counter perform only a few tasks and all are in the fish section (Figure 8.25):

1. Prepare and slice large fish
2. Cut and bag fish
3. Prepare the fish counter display with crates and individual fish
4. Package fish slices
5. Weigh and sell fish at the counter

In certain instances, the tasks entail different levels of risk for the left and right arms:

- Packing fish on trays puts the right arm at greater risk primarily due to both the high frequency of movements and stereotypy (OCRA checklist score: 24—purple). The other tasks feature OCRA checklist values of between 14 and 19 (red).
- The left arm is generally at only slight risk (OCRA checklist score: 10.5—yellow): However, preparing the fish counter with crates and individual fish and cutting large fish into slices are tasks classified as red (OCRA checklist scores: 17 and 18).

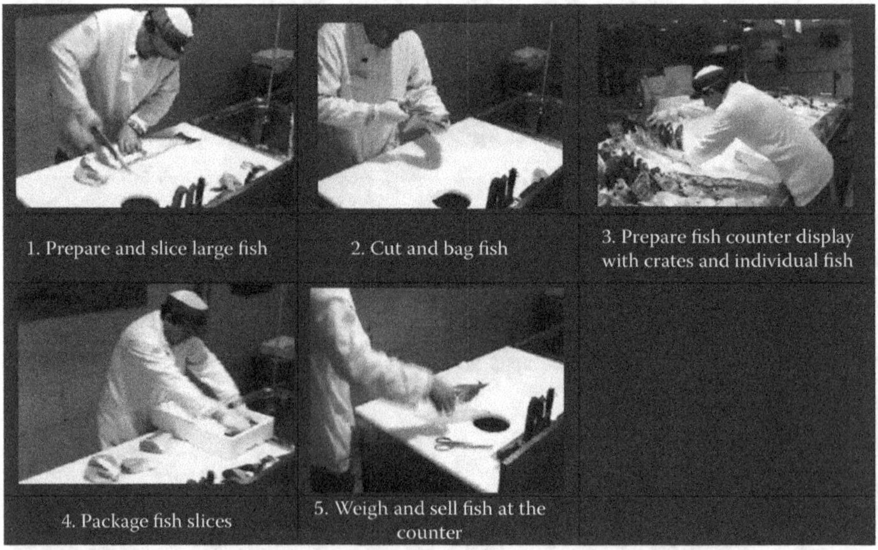

FIGURE 8.25 Various tasks performed at the fish counter.

8.3.4.8 Bakery

The following operations are performed in the bakery section (Figure 8.26):

1. Load frozen bread into oven
2. Sell small loaves at bakery counter
3. Package bread

All these operations are classified as red. The highest risk task is packaging bread, due to the high frequency of action and stereotypy. The OCRA checklist scores are between 12 and 20.5.

8.3.4.9 Cleaning

Most of the cleaning tasks involve cleaning the shelves and rotisserie; the risk is due mainly to the high frequency of action and stereotypy (Figure 8.27).

The OCRA checklist scores are between 19.5 and 22.5 for the dominant arm (red); cleaning shelves presents no risk for the nondominant arm.

8.3.4.10 Merchandise Handling Using Pallet Jacks

Figure 8.28 depicts several merchandise handling operations.

These operations require assessment due to the presence of *manual materials handling* rather than to evaluate risk for the upper limbs.

FIGURE 8.26 Various tasks performed in the bakery section.

FIGURE 8.27 Various tasks performed by the cleaning staff.

FIGURE 8.28 Merchandise handling: principal types of trolleys and carts.

The manual handling of empty pallets and the use of inadequate pallet jacks may, however, also place a strain on the upper limbs and therefore need assessing.

The OCRA checklist values are between 8 and 20.5 for the dominant side (yellow/red); the nondominant side is at slight risk (yellow; risk score 8.5).

8.3.5 Results of Organizational Analyses into Task Turnover and Identification of Homogeneous Groups with Respect to Risk Exposure

Having identified and assessed all the tasks performed by supermarket staff in our study, we then went on to calculate exposure indexes for the homogeneous groups identified and working in four different supermarkets (three medium sized and one large).

Homogeneous groups are formed by workers who perform the same tasks every week and have the same working hours. The exposure indexes were calculated using the OCRA checklist for exposure to weekly multitask shifts (see Chapter 6, Section 6.4).

Eight main work areas were found, each featuring the various tasks listed as follows:

1. Cashier
2. Stock clerk
3. Grocery clerk
4. Butcher
5. Grocery service counters/deli counter

6. Fruit and vegetables (fresh produce)
7. Fish counter
8. Bakery section

During the course of their work, supermarket staff may move from one area to another, spending variable amounts of time in each one, depending on the type of supermarket and/or the day of the week.

Most of the supermarkets in our study adopted similar shifts: 380 min for FT staff and 280 min for PT staff.

In some supermarkets, the shifts had a different duration on different days of the week (e.g., FT on certain days and PT on others); breaks might also be distributed differently. Therefore, in order to reconstruct shift duration and content on the various days of the week, and ensure an accurate and objective risk assessment, it is essential to carry out preliminary organizational studies.

Our results can thus be viewed as typical and representative but cannot be extended to assess risk in all supermarkets where shift length and task turnover may vary considerably.

In the supermarkets analyzed here, the percentage of time that workers spent on each task was calculated over a constant 480-min shift to ensure that the data were comparable (graphically and descriptively) and took PT workers into due account.

A description and analysis of exposure risk for the main homogeneous groups identified in the supermarkets follows, including details of the tasks and organizational setup (i.e., task turnover: which tasks and for how long).

8.3.5.1 Cashier

"Pure" cashiers, especially in the larger chains, seldom stock shelves, clean checkouts, or perform office jobs; however, they often work PT on different weekly schedules. At one large supermarket chain, for instance, there were 18 different shifts!

In the four supermarkets included in our study, both the PT and FT cashiers perform their tasks prevalently at the checkouts (Figure 8.29).

Supermarket	C	S	S	T	T	G	G
	Cashier full-time	Cashier full-time	Cashier part-time	Cashier full-time	Cashier part-time	Cashier full-time	Cashier part-time
Checkout	348	356	212	372	228	348	212
Stocking	20					12	
Cleaning		12	12				
Office				10	10		
Total shift duration min	368	368	224	382	238	360	212

FIGURE 8.29 Cashier, part-time (PT) and full-time (FT) with or without other task rotations.

Supermarket	G	G	T	T	S	S	S	G	T
	Stocking FT	Stocking PT	Stocking FT	Stocking-refilling shelves PT	Stocking FT	Stocking large items FT	Stocking PT	Stocking FT	Unloading merchandise and bottled water FT
Checkout	12	12			196	184	196	20	
Stocking	256	160	144	76	132	104	44	268	180
Cleaning			16	8		16		64	12
Handling	92	36	188	116	28	48		104	180
Office			20	20				16	
Total shift duration min	360	208	368	220	356	352	240	472	372

FIGURE 8.30 Stock clerk, part-time (PT) and full-time (FT) with rotations (i.e., different tasks and durations) including in different sections of the supermarket.

8.3.5.2 Stock Clerk

The work of the stock clerk is the most highly variable in terms of the variety of tasks performed. As required, stock clerks may also work at checkouts, load and move merchandise, clean, and prepare orders for variable lengths of time (Figure 8.30).

8.3.5.3 Grocery Clerk

Grocery clerks replenish stock (such as cheese, fresh pasta, and other prepared foods), unload and move merchandise, and clean and rearrange display cases. In terms of risk, grocery clerks are comparable with stock clerks; however, they seldom perform other tasks except for cleaning (Figure 8.31).

8.3.5.4 Meat Counter

Butchers are highly specialized and their tasks at the meat counter are largely limited to cutting, portioning, and packing meat; retrieving meat from the refrigerator; positioning meat in display cases; handling orders; and cleaning workbenches (Figure 8.32). In some supermarkets, the butchers also work at counters (meat counter).

8.3.5.5 Service Counter, Deli

Service counter staff serve customers cheese, cold cuts, and so on, over the counter and also work in the stockroom to prepare fresh food.

Deli counter staff may occasionally be assigned to the grocery and fresh food counter or the bakery section, and may also retrieve merchandise from the stockroom to place in display cases. They also clean the counter area and prepare orders (Figure 8.33).

Supermarket	Grocery counter FT	Grocery counter FT	Grocery counter PT
Stocking	12	24	8
Grocery counter	344	120	92
Cleaning	12	12	8
Handling		196	116
Total shift duration min	368	352	224

FIGURE 8.31 Grocery counter, part-time (PT) and full-time (FT) with rotations (i.e., different tasks and durations) including in different sections of the supermarket.

Supermarket	C	S	T	G	T	G
	Meat counter FT	Meat counter FT	Meat counter FT	Meat counter FT	Meat packaging	Meat packaging
Stocking			56		48	
Grocery counter					4	
Meat counter	412	344	196	360	136	208
Cleaning		44	88		20	
Handling	32				28	
Total shift duration min	444	388	340	360	236	208

FIGURE 8.32 Meat counter, part-time (PT) and full-time (FT) with rotations (i.e., different tasks and durations) including in different sections of the supermarket.

8.3.5.6 Fruit and Vegetable Section (Fresh Produce)

Workers in the fruit and vegetable section supply produce but from time to time may also man the checkouts, unload and move merchandise, clean the fresh produce area, restock the grocery counter, and perform stock clerk duties (Figure 8.34).

8.3.5.7 Fish Counter

As shown in Figure 8.35, the only tasks performed by fish counter staff are prepare and sell fish and clean their work counters.

8.3.5.8 Bakery

In all the supermarkets included in our study, the bakery staff almost always only prepared and sold bread, performing very few stocking or cleaning tasks in their section. Occasionally, they tidied up the merchandise (Figure 8.36).

Supermarket	C	S	S	T	T	G	G	T	G
	Service/deli counter FT	Service/deli counter FT	Service/deli counter PT	Service/deli counter FT	Service/deli counter PT	Service/deli counter FT	Service/deli counter PT	Stockroom PT	Packaging PT
Grocery	164			40	4	28	16	20	16
Service/deli counter	192	132	108	304	196	256	156	272	172
Bakery section		68	52						
Cleaning	32	20	20	12	4	64	28	20	20
Handling	4					12	12	56	
Office				10	10			10	
Total shift duration min	392	220	180	366	214	360	212	378	208

FIGURE 8.33 Grocery counter, part-time (PT) and full-time (FT) with rotations (i.e., different tasks and durations) including in different sections of the supermarket.

Supermarket	C	S	T	T	G	G
	Produce FT	Produce FT	Produce FT	Produce FT	Produce FT	Produce FT
Cashier		100				
Stocking		12				
Service/deli counter		12				
Fresh produce	316	192	320	124	316	212
Cleaning	32	32	44	100		
Handling	64		16	12	44	
Office			5	5		
Total shift duration min	412	348	385	241	360	212

FIGURE 8.34 Fresh produce, part-time (PT) and full-time (FT) with rotations (i.e., different tasks and durations) including in different sections of the supermarket.

8.3.6 Examples of Risk Assessment Results for Homogeneous Groups of Workers

Exposure risk is highly variable in terms of the shift duration and contents, that is, which tasks make up the turnover both qualitatively and quantitatively; we now offer some examples of risk assessments for certain exposure levels, specifying the duration and content of each.

Supermarket	T	G	G
	Fish counter FT	Fish counter FT	Fish counter PT
Fish counter	340	260	152
Cleaning	44	88	60
Total shift duration min	384	384	212

FIGURE 8.35 Fish counter, part-time (PT) and full-time (FT) with rotations (i.e., different tasks and durations) including in different sections of the supermarket.

Supermarket	T	G	G
	Bakery section FT	Bakery section FT	Bakery section PT
Stocking	76		
Bakery section	260	360	200
Cleaning	44		
Total shift duration min	380	360	200

FIGURE 8.36 Bakery section, part-time (PT) and full-time (FT) with rotations (i.e., different tasks and durations) including in different sections of the supermarket.

8.3.6.1 Cashiers

In our example, cashiers worked a 380-min shift over 6 days a week. They did not perform any other tasks, except for returning unwanted items from the checkout (10 min a day). All the data relating to the organizational aspects and concerning risk are reported in Tables 8.21 through 8.23.

For cashiers, the risk of biomechanical overload of the upper limbs was assessed over one FT 380-min shift with one 20-min break.

The work was over a 6-day week; therefore, the weekly risk was 17 for the right side and 16 for the left (Table 8.21). Since every day of the week featured the same exposure and the main job of the cashiers was to man the registers (their secondary tasks being minimal), the time-weighted average and the multitask complex methods both produced the same risk indexes.

However, the exposure index calculated with the time-weighted average method per day (15.6 on the right side and 15.8 with the multitask complex method) was slightly higher than per week (16.9 and 17). This slight difference is negligible.

TABLE 8.21
Example of Risk Assessment for Cashier, Full-Time (Shift Duration 380 min with One 20-Min Break). Time-Weighted Average: DX = 16.9; SX = 16.3; Multitask Complex: DX = 17; SX = 16.5

	Distribution of Shifts over the Week for a Homogeneous Group (Total Duration of Shift in Minutes, for Regular Working Hours and Overtime, Including Breaks)						
	Mon	**Tue**	**Wed**	**Thu**	**Fri**	**Sat**	**Sun**
Analysis of Work Organization in the First Week of a Representative Month							
Shift duration	380	380	380	380	380	380	0
Nonrepetitive task duration in minutes	0	0	0	0	0	0	0
Duration of actual breaks: 20 min, 1 break	20	20	20	20	20	20	0
Net duration of repetitive tasks	360	360	360	360	360	360	0
Duration of Repetitive Tasks in Different Shifts over 1 Week							
Cash register with flatbed scanner	349.2	349.20	349.20	349.20	349.20	349.20	
Return unwanted merchandise from checkouts	10.44	10.44	10.44	10.44	10.44	10.44	

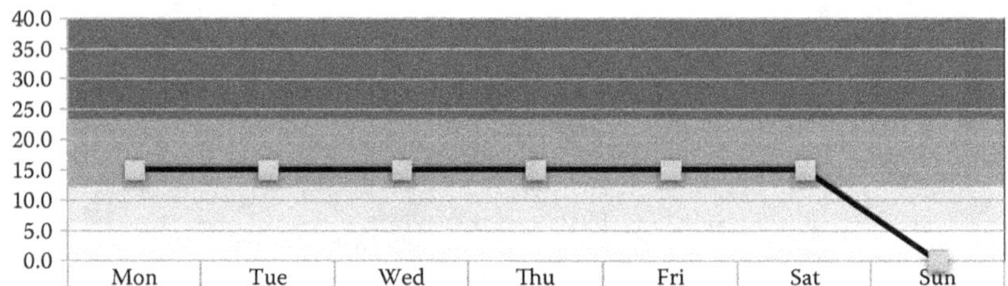

	Mon	Tue	Wed	Thu	Fri	Sat	Sun
Right	15.6	15.6	15.6	15.6	15.6	15.6	0.0
Left	15.1	15.1	15.1	15.1	15.1	15.1	0.0

If the shift duration is modified with PT workers doing 240 min a day with just one 15-min break, over a 6-day week, the daily risk drops to 10.1 (yellow) for the right side, and 11.5 (borderline, yellow/light red) for the week (Table 8.22). Again, there were no major differences between the two calculation methods. The last example, Table 8.23, describes a PT job (240 min a day) over a 3-day week. The risk for the week is 7.8–7.9, that is, yellow but borderline green (no risk). It can be concluded, therefore, that to assess the real exposure risk of cashiers, it is essential to carry out an in-depth organizational analysis that also covers any other tasks performed during the shift.

8.3.6.2 Stocking: Refilling Shelves

Tables 8.24 through 8.26 show three examples of three different organizational setups for stock clerks with relative risk assessments.

The examples refer to:

- An FT stock clerk working 380 min per day, with two 10-min breaks and only 3% of the shift at the checkout, over a 6-day week

TABLE 8.22
Example of Risk Assessment for Cashier, Part-Time (Shift Duration of Approx. 240 min with One 15-Min Break) over a 6-Day Week: Time-Weighted Average: DX = 11.5; SX = 11.1; Multitask Complex: DX = 11.6; SX = 11.2

	Distribution of Shifts over the Week for a Homogeneous Group (Total Duration of Shift in Minutes, for Regular Working Hours and Overtime, Including Breaks)						
	Mon	Tue	Wed	Thu	Fri	Sat	Sun
Analysis of Work Organization in the First Week of a Representative Month							
Shift duration	240	240	240	240	240	240	0
Nonrepetitive task duration in minutes	0	0	0	0	0	0	0
Duration of actual breaks: 15 min, 1 break	15	15	15	15	15	15	0
Net duration of repetitive tasks	225	225	225	225	225	225	0
Duration of Repetitive Tasks in Different Shifts over 1 Week							
Cash register with flatbed scanner	218.25	218.25	218.25	218.25	218.25	218.25	
Return unwanted merchandise from checkouts	6.53	6.53	6.53	6.53	6.53	6.53	

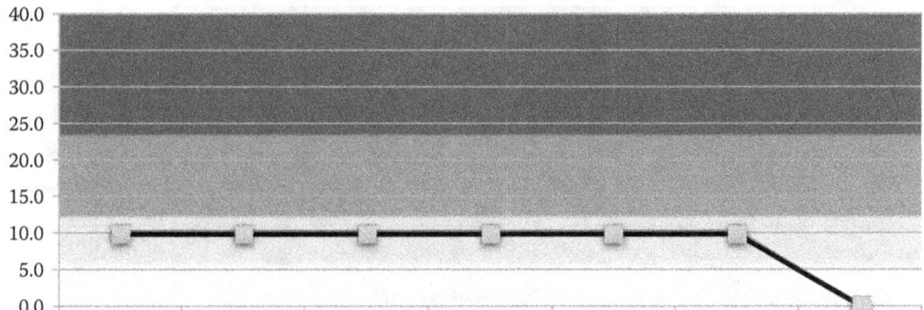

	Mon	Tue	Wed	Thu	Fri	Sat	Sun
Right	10.1	10.1	10.1	10.1	10.1	10.1	0.0
Left	9.8	9.8	9.8	9.8	9.8	9.8	0.0

- An FT stock clerk working 380 min/day, with two breaks (one 10 min and one 15 min) and 52% of the shift at the checkout, over a 6-day week
- A PT stock clerk working 240 min/day over 5 days and 180 min for 1 day, with no work at the checkout

In the first case (FT, 380 min, two 10-min breaks and only 3% of the shift at the checkout, over a 6-day week—Table 8.24), there are two different risk scores for each day of the week depending on whether the time-weighted average is used (12.9 on the right—light red) or the multitask complex (20.5 on the right—dark red).

The intrinsic risk indexes for the various tasks are quite different, reflecting a mismatch between the results obtained using the two formulas. The question is, which one is best?

TABLE 8.23
Example of Risk Assessment for Cashier, Part-Time (Shift Duration of Approx. 240 min with One 15-Min Break) over a 3-Day Week: Time-Weighted Average:DX=7.8; SX=7.6; Multitask Complex: DX=7.9; SX=7.6

	Distribution of Shifts over the Week for a Homogeneous Group (Total Duration of Shift in Minutes, for Regular Working Hours and Overtime, Including Breaks)						
	Mon	Tue	Wed	Thu	Fri	Sat	Sun
Analysis of Work Organization in the First Week of a Representative Month							
Shift duration		240			240	240	0
Nonrepetitive task duration in minutes		0			0	0	0
Duration of actual breaks: One 15-min break		15			15	15	0
Net duration of repetitive tasks		225			225	225	0

Duration of Repetitive Tasks in Different Shifts over 1 Week

		Mon	Tue	Wed	Thu	Fri	Sat
Cash register with flatbed scanner	X		218.2			218.25	218.25
Return unwanted merchandise from checkouts	X		6.53			6.53	6.53

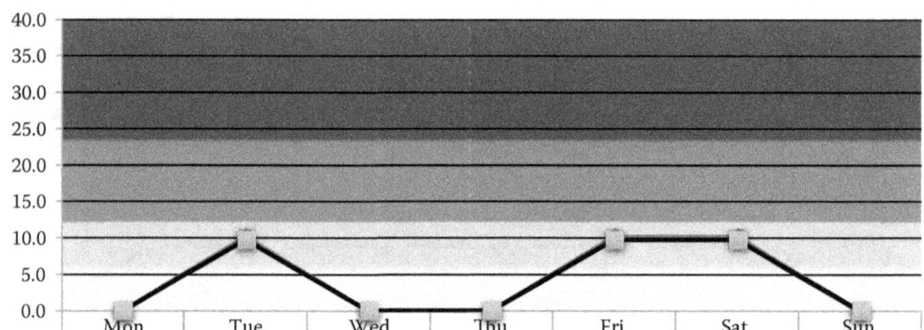

	Mon	Tue	Wed	Thu	Fri	Sat	Sun
Right	0.0	10.1	0.0	0.0	10.1	10.1	0.0
Left	0.0	9.8	0.0	0.0	9.8	9.8	0.0

If risk is calculated on the basis of each individual day and the pace of task rotation is less than 90 min, then the time-weighted average will be the more representative method.

If the rotations take place over a longer period, then exposure risk calculated using the multitask complex will be more representative. Generally speaking, when the work is organized in this way, tasks are rotated quite quickly and therefore the time-weighted average method produces the most reliable values.

Based on current understanding, the weekly risk score can be set midway between the score obtained with the time-weighted average and the one obtained with the multitask complex, which in our first example would be between 13.3 and 19 (medium red) for the right side.

TABLE 8.24
Stock Clerk/Shelf Filler, Full-Time (380 min with 2 × 10-min breaks)

	Distribution of Shifts over the Week for a Homogeneous Group (Total Duration of Shift in Minutes, for Regular Working Hours and Overtime, Including Breaks)					
	Mon	Tue	Wed	Thu	Fri	Sat
Analysis of Work Organization in the First Week of a Representative Month						
Shift duration	380	380	380	380	380	380
Nonrepetitive task duration in minutes	30	30	30	30	30	30
Duration of actual breaks: 20 min, 2 break	20	20	20	20	20	20
Net duration of repetitive tasks	330	330	330	330	330	330
Duration of Repetitive Tasks in Different Shifts over 1 Week						
Cashier						
Cash register with flatbed scanner	10.0	10.0	10.0	10.0	10.0	10.0
Stocking						
Move crates/boxes (6–12 kg) manually on to shelves	10	10	10	10	10	10
Stock shelves with items weighing up to 2 kg	161	181	181	181	181	181
Fill and transport cardboard boxes weighing between 6 and 12 kg using four-wheeled carts	10	10	10	10	10	10
Open cardboard boxes	9	9	9	9	9	9
Position boxes along aisles (weight 6–12 kg)	10	10	10	10	10	10
Position special offer signs	20					
Stock shelves with bottled water	15	15	15	15	15	15
Load/Unload/Handle Pallet Jacks and Empty Pallets						
Move merchandise with manual pallet jack (push and pull)	10	10	10	10	10	10
Unload trucks and move merchandise with electric pallet jack	10	10	10	10	10	10
Handle empty pallets	20	20	20	20	20	20
Other Transportation						
Flatbed trolleys	45	45	45	45	45	45

	Multitask Complex		Time-Weighted Average	
	Right	Left	Right	Left
Mon	18.5	17.1	13.0	12.2
Tue	20.5	18.8	12.9	12.0
Wed	20.5	18.8	12.9	12.0
Thu	20.5	18.8	12.9	12.0
Fri	20.5	18.8	12.9	12.0
Sat	20.5	18.8	12.9	12.0
Sun	0	0	0	0
Total over 1 week	19.0	17.4	13.3	12.3

In the second example of an FT worker on 380 min/day, with two breaks (one 10 min and one 15 min long) and 52% of the shift at the checkout, over a 6-day week, (Table 8.25), the daily scores range between 15.9 and 21.9. In this case, since there is no task rotation between stocking and working at the checkout (52% of every afternoon only at the checkout), the representative risk score will be 21.9 (dark red), using the multitask complex method.

As to the score defining weekly risk, based on current understanding, we may argue that the level will be midway between the results obtained with the time-weighted average and the multitask complex, so for the right side it will range between 16.3 and 20.8 (medium–dark red), although without suitable rotations it would be more plausible to choose the score generated by the multitask complex (20.8—dark red).

In the third case of a PT worker on 240 min/day over 5 days and 180 min for 1 day, and no work at the checkout (Table 8.26), there are again two different risk scores for each day of the week depending on whether the time-weighted average is used (10 on the right—yellow) or the multitask complex (16 on the right—medium red).

When the stock clerk's work is organized in this way, tasks are rotated quite quickly; therefore, the time-weighted average method produces the most reliable values.

As to the score defining weekly risk, based on current understanding, we may argue that the level will be midway between the results obtained with the time-weighted average and the multitask complex, so for the right side it will range between 10 and 14 (yellow–light red).

In short, the stock clerk's job is intrinsically at risk of biomechanical overload; however, the degree of risk depends largely on how the work is organized and can therefore be modified. As well, especially in smaller operations, stock clerks are often also called on to work at the checkouts.

8.3.6.3 Grocery

When the tasks performed at the grocery counter include stocking and filling display counters or shelves with fresh food, as far as risk is concerned, the job is entirely comparable with that of the stock clerk.

8.3.6.4 Meat Counter

Tables 8.27 and 8.28 show the high intrinsic risk associated with working at the meat counter; it comes as no surprise, therefore, that the risk scores are medium–high.

In the first organizational analysis (Table 8.27), the butchers' work shifts are 450 min/day, with two breaks, one 20 min long and one 10 over a 6-day week. In this analysis, the workers also work overtime: instead of the usual 2400 min/week (a constant 480 min a day over a 5-day week), they do 2700 (i.e., 450 min a day over 6 days a week). There are few breaks and no meal breaks at all (a fairly common situation among supermarket butchers).

There are two different risk scores for each day depending on whether the analysis uses the time-weighted average (18 on the right side—medium red) or

TABLE 8.25
Cashier, Full-Time (52% of the Time) and Stocking/Shelf Filler (380 min; 2 Breaks, one 10-Min and one 15-Min)

	Distribution of Shifts over the Week for a Homogeneous Group (Total Duration of Shift in Minutes, for Regular Working Hours and Overtime, Including Breaks)						
	Mon	Tue	Wed	Thu	Fri	Sat	Sun
Analysis of Work Organization in the First Week of a Representative Month							
Shift duration	380	380	380	380	380	380	0
Nonrepetitive task duration in minutes	30	30	30	30	30	30	0
Duration of actual breaks: 25 min	25	25	25	25	25	25	0
Net duration of repetitive tasks	325	325	325	325	325	325	0

Duration of Repetitive Tasks in Different Shifts over 1 Week						
	Mon	Tue	Wed	Thu	Fri	Sat
Cashier						
Cash register with flatbed scanner	170.0	170.0	170.0	170.0	170.0	170.0
Stocking						
Move crates/boxes (6–12 kg) manually onto shelves	30	30	30	30	30	30
Stock shelves with items weighing up to 2 kg	30	30	30	30	30	30
Open cardboard boxes	5	5	5	5	5	5
Stock shelves with bottled water	30	30	30	30	30	30
Cleaning						
Clean shelves	15	15	15	15	15	15
Load/Unload/Handle Pallet Jacks and Empty Pallets						
Move merchandise with manual pallet jack (push and pull)	30	30	30	30	30	30
Unload trucks and move merchandise with electric pallet jack	10	10	10	10	10	10
Handle empty pallets	5	5	5	5	5	5

	Multitask Complex		Time-Weighted Average	
	Right	Left	Right	Left
Mon	20.3	18.4	15.9	14.3
Tue	21.9	19.9	15.9	14.3
Wed	21.9	19.9	15.9	14.3
Thu	21.9	19.9	15.9	14.3
Fri	21.9	19.9	15.9	14.3
Sat	21.9	19.9	15.9	14.3
Sun	0	0	0	0
Total over 1 week	20.8	18.9	16.3	14.7

TABLE 8.26
Stock Clerk/Shelf Filler, Part-Time (240 min over 5 days, 180 on One Day and 4% at Checkout); 1 × 20-Min Break

	Distribution of Shifts over the Week for a Homogeneous Group (Total Duration of Shift in Minutes, for Regular Working Hours and Overtime, Including Breaks)						
	Mon	Tue	Wed	Thu	Fri	Sat	Sun
Analysis of Work Organization in the First Week of a Representative Month							
Shift duration	240	180	240	240	240	240	0
Nonrepetitive task duration in minutes	20	20	20	20	20	20	0
Duration of actual breaks: One 20-min break	20	20	20	20	20	20	0
Net duration of repetitive tasks	200	140	200	200	200	200	0

Duration of Repetitive Tasks in Different Shifts over a Week	Mon	Tue	Wed	Thu	Fri	Sat
Cashier						
Cash register with flatbed scanner	10.0	5,0	10.0	10.0	10.0	10.0
Stocking						
Move crates/boxes (6–12 kg) manually onto shelves	5	5	5	5	5	5
Stock shelves with items weighing up to 2 kg	116	86	116	116	116	116
Fill and transport cardboard boxes weighing between 6 and 12 kg using four-wheeled carts	10	5	10	10	10	10
Open cardboard boxes	9	4	9	9	9	9
Position boxes along aisles (weight 6–12 kg)	5	5	5	5	5	5
Stock shelves with bottled water	10	5	10	10	10	10
Load/Unload/Handle Pallet Jacks and Empty Pallets						
Move merchandise with manual pallet jack (push and pull)	5	5	5	5	5	5
Unload trucks and move merchandise with electric pallet jack	5	5	5	5	5	5
Other Transportation						
Flatbed trolley	25	15	25	25	25	25

	Multitask Complex		Time-Weighted Average	
	Right	Left	Right	Left
Mon	14.1	13.0	10.0	9.2
Tue	13.9	12.8	8.1	7.5
Wed	16.2	14.8	10.0	9.2
Thu	16.2	14.8	10.0	9.2
Fri	16.2	14.8	10.0	9.2
Sat	16.2	14.8	10.0	9.2
Sun	0	0	0	0
Total over 1 week	14.1	13.0	10.0	9.2

the multitask complex (22.5–22.7 on the right side—purple). As emphasized previously, if the task rotation schedule is below 90 min, in terms of daily risk, the time-weighted average method will produce the most representative results; if above 90 min, it will be the multitask complex method. If the butchers alternate between multiple tasks quite frequently, the time-weighted method should be

TABLE 8.27
Butcher, Full-Time (450 min, 2 Breaks, One 20-Min and One 10-Min, over a 6-Day Week)

	Distribution of Shifts over the Week for a Homogeneous Group (Total Duration of Shift in Minutes, for Regular Working Hours and Overtime, Including Breaks)						
	Mon	Tue	Wed	Thu	Fri	Sat	Sun
Analysis of Work Organization in the First Week of a Representative Month							
Shift duration	450	450	450	450	450	450	0
Nonrepetitive task duration in minutes	10	10	10	10	10	10	0
Duration of real breaks: one 20-min break and one 10-min break	30	30	30	30	30	30	0
Net duration of repetitive tasks	410	410	410	410	410	410	0

Duration of Repetitive Tasks in Different Shifts over a Week

	Mon	Tue	Wed	Thu	Fri	Sat
Service Counter: Grocery Counter						
Restock cheese and dairy counter				10		
Meat Counter: Prepare, Pack, and Restock Counters						
Halve chickens with cleavers and place them on trays	15	15	15	15	15	15
Remove chicken breasts and place on trays	15	15	15	15	15	15
Remove chicken thighs and place on trays	10	10	10	10	10	10
Prepare and clean pieces of meat	60	50	60	50	60	50
Cut meat into large portions and place on trays	40	40	40	40	40	40
Slice steaks using electric slicer and place on trays	90	70	90	60	90	70
Cut pork chops using a knife and place on trays	20	20	20	20	20	20
Cut through backbones using a cleaver and slice meat using a knife	2	2	2	2	2	2
Pack trays of precut meat	28	28	28	28	28	28
Pack sausages on trays	20	20	20	20	20	20
Wrap, weigh, and price full trays	20	20	20	20	20	20
Restock meat counter	60	60	60	60	60	60
Restock meat refrigerator		30		30		30
Other Transportation						
Roller and trolley	30	30	30	30	30	30

	Multitask Complex		Time-Weighted Average	
	Right	Left	Right	Left
Mon	21.1	20.7	18.0	17.4
Tue	22.7	22.4	18.3	17.7
Wed	22.5	22.2	18.0	17.4
Thu	22.7	22.1	18.4	17.2
Fri	22.5	17.4	18.0	17.4
Sat	22.7	22.4	18.3	17.7
Sun	0	0	0	0
Total over 1 week	33.5	32.7	28.7	27.6

TABLE 8.28
Butcher, Full-Time (450 min, 2 × 10-Min Breaks and One Unpaid 30-Min Meal Break, over a 5-Day Week)

	Distribution of Shifts over the Week for a Homogeneous Group (Total Duration of Shift in Minutes, for Regular Working Hours and Overtime, Including Breaks)						
	Mon	Tue	Wed	Thu	Fri	Sat	Sun
Analysis of Work Organization in the First Week of a Representative Month							
Shift duration	450		450	450	450	450	0
Nonrepetitive task duration in minutes	10		10	10	10	10	0
Duration of actual breaks: one 20-min break and one 10-min break	20		30	30	30	30	0
Net duration of repetitive tasks	410		410	410	410	410	0

Duration of Repetitive Tasks in Different Shifts over a Week

	Mon	Tue	Wed	Thu	Fri	Sat
Service Counter: Grocery Counter						
Restock cheese and dairy counter				10		
Meat Counter: Prepare, Pack, and Restock Counters						
Halve chickens with cleavers and place them on trays		15	15	15	15	15
Remove chicken breasts and place on trays		15	15	15	15	15
Remove chicken thighs and place on trays		10	10	10	10	10
Prepare and clean pieces of meat		50	60	50	60	50
Cut meat into large portions and place on trays		40	40	40	40	40
Slice steaks using electric slicer and place on trays		70	90	60	90	70
Cut pork chops using a knife and place on trays		20	20	20	20	20
Cut through backbones using a cleaver and slice meat using a knife		2	2	2	2	2
Pack trays of pre-cut meat		28	28	28	28	28
Pack sausages on trays		20	20	20	20	20
Wrap, weigh, and price full trays		20	20	20	20	20
Restock meat counter		60	60	60	60	60
Restock meat refrigerator		30		30		30
Other Transportation						
Roller and trolley	40		40	40	40	40

	Multitask Complex		Time-Weighted Average	
	Right	Left	Right	Left
Mon				
Tue	16.2	16.0	13.1	12.6
Wed	20.4	20.1	16.3	15.7
Thu	20.5	20.0	16.6	15.5
Fri	20.4	15.7	16.3	15.7
Sat	20.5	20.2	16.6	16.0
Sun				
Total over 1 week	19.5	19.0	16.5	15.8

used, otherwise, the values generated using the multitask complex method will be more representative.

As to the score defining weekly risk, based on current understanding, we may argue that the level will be midway between the results obtained with the time-weighted average and the multitask complex, so for the right side it will range between 28 and 33.5 (purple—extremely high risk). The high weekly risk scores are also due to overtime (12%) versus a constant 2400 h/week.

In the second organizational analysis (Table 8.28), the butchers work 5 days a week and in addition to the previous two daily breaks also have one 30-min meal break (unpaid, therefore not included in the shift). In this case, the score defining weekly risk will be midway between the results obtained with the time-weighted average and the multitask complex, so for the right side it will range between 16 and 19.5 (medium red).

Based on these examples, it appears that for meat counter staff, exposure levels may vary considerably in terms of job description and duration. Overtime should be avoided and breaks should be adequate in length and number (for instance, there should be a meal break during a 7–8 h shift).

8.3.6.5 Deli Counter

Depending on the type of supermarket and whether the work is FT or PT, the risk scores obtained with the OCRA checklist will vary considerably.

Table 8.29 shows a very complex setup for the deli counter where staff cut and pack food, restock cheeses and deli meats, and also perform other tasks in the bakery section. The work is PT 4 days at 240 min/day and FT 1 day at 450 min/day (with the usual two 15-min breaks).

Figure 8.37 shows the risk levels obtained using the time-weighted average (first graph) versus the multitask complex method (second graph).

The first graph depicts a schedule where tasks with frequent rotations (every 90 min or less) are performed, over a 240-min shift; the daily risk score for the right side is 13.3 (light red). In the second graph, where rotations between tasks is less frequent, with a 240-min shift, the daily risk score for the right side rises to 35.9 (purple). However, since there are 15 tasks, it is likely that, with a good turnover rate, the time-weighted average method would be more appropriate. The difference between the two values underscores the presence of tasks featuring both low and extremely high intrinsic risk. On Saturday, when the shift is 450 min long, the risk score is higher at 22.6 (time-weighted average on the right side).

As to the score defining weekly risk, based on current understanding, we may argue that the level will be midway between the results obtained with the time-weighted average and the multitask complex, so for the right side it will range between 15 and 33.4 (red high/purple extremely high).

8.3.6.6 Fresh Produce

In the fresh produce section in this example, the workers are mainly assigned to loading and unloading and stocking tasks (Table 8.30). Table 8.31 shows the results of the risk assessment for situations with different shift durations and breaks:

TABLE 8.29
Service/Deli Counter, Part-Time, 4 days at 240 min/day and Full-Time and 1 Day at 450 Min/Day (with Two 15-Min Breaks)

	Distribution of Shifts over the Week for a Homogeneous Group (Total Duration of Shift in Minutes, for Regular Working Hours and Overtime, Including Breaks)						
	Mon	Tue	Wed	Thu	Fri	Sat	Sun
Analysis of Work Organization in the First Week of a Representative Month							
Shift duration	240		240	240	240	450	0
Nonrepetitive task duration in minutes	5		5	5	5	5	0
Duration of actual breaks: 20 min with only 1 break	30		30	30	30	30	0
Net duration of repetitive tasks	205		205	205	205	415	0

Duration of Repetitive Tasks in Different Shifts over a Week	Mon	Tue	Wed	Thu	Fri	Sat
Deli Counter: Prepare, Pack, and Restock Counters						
Cut hard cheeses into large pieces using a knife	5		5	5	5	10
Pack and wrap cheese pieces (low machine)	4		4	4	4	8
Open large packages of cheese	4		4	4	4	8
Weigh and price cheeses	4		4	4	4	8
Prepare prosciutto hams for slicing on electric slicer	4		4	4	4	8
Wrap prosciutto ham	4		4	4	4	8
Move and position deli meats and cheeses on shelves (weight: over 8–10 kg)	4		4	4	4	8
Slice deli meats on manual slicer and prepare (slice, package, weigh, label)	73		73	73	73	140
Slice deli meats on electric slicer and prepare (slice, package, weigh, label);	18		18	18	18	36
Cut small rounds of soft and semi-soft cheeses	4		4	4	4	14
Move and position smaller deli meats and cheeses on shelves	2		2	2	2	9
Bakery: Prepare, Pack, and Restock Counters						
Load frozen loaves into oven	18		18	18	18	36
Sell small loaves at bakery counter	18		18	18	18	36
Package	28		28	28	28	56
Cleaning						
Clean shelves	15		15	15	15	30

- For the FT shift (380 min over a 6-day week) with one 20-min break, the weekly risk index for the right side is between 24.6 (time-weighted average) and 27.2 (multitask complex), placing it in the purple/high area.
- For the FT shift (380 min over a 6-day week) with two 10-min breaks, the weekly risk index for the right side is between 22.2 (time-weighted average) and 24.9 (multitask complex), placing it in the red/purple area.
- For the PT shift (240 min over a 6-day week) with one 15-min break, the weekly risk index for the right side is between 17.7 (time-weighted average) and 17.9 (multitask complex), placing it in the medium red area.

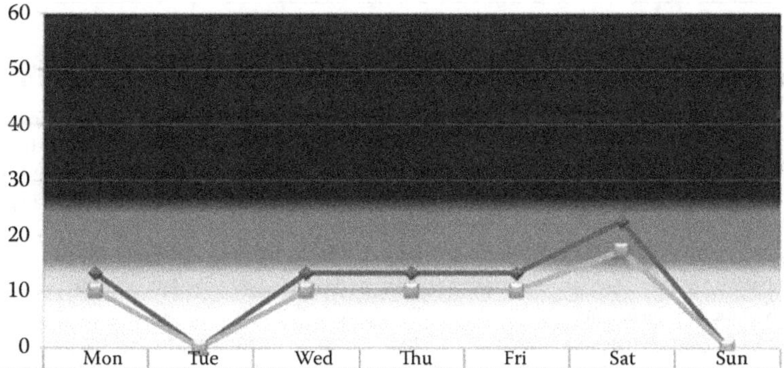

	Mon	Tue	Wed	Thu	Fri	Sat	Sun
Right	13	0	13	13	13	23	0
Left	10	0	10	10	10	17	0

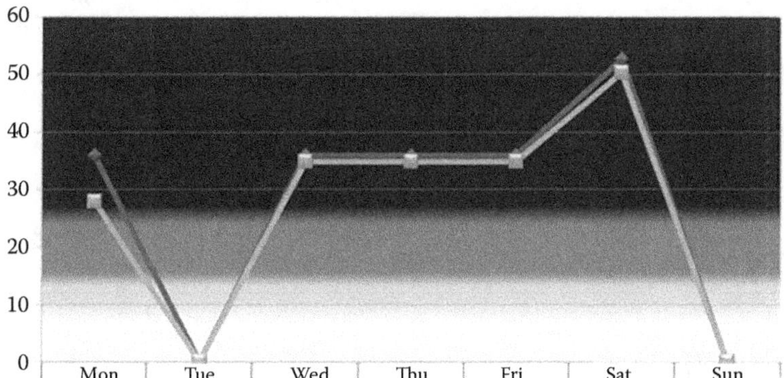

	Mon	Tue	Wed	Thu	Fri	Sat	Sun
Right	36	0	36	36	36	53	0
Left	28	0	35	35	35	50	0

FIGURE 8.37 Service/deli counter, part-time 4 days a week on a 240-min shift and 1 day full-time on a 450-min shift (with 2 × 15-min breaks). Weekly risk indexes using the time-weighted average vs the multitask complex: between 15 and 33.4).

8.3.6.7 Fish Counter

Table 8.32 shows the organizational data and results of the risk assessment for a group of workers at a fish counter. Fish counter staff generally work only in their own section.

For an FT shift (380 min over a 6-day week) with one 20-min break, the weekly risk index for the right side is between 14.8 (time-weighted average) and 18.3 (multi-task complex), placing it in the medium red area.

8.3.6.8 Bakery Section

Table 8.33 shows the organizational data and results of the risk assessment for a group of workers in a bakery section. These workers generally work only in their own section but from time to time may rotate between bakery and the service counter/deli counter.

TABLE 8.30
Fresh Produce Section, Full-Time, 380-Min Shift: Main Organizational Data

	Distribution of Shifts over the Week for a Homogeneous Group (Total Duration of Shift in Minutes, for Regular Working Hours and Overtime, Including Breaks)						
	Mon	Tue	Wed	Thu	Fri	Sat	Sun
Analysis of Work Organization in the First Week of a Representative Month							
Shift duration	380	380	380	380	380	380	0
Nonrepetitive task duration in minutes	30	30	30	30	30	30	0
Duration of actual breaks: 20 min with only 1 break or 2 × 10-min breaks	20	20	20	20	20	20	0
Net duration of repetitive tasks	330	330	330	330	330	330	0

Duration of Repetitive Tasks in Different Shifts over a Week	Mon	Tue	Wed	Thu	Fri	Sat
Fruit and Vegetable Section (Fresh Produce) Prepare, Pack, and Restock Section						
Price wrapped produce	10	10	10	10	10	10
Position prices (e.g., position special offer labels)	15	15	15	15	15	15
Move and position hampers on display units	265	265	265	265	265	265
Load/Unload/Handle Pallet Jacks and Empty Pallets						
Unload trucks and move merchandise with electric pallet jack						
Handle empty pallets	40	40	40	40	40	40

For an FT shift (380 min over a 6-day week) with one 20-min break, the weekly risk index for the right side is between 18.7 (time-weighted average) and 20.4 (multitask complex), placing it in the medium red area.

8.3.7 Conclusions

In conclusion, in light of the risk assessment results outlined so far, the following observations can be made with reference to open issues that still require further research:

- Intrinsic risk scores for various tasks: Tasks are performed in much the same way in the various supermarkets: The intrinsic risk values assigned to the various tasks can, for the time being, be regarded as representative of all supermarkets (thus providing nonexperts with libraries containing lists of preassessed tasks and making it easier for them to assess risk).
- Studies on the duration and content of weekly shifts: In terms of assessing risk in the various supermarkets, the differences stem primarily from the way the work is organized (i.e., duration of shift(s) over the week, duration and distribution of breaks, task assignment, and turnover). All these aspects must be analyzed in the utmost detail in each individual situation.

TABLE 8.31
Fresh Produce Section, Full-Time: 380-Min Shift with One 20-Min Break; 380-Min Shift with 2 × 10-Min Breaks; Part-Time: 220-Min Shift with One 15-Min Break

	Multitask Complex		Time-Weighted Average	
	Right	Left	Right	Left
380-Min Shift with One 20-Min Break over a 6-Day Week				
Mon	26.3	26.3	23.9	23.8
Tue	27.5	27.5	23.9	23.8
Wed	27.5	27.5	23.9	23.8
Thu	27.5	27.5	23.9	23.8
Fri	27.5	23.8	23.9	23.8
Sat	19.7	19.7	17.1	17.0
Sun				
Total over 1 week	27.2	27.2	24.6	24.5
380-Min Shift with Two 10-Min Breaks over a 6-Day Week				
Mon	24.0	24.0	21.6	21.5
Tue	24.9	24.9	21.6	21.5
Wed	24.9	24.9	21.6	21.5
Thu	24.9	24.9	21.6	21.5
Fri	24.9	21.5	21.6	21.5
Sat	19.7	19.7	17.1	17.0
Sun				0.00
Total over 1 week	24.9	24.8	22.2	22.1
240-Min Shift with One 15-Min Break over a 6-Day Week				
Mon	17.9	17.9	17.7	17.7
Tue	14.7	14.7	14.3	14.2
Wed	18.1	18.1	17.7	17.7
Thu	18.1	18.1	17.7	17.7
Fri	18.1	17.7	17.7	17.7
Sat	18.1	18.1	17.7	17.7
Sun				0.00
Total over 1 week	17.9	17.9	17.7	17.6

- IT support: Each situation is so complex that risk assessments call for the use of more than mere spreadsheets: Specifically designed software programs are thus a necessity (reference can be made to http://www.petrasofware-weekly/monthly assessment cycles with preassessed libraries prepared by the authors using the OCRA method applied to supermarket tasks).

TABLE 8.32
Fish Counter, Full-Time (380 min; One 20-Min Break)

	Distribution of Shifts over the Week for a Homogeneous Group (Total Duration of Shift in Minutes, for Regular Working Hours and Overtime, Including Breaks)						
	Mon	Tue	Wed	Thu	Fri	Sat	Sun
Analysis of Work Organization in the First Week of a Representative Month							
Shift duration	380	380	380	380	380	380	0
Nonrepetitive task duration in minutes	30	30	30	30	30	30	0
Duration of real breaks: 20 min with only 1 break	20	20	20	20	20	20	0
Net duration of repetitive tasks	330	330	330	330	330	330	0

Duration of Repetitive Tasks in Different Shifts over a Week Fish Counter: Prepare, Pack, and Restock Counters						
	Mon	Tue	Wed	Thu	Fri	Sat
Gut and bag fish	70	70	70	70	70	70
Prepare fish counter display with crates and individual fish	90	90	90	90	90	90
Package fish slices	90	90	90	90	90	90
Cleaning						
Clean shelves	80	80	80	80	80	80

	Multitask Complex		Time-Weighted Average	
	Right	Left	Right	Left
Mon	17.8	13.0	14.4	7.0
Tue	19.6	13.0	14.4	7.0
Wed	19.6	13.0	14.4	7.0
Thu	19.6	13.0	14.4	7.0
Fri	19.6	13.0	14.4	7.0
Sat	19.6	13.0	14.4	7.0
Sun				0.00
Total over 1 week	18.3	11.3	14.8	7.2

- General level of risk scores: Based on the evaluations, risk is often "medium" and sometimes "high" (e.g., deli counter and butcher) among FT workers; for PT workers, risk is often medium/low red or borderline (yellow).
- Estimating unsaturation: To avoid overestimating the final outcome, it is essential to investigate and quantify the amount of "downtime," that is, estimate the percentage of unsaturation so as to reduce the *net duration of repetitive work*. For example, *time between customers* at the deli counter or fish counter during quiet times of the day, walking around empty-handed (stock clerks), waiting for customers at the checkout (cashiers), and so on.

This study provides data for other work areas, indicating the following unsaturation levels:

TABLE 8.33
Bakery Section, Full-Time (380 min; Two 10-Min Breaks)

	Distribution of Shifts over the Week for a Homogeneous Group (Total Duration of Shift in Minutes, for Regular Working Hours and Overtime, Including Breaks)						
	Mon	Tue	Wed	Thu	Fri	Sat	Sun
Analysis of Work Organization in the First Week of a Representative Month							
Shift duration	380	380	380	380	380	380	0
Nonrepetitive task duration in minutes	30	30	30	30	30	30	0
Duration of real breaks: 20 min with only 1 break	20	20	20	20	20	20	0
Net duration of repetitive tasks	330	330	330	330	330	330	0

Duration of Repetitive Tasks in Different Shifts over a Week	Mon	Tue	Wed	Thu	Fri	Sat
Bakery: Prepare, Pack, and Restock						
Load frozen loaves into oven	30	30	30	30	30	30
Sell small loaves at bakery counter	140	140	140	140	140	140
Package	160	160	160	160	160	160

	Multitask Complex		Time-Weighted Average	
	Right	Left	Right	Left
Mon	19.7	12.6	18.2	12.2
Tue	20.4	12.7	18.2	12.2
Wed	20.4	12.7	18.2	12.2
Thu	20.4	12.7	18.2	12.2
Fri	20.4	12.2	18.2	12.2
Sat	20.4	12.7	18.2	12.2
Sun				0.00
Total over 1 week	20.4	12.9	18.7	12.6

- 8%–10% average unsaturation for all tasks
- 39% unsaturation for the deli counter
- 10% unsaturation for the fresh produce section and stocking/refilling shelves
- 0% for the meat counter

To increase the reliability of this preliminary unsaturation data in the future, further research needs to be carried out in many more operational areas. This would generate representative average unsaturation times that could then be extended to all supermarkets.

- Weekly risk: Mathematical models and their predictive accuracy: The decision as to whether to use the time-weighted average or the multitask complex formula to calculate the final weekly risk score should consider more statistical data looking at the association between the risk indexes and the clinical data deriving from on-site assessments, and also to test which method produces the most predictive results.

8.4 COMMERCIAL KITCHENS: MONTHLY CYCLE

8.4.1 Introduction and Main Issues Relating to Risk Assessment

When one considers the way large commercial kitchens are organized to prepare and distribute meals, there are obviously going to be difficulties in evaluating risk due to biomechanical overload, especially of the upper limbs.

This work entails:

- Multitask exposure on a monthly basis with numerous variables determined by changing monthly menus.
- Two shifts to prepare lunches and dinners.
- Different locations including hospitals, schools, self-service cafeterias, and so on, of different sizes.
- The use of both manual and automatic equipment.
- Ingredients of different sizes and weights.
- Different procedures for allocating tasks and variable execution times.

Such extreme variability makes it challenging even for experts to assess risk objectively.

The approach used to tackle yet another "mission impossible" has been to create a software program (*Petrasoftware* for meal preparation) featuring predefined and preassessed variables, which even people with little experience in ergonomics will find easy to use. The following operational strategies have made that possible:

- Preliminary guide to collecting organizational data
- Identification of homogeneous groups of workers exposed to the same risk
- Preassessment using the OCRA checklist of all possible tasks and methods of execution, including filing the relevant video clips
- Calculation of exposure index for every homogeneous group, by entering into the software only the technical data generally required to define the menu, shift duration, and tasks performed by the homogeneous group

8.4.2 Classification of Foods, Operating Areas, and Utensils

Before actually listing the assessment criteria, it is crucial to define the structural and organizational elements required to evaluate risk.

Table 8.34 shows a preliminary classification of foods handled in a commercial kitchen which is an example of a short representative list of foods used to prepare meals:

Figure 8.38 briefly describes the work areas, macrophases, and food flows.

Food is transported using manual carts and trolleys of different sizes and capacities: There is a considerable amount of manual pushing and pulling between the prep areas and also to and from the final distribution points (Figure 8.39). Figure 8.40 describes the main utensils and equipment used to prepare, cook, and distribute food. Figure 8.41 describes the main types of carts and trolleys

TABLE 8.34
Classification of Foods Handled in a Commercial Kitchen

Food Group	Nutrients	Subgroups	Food Examples
Vegetables	Vitamin C, carotenes, folate, fibre, carbohydrate	Starchy vegetables	Potato, kumara, taro, parsnip
		Nonstarchy vegetables	Lettuce, tomato, carrot, pumpkin, capsicum, spinach, green beans, cabbage
Fruits	Vitamin C, carotenes, folate, fibre, carbohydrate		Apple, orange, banana, apricot, peach, pear, plum, melon, pineapple
Breads and cereals	Carbohydrate, fibre, B vitamins, calcium, iron	Bread and flour-based products excluding pasta	Bread, bagels, muffins, crispbreads
		Rice, pasta, cereal, grains	Breakfast cereal, rice, pasta, couscous, bulghur
Milk, yoghurt, cheese	Protein, calcium, Vitamin B12, Vitamin B2, Vitamin A		Milk, yoghurt, cottage cheese, soy milk
Meat and alternatives	Protein, iron, B vitamins, Vitamin D, zinc, magnesium, carbohydrate (alternatives)		Lamb, beef, chicken, pork, fish and seafood, tofu, dried beans, lentils
Healthy fats and oil			Avocado, canola oil, olive oil, walnuts, sunflower seeds
Others			Butter, high fat cheeses, potato crisps, chocolate, cakes, sweet biscuits
Condiments			Jam, peanut butter, sugar, Vegemite, chutney, tomato sauce
Free foods			Herbs and spices, tea, coffee, plain soda, diet drinks

used to transport food from one work area to another during different stages of preparation.

8.4.3 Identifying and Classifying Operations

Given the extreme variability and scope of these jobs, in order to define worker exposure, it has been necessary to begin by classifying work into macrophases, phases, and tasks. This preliminary classification is essential to identify the *repetitive*

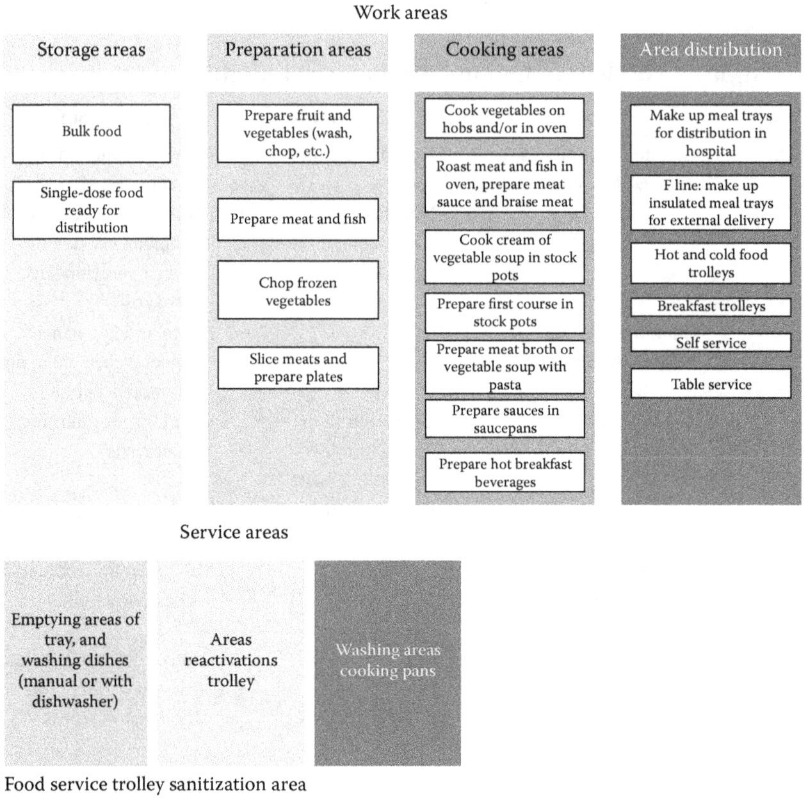

FIGURE 8.38 Classification of work and service areas and macrophases.

tasks that the exposure risk analysis will focus on using the OCRA checklist (see Figure 8.42).

Table 8.35 lists the numerous macrophases and phases that have been identified and broken down into work areas and types of food.

Since they are many and complex, they will be described schematically; only the macrophases and phases prior to identifying the actual tasks are reported here.

8.4.4 Identifying Tasks and Estimating Relative Intrinsic OCRA Checklist Values

The tasks making up each macrophase have been identified.

Table 8.36 summarizes the number of tasks per phase: All together a total of 288 tasks were analyzed!

The intrinsic risk indexes for the right and left sides were calculated for each task.

As already indicated in Chapter 6, the *intrinsic risk value* or *score* is the result of evaluating each task as if it were the only task performed by the worker for an entire shift (approximately 420–440 net min of repetitive work: that is, duration

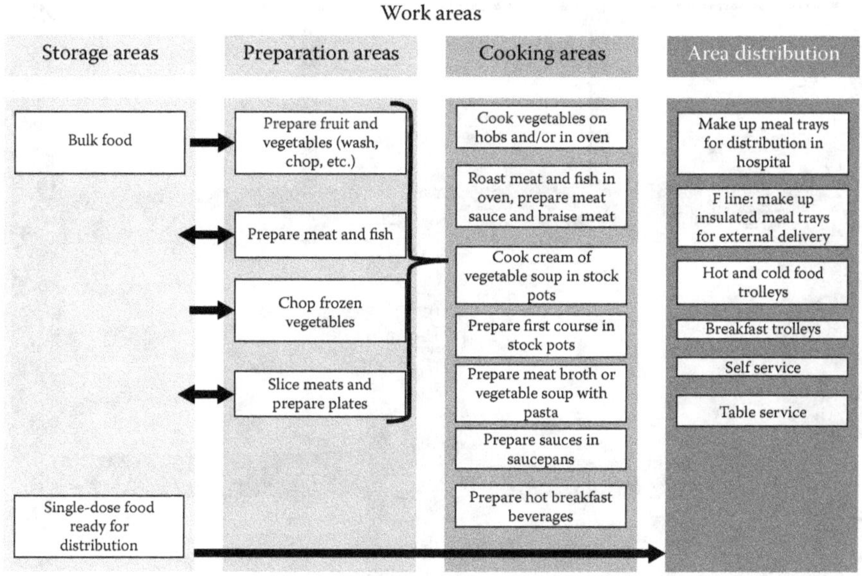

FIGURE 8.39 Classification of work areas and food flows.

multiplier = 1, with one meal break lasting at least 30 min and two breaks lasting at least 8 min, making for recovery multiplier = 1.33.

With regard to analyzing the risk of biomechanical overload associated with the manual pushing and pulling of trolleys, given the extreme variability among the trolleys in the study, a representative average of the intrinsic OCRA checklist was estimated, based on the weight of the loads transported using a manual trolley and the type of wheels and floors, as shown in Table 8.37.

8.4.5 Identifying Homogeneous Groups Based on Risk Exposure: Shift, Task, and Duration

8.4.5.1 Identification of Homogeneous Groups

Before assessing risk, it is necessary to identify homogeneous groups of workers; groups are homogeneous if the workers perform the same tasks over the period (i.e., same tasks, same shifts, and same duration of exposure).

The following therefore need to be analyzed:

- How many workers are involved and who are they?
- What tasks do they perform?
- How long to they perform them for?

8.4.5.2 Task Rotation Study

Tasks are often rotated on a monthly basis (if meal menus are changed every 4 weeks), over two daily shifts.

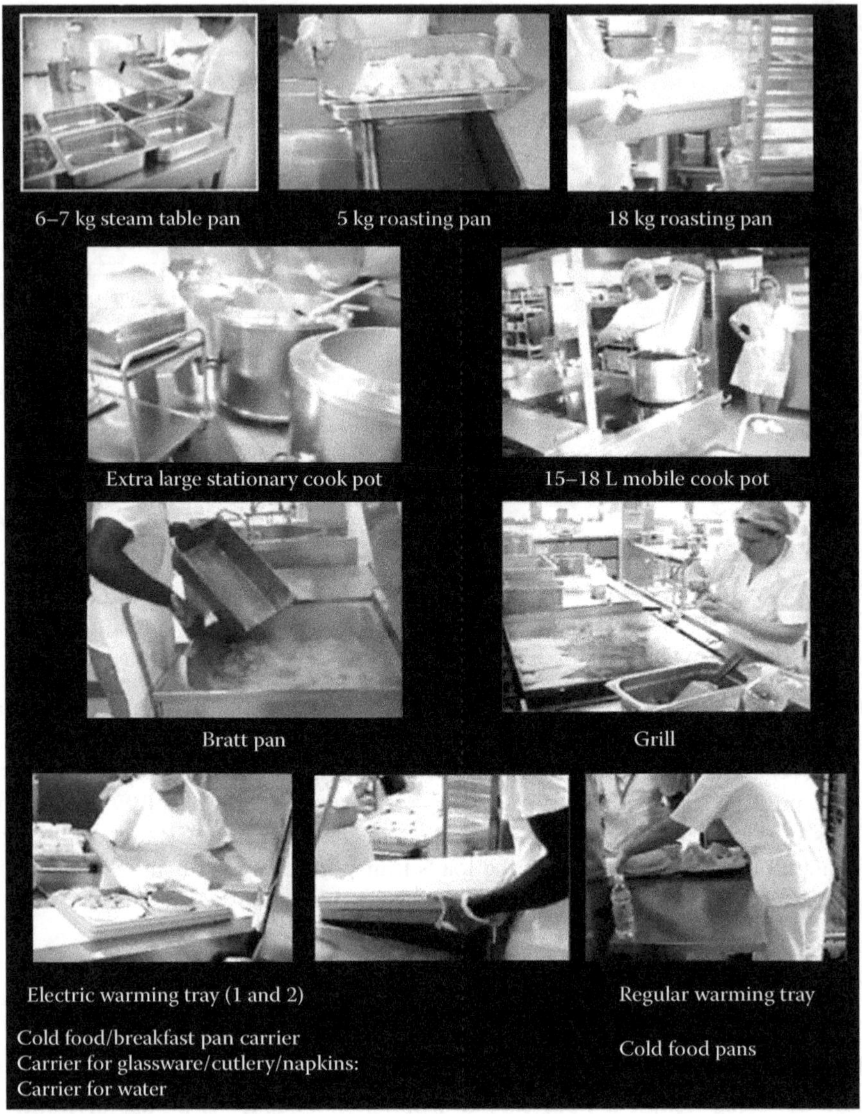

FIGURE 8.40 Classification of most common containers for food preparation, cooking, and distribution.

The preliminary organizational analysis for assessing the exposure index and determining the *net duration of repetitive work* and *distribution of recovery times* must be carried out for every working day of the 4-week month. The various days of the week might differ in terms of which tasks are performed and also their duration, as well as the duration of the shift itself—often when there are two daily shifts the evening shift is shorter.

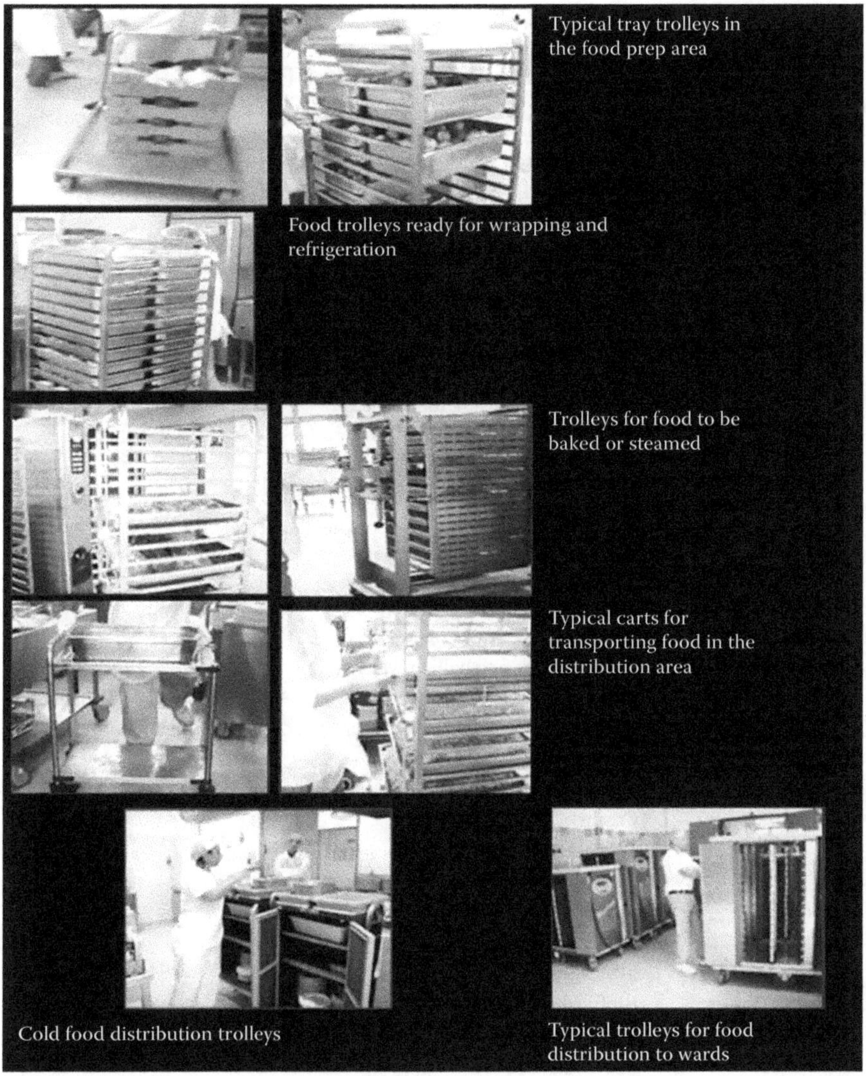

FIGURE 8.41 Typical carts and trolleys used to transport food from one work area to another during different stages of preparation.

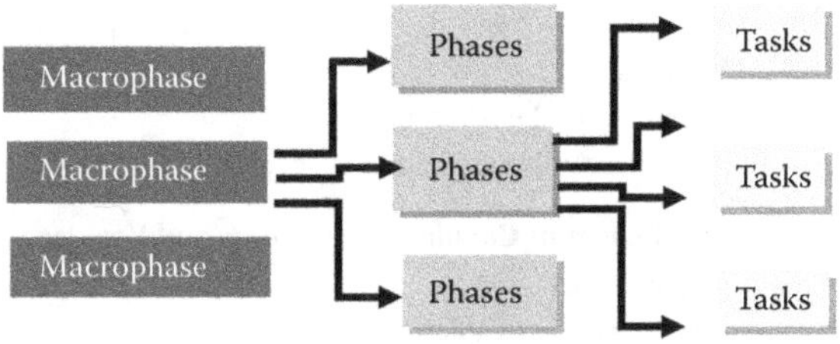

FIGURE 8.42 Main steps in the organizational analysis to identify and classify tasks.

TABLE 8.35
Macrophases and Phases Listed by Area and Food Type

	Macrophases in the Prep Area
Prepare vegetables	Green or mixed salad
	Tomato salad
	Transport fruit/vegetables for washing
	Rinse fruit
	Wash and chop other vegetables
	Chop carrots using food processor
	Wash and chop herbs
	Transport fruit/vegetables that do not need washing
	Chop capsicum using food processor
Prepare meat	Prepare chicken portions
	Wrap prepared meat trays and trolley
	Prepare meat in large pieces for baking in oven
	Supply empty oven trays for cooking meat
Prepare fish	Prepare fish fillets or slices
	Macrophases and Phases in Cooking Area: Boilers and Hobs
Boil pasta	Get empty oven trays to prepare pasta
	Transfer pasta into boiling water
	Strain pasta using automatic strainer and add sauce in large steam table pans
	Strain pasta using manual strainer and add sauce in smaller pans
	Prepare lasagne
Stock pots	Get stock pots for cooking cream of vegetable soup
	Cream of vegetable soup
	Béchamel sauce
	Tomato sauce
	Meat broth
	Minestrone with or without grains
	Mashed potato
	Broth with or without pasta and grains
	Semolina soup
	Egg drop soup
	Macrophases and Phases in Cooking Area: Grills and Bratts
Grill	Cook various types of meat on grill
	Clean grill
Bratt pan	Meat sauce
	Meat balls
	Risotto cooked in Bratt pan
	Risotto cooked in saucepan
	Macrophases and Phases in Cooking Area: Steam and Regular Ovens
Bake pizza and focaccia	Prepare pizza
Steamer	Get rice and vegetables from storage area
	Rice salad

TABLE 8.35 (CONTINUED)
Macrophases and Phases Listed by Area and Food Type

	Boil rice
	Steam vegetables and rice
	Roast or steam potatoes
	Vegetable gratin
	Creamed spinach
	Vegetable miniflan
Traditional oven	Roast chicken portions
	Roast large pieces of meat
	Place/remove whole trays of meat in oven
	Bake fish
	Place/remove whole trays of fish in oven
Macrophases and Phases in Cooking Area: Other Food Preparation	
Prepare vegetables for cream of vegetable soup	
Machine-slice prosciutto or other deli meats	
Prepare thermos with hot liquids	
Macrophases and Phases in Cooking Area: Wash Pots and Pans	
Wash pots and pans	Wash kitchen pans
Macrophases and Phases in Cold Trolley Prep Area	
Prepare cold trolleys with identical loads	Load trolleys with crates of drink bottles or breadsticks, and so on.
Prepare cold trolleys with mixed loads	Load trolleys with meals for distribution (without trays)
Macrophases and Phases in Distribution Area	
Make up single course meal trays	Fill one-course meal trays
Macrophases and Phases in Distribution Area	
Distribute meals	Prepare insulated crates for transport
Self-service	Meal distribution line for employee self-service
Table service	School lunch distribution (table service)
Macrophases and Phases in Tray Emptying and Washing Area	
Empty trays and wash dishes	Empty trays
	Rinse dishes and load dishwasher
	Rinse dishes by hand
	Fill and empty dishes in small dishwasher
Macrophases and Phases: Other Prep Areas	
Sanitize trolleys clean trolleys before transporting trays	
Prepare disposable glasses and cutlery	

TABLE 8.36
Summary of Tasks per Phase

Phase	No. of Tasks Identified in Each Phase
Prepare vegetables	28
Prepare meat	11
Cook pasta	11
Grill meat	6
Bake pizza and focaccia	17
Cook soups	39
Chop vegetables and slice deli meats	9
Bake and steam first courses and vegetables	38
Roast meats	7
Bake fish	10
Cook in bratt pan (risotto, meat sauce, stews)	13
Make up one-course meal trays	13
Pack one-course meals on belt for insulated crates	6
Empty trays and wash dishes	14
Clean trolleys, other storage and breakfast services	10
Prepare trolleys with cold food	12
Wash pots and pans	5
Self-service distribution	11
Table service and other supplies	14
Pushing and pulling trolleys	14
Total tasks analysed	288

Therefore, the following information needs to be acquired for each day of the sample month:

- Shift duration
- Total net duration of repetitive work
- Duration of each repetitive task
- Duration and distribution of actual breaks

These data are necessary to recalculate the intrinsic OCRA checklist scores based this time on the actual net duration of the repetitive tasks performed during the shift and the actual distribution of recovery times for each individual exposure day.

We will show a some examples of the kinds of preliminary organizational studies that must be carried out before a risk assessment.

Figure 8.43 summarizes the shifts assigned to a homogeneous group of workers during the sample month.

Figure 8.44 describes the duration of shifts and breaks on each day of the week.

TABLE 8.37
Estimated OCRA Checklist Scores for Pushing/Pulling Trolleys Carrying Different Load Weights

Weight Carried by Trolley (kg)	Force Score	Frequency Score (Static)	Recovery Multiplier	OCRA Checklist Index
Pushing/Pulling Trolley with Wheels in Good Working Order over Smooth Surfaces				
8–25	1	4.5	1.33	7.3
26–50	1.5	4.5	1.33	8.0
51–75	2	4.5	1.33	8.6
76–100	2.5	4.5	1.33	9.3
100–120	3	4.5	1.33	10.0
121–150	3.5	4.5	1.33	10.6
More than 150	4	4.5	1.33	11.3
Pushing/Pulling Trolley with Wheels in Poor Working Order over Smooth Surfaces				
8–25	2	4.5	1.33	8,6
26–50	3	4.5	1.33	10.0
51–75	4	4.5	1.33	11.3
76–100	5	4.5	1.33	12.6
100–120	6	4.5	1.33	14.0
121–150	7	4.5	1.33	15.3
More than 150	8	4.5	1.33	16.6

Week 1	Start and end of shift	Mon	Tue	Wed	Thu	Fri	Sat	Sun
1st shift	7a.m.–1.40p.m.	×	×	×	×			
2nd shift	1.40p.m.–7p.m.						×	×
Week 2	**Start and end of shift**	**Mon**	**Tue**	**Wed**	**Thu**	**Fri**	**Sat**	**Sun**
1st shift	7a.m.–1.40p.m.				×	×	×	×
2nd shift	1.40p.m.–7p.m.	×	×					
Week 3	**Start and end of shift**	**Mon**	**Tue**	**Wed**	**Thu**	**Fri**	**Sat**	**Sun**
1st shift	7a.m.–1.40p.m.	×	×	×	×			
2nd shift	1.40p.m.–7p.m.						×	×
Week 4	**Start and end of shift**	**Mon**	**Tue**	**Wed**	**Thu**	**Fri**	**Sat**	**Sun**
1st shift	7a.m.–1.40p.m.				×	×	×	×
2nd shift	1.40p.m.–7p.m.	×	×					

The same analysis is repeated for the other three weeks of the month.

FIGURE 8.43 Example of preliminary description of shift distribution over a representative month.

Distribution of shifts in the week for a homogeneous group (total duration of shift in minutes, for working hours and overtime, including breaks)	Mon	Tue	Wed	Thu	Fri	Sat	Sun
Shift duration (minutes)	400	400	400	400		360	360
No. of actual breaks during the shift lasting at least 8 minutes (except meal breaks) that can be considered as recovery periods	2	2	2	2		1	1
Overall duration of all actual breaks (excluding meal break) in minutes	20	20	20	20		16	16
Duration of meal break if included in shift duration (minutes)	30	30	30	30		30	30
No. of other breaks (i.e. meal break before or after working time; travel time to/from different company locations). Enter a number only if these breaks last at least 30 minutes							
Total duration of breaks (If included in shift duration)	50	50	50	46	0	46	46

FIGURE 8.44 Example of analysis of shift and break durations in week 1 (morning shift on the first 4 days and afternoon shift on Saturday and Sunday).

8.4.5.3 Analysis of Tasks Performed by the Homogeneous Group on Each Day of the Month and Each Shift, Including Durations

The next step is to examine the tasks performed by the specific homogeneous group (or individual worker) exposed to risk from the qualitative and quantitative standpoint.

The qualitative analysis does not present major difficulties, while the quantitative analysis may be complex, especially when there is a very large number of tasks.

The *qualitative* analysis requires the following aspects to be observed:

- Tasks actually performed by the homogeneous group over the entire period under examination, for the time being without breaking down by the week or month
- Tasks actually performed by the homogeneous group every day over the month

The *quantitative* assessment describes the duration of each task as a percentage of each day of the month.

The sum of the percentages for each day of the week must always add up to 100%.

Extreme accuracy is not essential for estimating the proportional duration of the various tasks. The employer, or even the members of the homogeneous group, should be able to provide this information. Once the proportional duration of the tasks performed on a rotating basis in a small commercial kitchen has been provided, it will be possible to carry out the risk analysis using the software program described in Chapter 6, Section 6.4.

Unfortunately, experience has shown that for larger commercial kitchens (2000–3000 meals), it is difficult if not impossible to obtain reliable and objective information about the net duration of many tasks, for the following reasons (Figure 8.45):

- Some workers perform their work faster or slower than others.
- Certain preparations call for lengthy cooking times and testing for doneness, without necessarily using the upper limbs.

Duration in minutes of each task performed by a worker (or by a homogeneous group of workers) during each day of the week							
Tasks	Mon	Tue	Wed	Thu	Fri	Sat	Sun
A	40	100	50	30		100	
B	60	120	100	30		50	50
C	20	30	50	30		50	50
D	10	25	25	50		25	25
E	120	25	25	100		25	25
F	100	50	100	110			100
Total net duration of repetitive tasks (in minutes) on each day of the week	350	350	350	350	0	250	250

FIGURE 8.45 Identification of the actual duration of each task in the shift over the period: A difficult calculation.

- Workers are often empty-handed when returning trolleys, thus no strain is placed on the upper limbs.
- Workers are often required to retrieve equipment such as empty pots, oven trays, and so on; these tasks can be classified as nonrepetitive and are difficult to quantify.

8.4.6 Simplifying and Conducting an Analysis to Assess Risk due to Repetitive Movements: Estimating Task Duration

There is obviously the issue of objectively estimating the net duration of repetitive work involving the upper limbs, as it is hard enough to measure the actual duration of each individual repetitive task performed in the shift. It is quite easy to over- or underestimate the time since, workers often have to wait for food to cook or return empty-handed after transporting trolleys.

We shall now explain the rationale for estimating the duration of repetitive work and overcoming excessively subjective assessments due to the different speed with which workers perform their tasks. The rationale is based on the observation of objective organizational elements such as actual amounts of food or other materials processed by the workers, when the average time required to handle them, by unit of measurement, is a known quantity. The definition might sound somewhat complex, but we will now provide some examples to make it more comprehensible.

We began by gathering the main general quantitative data relative to each day of the four representative weeks of the month (corresponding to the monthly menu of a large hospital kitchen):

- Number of meals per shift
- Kilogram of food processed per task and per shift
- Average time required to cook/process one predefined unit of measurement

The average time per unit of measurement could be regarded as a "times and methods analysis," serving a twofold purpose:

- It allows the company to more effectively build a typical work day.
- It identifies the actual net repetitive work performed by the upper limbs, which is essential for defining levels of exposure risk due to biomechanical overload.

This produces the *average duration* of each task required to fully process one unit of product. The average durations identified and used here need further and more in-depth verification. Next, we illustrate the procedure adopted to obtain these net average durations/unit of product.

8.4.6.1 Phase A: Determine Quantity of Food (or Better, Groups of Similar Foods) Processed per Day of the Week and per Shift

- Weight in grams/portion of the main foods needs to be known and generally applies across the entire company (this information can always be provided by admin).

- For each food, there will be an indication of how many portions must be prepared on a specific day and shift (it should be noted that this will vary depending on the location and number of meals to be prepared).
- The following formula is used to calculate the kilograms of total food processed during a specific shift:

$$\text{Grams per portion} \times \text{No. of portions} = \text{Kilograms of food processed in the shift}$$

To calculate how many kilograms are processed by each member of a homogeneous group, the total weight is divided by the number of workers in the group: the result is the *kilograms processed/worker* on that specific day and in that specific shift (Figure 8.46).

This, and all other calculations that will be illustrated later, require the use of a software program developed specifically to enable risk to be assessed even by non-experts in ergonomics (see Table 8.38).

8.4.6.2 Phase B: Determine, for Each Main Food (or Better, Group of Similar Foods), the Unit of Measurement and Average Processing Time per Unit of Measurement within the Task

The example shown in Table 8.38 illustrates the concept of time per unit of measurement within a certain task.

For example, the unit of measurement for a crate of salad is 7.5–8 kg (salad is already delivered in crates of this weight). For the task defined as "Rinse salad

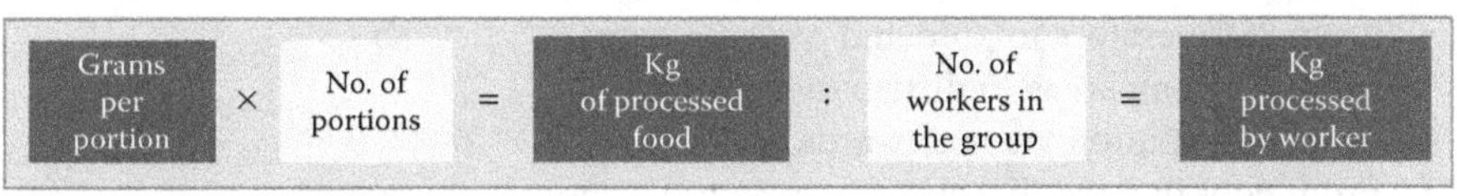

FIGURE 8.46 Estimated amount of food (in kilograms) processed by each member of a homogeneous group.

TABLE 8.38
Summary of Calculation of Unit Times per Task

Task	Unit of Measurement	Intrinsic Active Time (in seconds) per Unit of Measurement	Unit of Measurement (in kg) per Relative Unit of Time
Rinse salad ingredients (and other vegetables such as fennel, radishes, etc.,) at tall sink	7.5–8 kg crate	66	7.5
Manually chop and sort salad ingredients (and other vegetables such as fennel, radishes, etc.,) more thoroughly		396	7.5

ingredients (and other vegetables including fennel, radishes, etc.) at tall sink," the average time required in the estimation of company experts is 66 s. The second task indicated in the table is "Manually chop salad ingredients and sort other vegetables (such as fennel, radishes, etc.,) more carefully." The same unit of measurement as the crate of salad is used, but to manually chop 7.5 kg of salad a single operator will need 396 s.

8.4.6.3 Phase C: Determine, for Each Main Food Type (or Better, Group of Similar Foods), the Actual Average Processing Time to Prepare the Necessary Quantities per Task and per Shift

For example, once the amount of food to be rinsed by an operator during the shift is known, it will be possible to calculate the average time required for the operator to complete the task (Figure 8.47).

Table 8.39 shows a calculation of the actual duration of repetitive work, per task, based on the quantities processed, units of measurement, and average times per unit of measurement.

8.4.6.4 Phase D: Transport with Trolleys. Determine Actual Average Time Spent Transporting Trolleys (i.e., Manually Pushing and Pulling)

In commercial kitchens, personnel spend a great deal of time manually pushing and pulling trolleys:

- From storage areas to prep areas
- From prep areas to refrigerators for cooling
- From refrigerators to prep and cooking areas
- From cooking areas to distribution areas
- From distribution areas to wards
- From cooking and distribution areas to various washing areas

Operators often transport trolleys to certain areas, leave them there and return "empty-handed." Hence the need to factor in not only the duration of active transport tasks (to be assessed for risk) but also passive periods when the upper limbs are not exposed to the risk of biomechanical overload.

With reference to calculating the duration of cooking tasks, here too it was necessary to estimate the average duration of the transport tasks observed. Table 8.40 shows how *active transport tasks* and *passive return from transport tasks* are calculated. The following data is used: distance covered, duration of task by "unit of distance" (i.e., one step entailing active transport = 0.8 s; one step entailing no

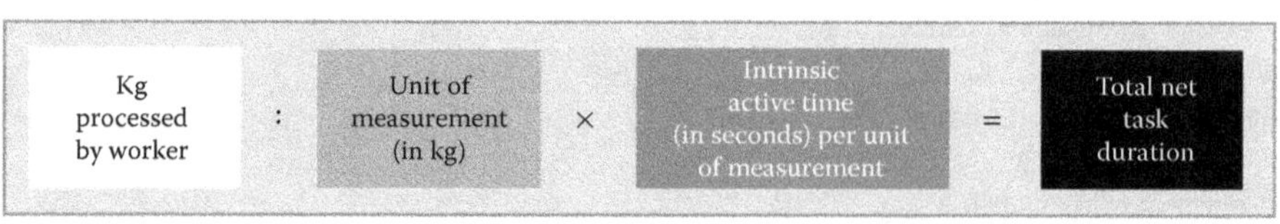

FIGURE 8.47 Calculation of total task duration.

TABLE 8.39

Example of Calculation of the Time Required to Complete Tasks Based on the Quantity of Food and Average Preparation Time per Unit of Food

Task	Intrinsic Active Time (in seconds) per Unit of Measurement	Unit of Measurement (in kg) per Time Unit	Monday	Kilogram Processed in First Shift per Operator	Time Required to Perform the Task
Rinse salad ingredients (and other vegetables such as fennel, radishes, etc.) at tall sink	66	7.5	X	84.0	12.3
Manually chop and sort salad ingredients (and other vegetables such as fennel, radishes, etc.) more thoroughly	396	7.5	X	84.0	73.9
Manually chop and sort salad ingredients (and other vegetables such as fennel, radishes, etc.) more quickly	264	7.5	X	84.0	49.3
Transfer salad (with or without other vegetables ingredients) manually from colander to spinner after machine washing	100	7.5	X	84.0	18.7
Prepare single portions of green or mixed salad: fill single-portion containers	84	7.5	X	121.5	22.7

TABLE 8.40
Example of Calculation of Time Required to Transport Salads Manually and/or Using Manual Trolleys: Estimated Active Transport Time and Return Time (Empty-Handed) Based on Distance Covered and Weight (in Kg) of Load

Task	Technical Data	Distance Covered (No. of Steps)	Unit of Measurement	Intrinsic Active Time (in seconds) per Unit of Measurement	Intrinsic Passive Time (in seconds) per Unit of Measurement	Unit of Measurement (in kg) per Time Unit	Kilogram Performed in the First Shift per Person	Minutes of Active Transport (Carrying Objects) in the First Shift	Minutes of Passive Transport (Walk Empty-Handed) in the First Shift
Manually transport 7 kg crate of salad (green or mixed) packed in various single-portion containers (approx. 45) in storage areas	Approx.7.5 kg crate with 45 single-portion salads in which the salad alone without tare plus a few other vegetables would weigh 4.5 kg	20	4.5 kg of salad	16	10	4.5	121.5	7.2	4.5
Transport salad (green or mixed) ready for consumption using low trolley in distribution area (trolley plus approx. 12 kg of salad)	Transport low trolley with two 7.5 kg crates (net weight 4.5 kg of salad each)	60	1 trolley = 9.5 kg of salad	48	30	9.5	121.5	10.2	6.4

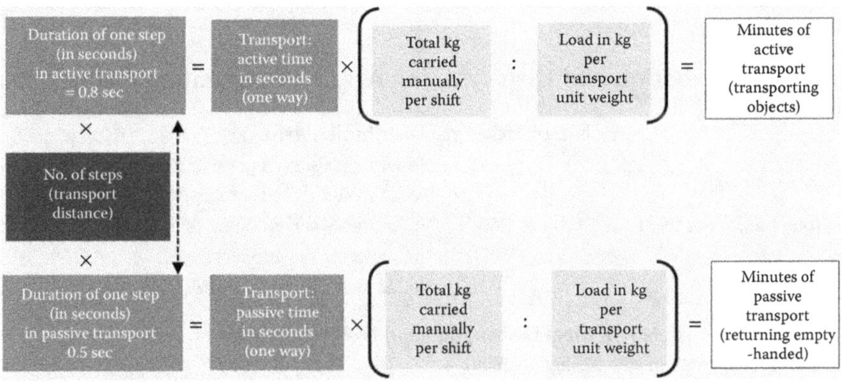

FIGURE 8.48 Criteria for calculating active and passive transport times based on distance covered.

transport = 0.5 s), weight transported (in kg) per "unit of transport," and total weight transported (in kg) per shift and per person. This data generates the total duration of active transport tasks and the total duration of passive returns. In Figure 8.48, the flowchart indicates how these values are calculated.

Table 8.40 shows an example of the total net repetitive work involved in transporting objects.

8.4.6.5 Phase E: Distribute Meals on Trays and Wash Trays. Determine Time to Distribute Meals or Wash Dishes on Trays. Wash Pots and Pans

Here, the organizational data required to calculate task duration is easier to acquire since it consists of the total number of meals served per distribution area (i.e., individual meal tray preparation, self-service, outside catering meal preparation, table service, etc.). Similar data is also required for the washing areas. Of course, it is necessary to know how many meal distribution and washing areas there are (Table 8.41).

Figure 8.49 shows the flowchart used to estimate the duration of meal preparation and tray washing tasks.

The same rationale used for calculating the duration of tray preparation and washing tasks is employed for estimating the duration of other tasks, such as manually washing large pots and pans. It is necessary to know the number of pots and pans washed by the homogeneous group per shift (see Figure 8.50).

8.4.6.6 Phase F: Prepare Chilled Trolleys Loaded with Boxes

The rationale underlying the calculation is as described for Phases E and F (see Figure 8.51).

8.4.6.7 Phase G: Sanitize Trolleys

This calculation is also based on the number of meals distributed on trays. In this case, the number of trays per trolley is fixed (and known): approximately 15 (see Figure 8.52).

TABLE 8.41
Meal Distribution and Dish Washing Areas: Organizational Data

			No. of Meals per Distribution Area
		500	No. of meals served per outside catering line
			No. of meals served at tables
Total number of meals	4000	2500	No. of meals served on trays in wards
		1000	No. of meals for self-service
			No. of meals served in wards using trolleys

No. of Meal Distribution and Dish Washing Lines	
No. of meals with dishes washed by hand	
No. of lines for washing dirty dishes on trays	1
No. of preparation lines for meals on trays	3
No. of meal distribution lines for self-service	1

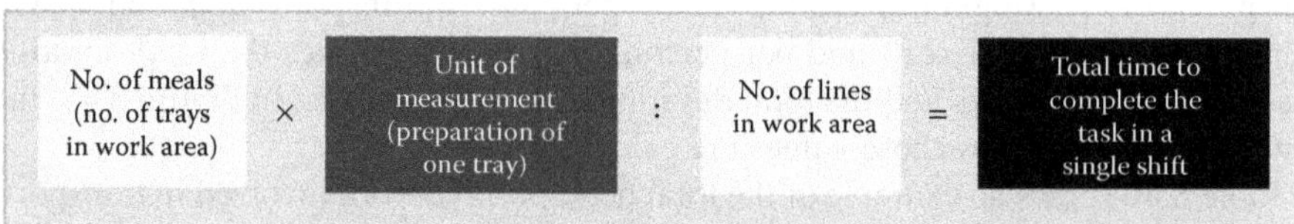

FIGURE 8.49 Criteria for calculating task duration: prepare and wash trays.

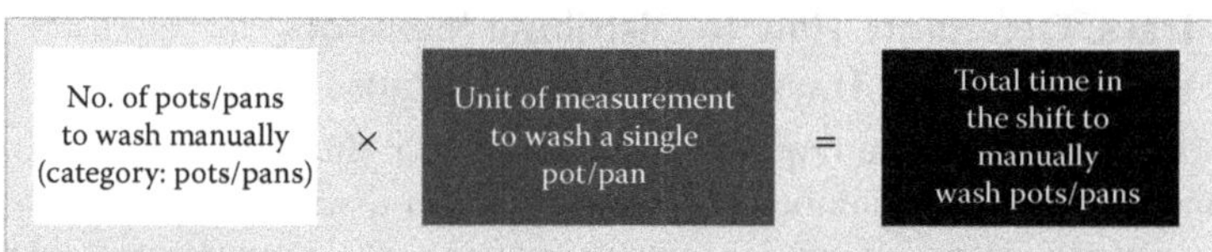

FIGURE 8.50 Criteria for calculating task duration: manually wash pots and pans.

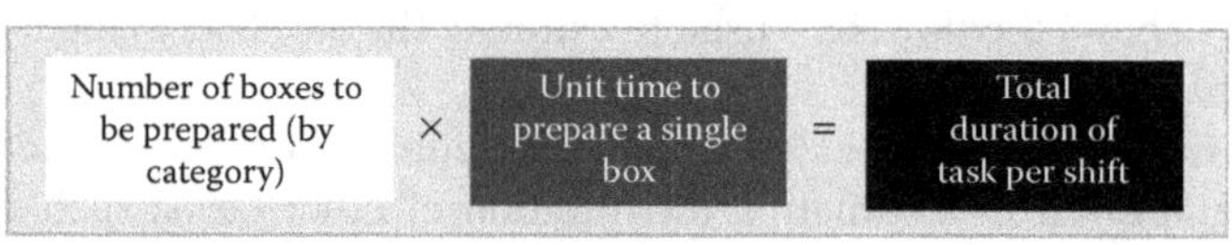

FIGURE 8.51 Criteria for calculating task duration: load chilled trolleys (bread, beverages, puddings, etc.).

FIGURE 8.52 Criteria for calculating task duration: sanitize trolleys for meal tray distribution.

8.4.7 Application of Mathematical Models

Once the type and number of tasks performed in each shift is known (for every day of the week and over the 4 weeks of the month), the criteria and formulas described in Chapter 6, Section 6.4, will be applied to determine the exposure index according to the general flowchart illustrated in Figure 8.53.

It is descriptively interesting to store and compare the daily risk index curves for the right and left limb over the entire month, especially when there are significant variations between days. A graphic depiction of risk indexes over the month (see Figure 8.54) can be an extremely useful tool for indicating which days require urgent remedial action. "0" means "nonoperational" days.

8.4.8 Conclusions

All of the data presented here, especially the units of measurement (and units of time) required to perform tasks and the relevant "intrinsic" OCRA checklist scores, led to the creation of a dedicated software program featuring predetermined and preassessed variables that even individuals without any experience in ergonomics can easily use.

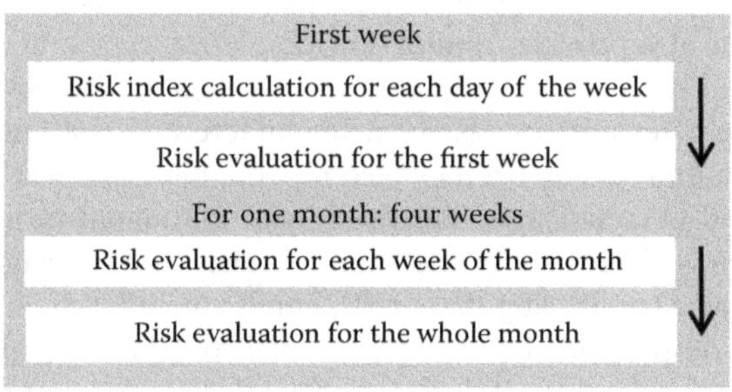

FIGURE 8.53 Flowchart utilized to calculate the final risk index (Chapter 6, Section 6.4).

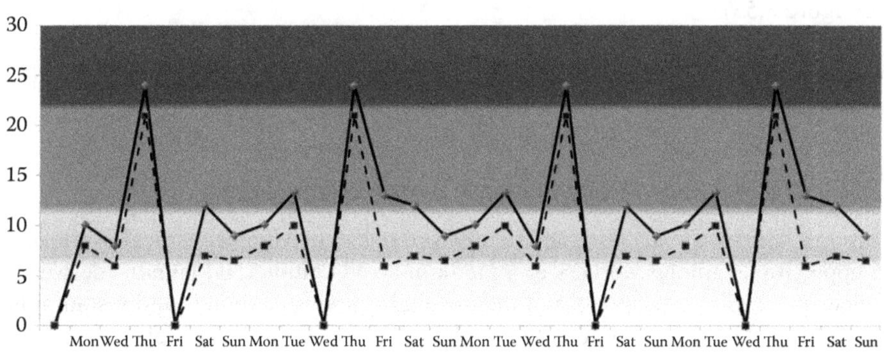

FIGURE 8.54 Graphic representation of risk index trends over the month (with right and left limbs indicated separately).

The main aim of the software (Petrasoftware for meal preparation) is to

- Allow people, even with no expertise in ergonomics, to easily but accurately assess risk; all they need to do is enter the specific organizational data, since the software already supplies the intrinsic risk indexes for 288 preanalyzed tasks.
- Evaluate exposure in the preliminary stages of the task assignment process.

As more and more experience is acquired, the software will be extended to include many more tasks than those presented here.

8.5 INDUSTRIAL LAUNDRIES

8.5.1 Introduction

Operations in this sector range from small local businesses to large industrial facilities.

Small dry-cleaning businesses are generally found in local neighborhoods or shopping malls and serve private customers, cleaning residential clothes and household linens. They provide an essential service, especially for city-dwellers who may find it difficult to clean certain types of garments or linens at home. Small dry-cleaning businesses also provide ironing services.

Industrial laundries clean large quantities of linens or clothes for commercial users such as hotels, restaurants, garment manufacturers, communities, and other laundry businesses. In some cases, they also rent out table and bed linens.

Italy has around 600 industrial laundries with an annual turnover of approximately €1.3 billion. It has roughly the same number of employees as the oil and pharmaceuticals industry. During the 1990s, employment levels in this sector grew by 9%, and today, despite the global financial crisis that has also hit Italy extremely hard, jobs are still on the rise. Italian industrial laundries employ around 15,000 workers including 93% on a permanent employment contract a large proportion of whom are women (65%). About 35% of these 600 laundries have less than 10 employees; 55% have between 10 and 49; and just 10% employ more than 50 workers (see Figure 8.55).

Half of the workers are located in facilities in Northern Italy (80%), and 20% are in the South. Lombardy alone has the largest number of workers (15%) (see Table 8.42).

8.5.2 Organizational Data and Laundering Technology

We report here on a study designed to measure risk due to biomechanical overload of the upper limbs among workers at a medium-sized commercial laundry delivering services primarily to hospitals and the hospitality industry (hotels and restaurants). The laundry is not highly automated therefore such risks are very high.

Traditionally, industrial laundries wash, recondition, and sterilize fabrics (table, bed and bath linens, work clothes, mattresses, and so on) for the owners of the articles. Today, many laundries also offer the following services (see Figure 8.56):

FIGURE 8.55 Industrial laundries broken down by size: 2001 data. (From Industrial Laundries Observatory.)

TABLE 8.42
Distribution of Industrial Laundries by Region. Facilities and Workers (ISTAT—Italian Statistics Bureau Data)

	No. of Facilities	% of Facilities	No. of Workers	% of Workers
North	319	47.9	7,936	53.2
Centre	185	27.8	4,049	27.1
South	162	24.3	2,938	19.7
Total Italy	666	100.0	14,923	100.0

- Linen rentals
- Logistics at the client's premises
- Transport (soiled–clean)
- Sterilization and disinfection
- Associated services such as management and sterilization of surgical instruments, supply of disposables, mattress management, wardrobe management at the customer's premises, sale of cleaning products, and so on

Only half of the laundries regularly rely on subcontractors or outsourcing for services they offer. The activities that tend to be decentralized are transportation and washing. Most laundries outsource transportation, but outsource less than half of their other services, and even then, only very occasionally.

8.5.2.1 General Production Cycle in Industrial Laundries and the Sample Laundry Described Here

Although the classification of economic activity used by the National Institute for Statistics (ISTAT) for official purposes lists industrial laundries as a personal-care service, this activity is organized along industrial lines, featuring capital-intensive production processes and highly automated machinery and equipment.

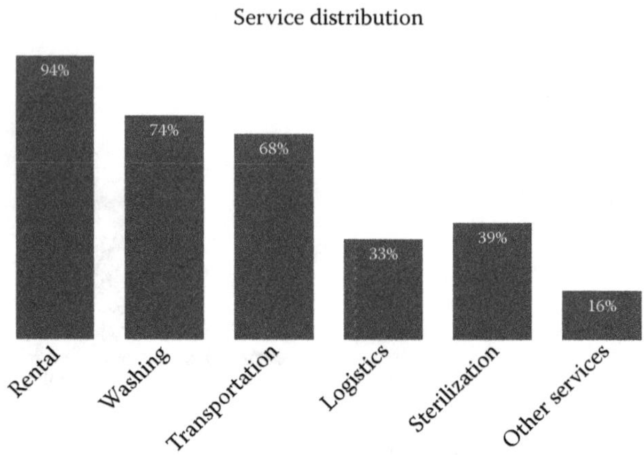

FIGURE 8.56 Range of services offered (percentage of total laundries): 2004 data. (From Industrial Laundries Observatory.)

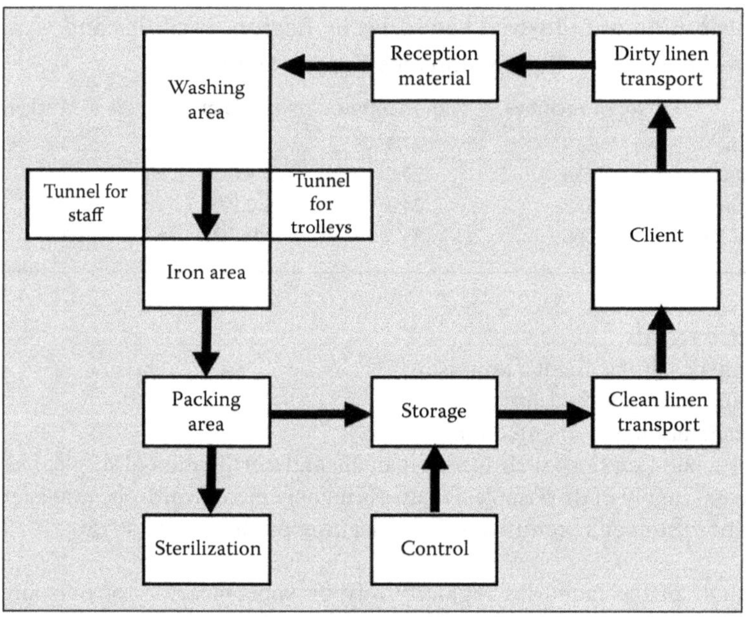

FIGURE 8.57 Industrial laundries: production cycle.

The production cycle consists of three main activities (see Figure 8.57):

- Receive incoming goods: Examine linens, check quality of materials, separate by type, combine batches of articles based on chemical and physical care codes.
- Mechanical, chemical, and thermal treatments: using largely computerized continuous or noncontinuous batch processing systems.

- Pressing and packaging: Dry, unload linens, inspect articles and quality of chemical finishing, iron, fold, and package.

In the industrial laundry examined here, the production cycle begins with receiving incoming goods and unloading goods collected from various customer premises with the laundry's vans. The drivers unload bags of soiled linens to special carts that then go to the "soiled goods department," where they are emptied into hampers and then sorted by workers.

Thus separated, the soiled laundry goes to the washing area where the same workers load up the machines, which have two openings, one on the soiled side and the other on the clean side, and start the appropriate washing cycle depending on the type of product. Operators (mainly female) take the clean laundry out of the washing machines in special carts in the "clean area" and load them into dryers. Once dried, the laundry is manually unloaded from the dryers, loaded onto special carts, and transported either to the manual folding table (i.e., mattress covers, noniron garments), to the automatic steam press (sheets, tablecloths, pillow cases, mattress covers), or to the manual ironing area (mainly shirts). Garments received from nursing homes are folded and placed on special shelves divided into pigeon holes marked with the name of the resident so that they can be individually packaged. They are then transported by cart to the packaging machines. The operators place one or more garments into the machine and the plastic wrap is heat-sealed and cut to the appropriate size by lowering a lever. The bags are then taken to a storage area where they are collected by drivers and loaded onto vans for transportation to the customer's premises. The laundry also has a small sewing department that stitches on name tags and does minor repairs for nursing home residents.

8.5.2.2 Classification of Operations Broken Down into Macrophases, Phases, and Tasks, and Calculation of Intrinsic Risk Scores: Construction of the "Basic" Assessment for an Industrial Laundry

The first step in assessing the risk of biomechanical overload of the upper limbs was to analyze the work organization and identify the macrophases, phases, and tasks involved. Figure 8.58 shows the macrophases, phases, and tasks at the industrial laundry under examination.

Repetitive tasks were identified in each phase. Repetitive tasks included frequent pushing and pulling of carts within the facility, often covering quite long distances, and with the hands held occasionally in awkward postures.

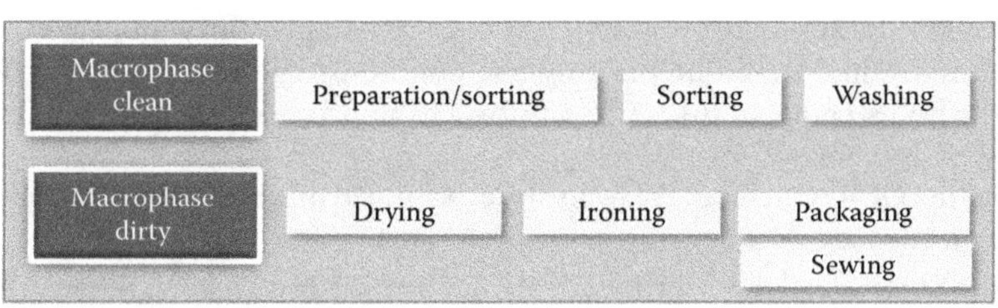

FIGURE 8.58 Principal macrophases and phases identified in an industrial laundry.

TABLE 8.43
Identification and Analysis of Phases and Tasks

Prepare Linen for Sorting	**Pressing**
Empty soiled linen bags into wheeled hamper	Transfer loaded cart to steam press iron
Transfer soiled linen hamper to sorting area	Press sheets and table cloths on steam press and fold manually
Transfer empty hamper to soiled linen storage area	Press mattress covers on steam press and fold manually
	Press pillow cases on steam press iron and fold manually
Linen Sorting	
Manually sort linens from hamper to carts	Transfer empty cart to drying area
Transfer loaded carts to washing machines	Transfer pressed and folded laundry to packaging unit
Transfer empty carts to sorting area	Manually iron shirts
Washing	**Packaging**
Manually load washing machines	Transfer loaded cart to folding table
Transfer empty carts to washing machines for loading	Fold bath towels
Manually unload washing machine to cart	Fold sweaters
Transfer loaded cart to drying area	Fold trousers
Drying	Transfer empty cart to drying area
Transfer loaded cart to dryers	Separate and stow garments
Load dryers	Load cart for packaging unit
Transfer partially loaded cart to area adjacent to dryers	Transfer loaded cart to packaging unit
Transfer empty cart to dryers for unloading	packaging unit
Empty dryers on to cart	**Sewing Department**
Transfer loaded cart to storage area before packaging	Stitch name tags
Pressing	Minor repairs

Table 8.43 lists the seven phases and their respective repetitive tasks; Figures 8.59 through 8.61 show pictures of these tasks.

8.5.3 Calculation of Intrinsic Risk Scores for Each Individual Task

A separate OCRA checklist was assigned to each task to estimate its intrinsic risk, as if that individual task were the only one performed by the same worker for the whole shift, this being an 8-h shift with two 10-min breaks and one 30-min meal break (see Chapter 6.4).

As per the OCRA method, the tasks were examined by viewing movies shot at the premises while they were actually being performed. Thirty-four films representative of the tasks performed at the laundry were analyzed, and 34 OCRA checklists were obtained with their respective intrinsic risk scores for the right and the left side.

For the right side, the estimated intrinsic risk indexes showed that:

FIGURE 8.59 Principal tasks: Presorting, sorting, and washing.

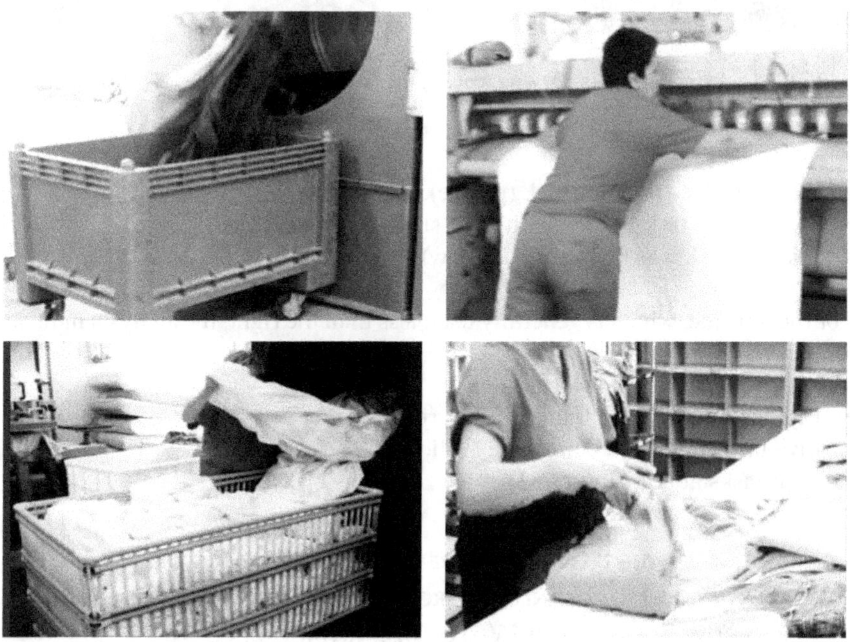

FIGURE 8.60 Principal tasks: Drying and packaging (loading and emptying dryer, steam press, and folding).

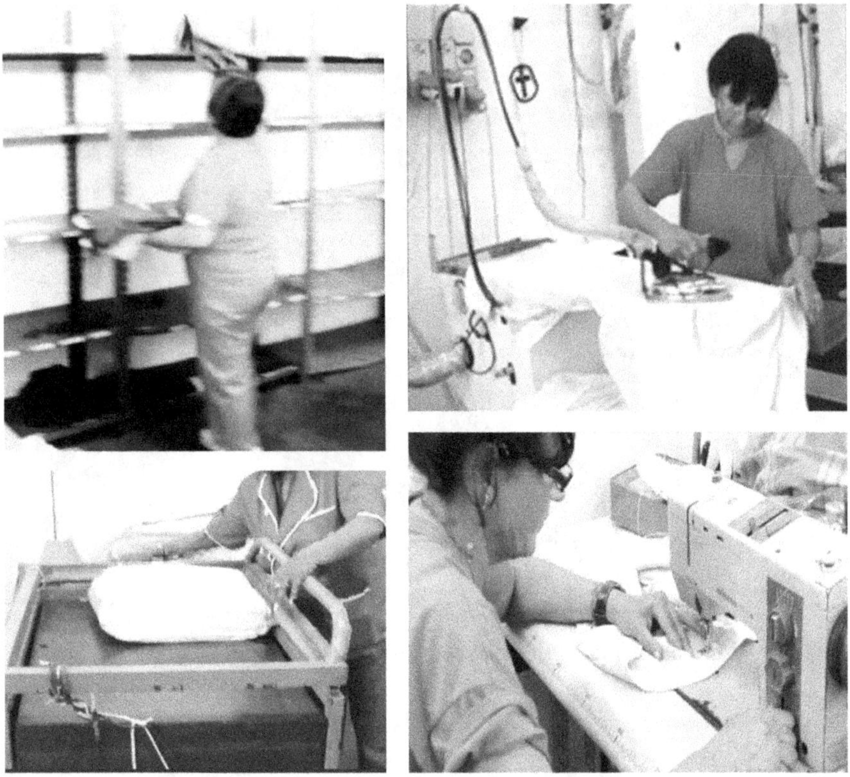

FIGURE 8.61 Principal tasks: Drying, packaging and sewing (placing on shelves, pressing, packaging, and sewing on name tags).

- Eleven tasks were high-risk (purple).
- Twenty-two tasks were medium or slight risk (red).
- Just one task was borderline (yellow).

For the left side, which is generally used less than the right, the analysis indicated:

- Seven tasks at high risk (purple).
- Eighteen tasks at medium or slight risk (red).
- Five tasks that were borderline (yellow).
- Four tasks at no risk (green).

8.5.4 Setting up the Basic Software for Industrial Laundries; Identifying Homogeneous Groups and Calculating Multitask Exposure in Daily and Weekly Task Rotation Scenarios

The workers at the industrial laundry in our study performed both daily and weekly task rotations. The ERGOepmCHECKOCRAmultiMONTH-EN software was used for the weekly task turnovers; this program assesses biomechanical overload risk in workplaces with a weekly or monthly work cycle (see Chapter 6.4).

First, the organizational data was entered (i.e., macrophases, phases, and repetitive tasks). Next, the scores for the various risk scores relative to the relevant tasks, as per the OCRA method, along with the final intrinsic risk index, were entered. The end result was the basic ERGOepmCHECKOCRAmultiMONTH-EN file for an industrial laundry. The software could then calculate the final risk indexes for exposed workers. It is only after homogeneous groups of workers performing the same tasks and shifts have been identified that risk can be assessed relative to the specific exposure of each homogeneous group and individual worker.

Three main jobs were singled out in the course of the evaluation:

- Drying/packaging
- Sewing
- Sorting

In terms of exposure, these three job areas included the following five homogeneous groups of workers:

1. Drying/packaging over a 6-h shift
2. Drying/packaging over a 5-h shift
3. Drying/packaging over a 3-h shift
4. Sewing
5. Sorting

Every different homogeneous group needs a specific risk evaluation: for each of them (present in industrial laundry under study), open a separate BASIC ERGOepmCHECKOCRAmultiMONTH-EN, proceeding with risk assessment. Next are some examples of the dedicated software files for each of the groups, including the respective organizational data; the program "automatically" evaluates the relevant exposure risk.

8.5.5 Homogeneous Group: "Drying and Packaging over a 6-h Shift"

We shall now illustrate each individual phase of the risk assessment process applied to a homogeneous group: *drying and packaging over a 6-h shift.*

First there is a description of the facility and the job carried out by the homogeneous group, as shown in Table 8.44.

Next comes an analytical description of breaks and nonrepetitive tasks performed on every day of a *working week* that is representative of the whole year. As in the case of a standard OCRA checklist, the duration of the shift is then indicated, along with the duration of breaks and nonrepetitive tasks, so as to obtain the net duration of the repetitive work.

Automatic scores can also be obtained for the lack of recovery times (i.e., calculation of the new *recovery multiplier* and thus obviously also of the new *duration multiplier*), which will adjust the intrinsic checklist values so as to define the actual new exposure score for the specific homogeneous group (see Table 8.45). As can be

TABLE 8.44
Definition of Homogeneous Group: Drying-Packaging over a 6-H Shift

Company Name	Industrial Laundry
Name and address of company head office and branches	YYYYYYY
Name of company employer	ZZZZZZZZ
Total number of employees	13
Name of worker or homogeneous group	Workers to drying-packaging over 6-h shift
No. of workers in the homogeneous group (persons performing the same work with the same work schedule)	3
Brief description of the tasks performed by the employee or the homogeneous group	Load and empty dryer, iron and press (manually); fold and package clothes and linens (sheets, sleepers, pillow cases, etc.)

TABLE 8.45
Shift Distribution in a Representative Week for a Homogeneous Group: Drying-Packaging over a 6-H Shift

	Distribution of Shifts in the Week for a Homogeneous Group (Total Duration of Shift in Minutes, for Working Hours and Overtime, Including Breaks)						
	Mon	Tue	Wed	Thu	Fri	Sat	Sun
Organizational Study of 1 Week, in a Representative Month							
Shift duration (minutes)	360	360	360	360	360	360	0
Minutes of nonrepetitive tasks in the shift	20	20	20	20	20	20	0
Total duration of breaks	15	15	15	15	15	15	0
Net duration of repetitive tasks in the shift B	325	325	325	325	325	325	0
Recovery multiplier	1.33	1.33	1.33	1.33	1.33	1.33	

seen, the homogeneous group performing "drying-packaging" tasks works 6-h shifts over a 6-day week (Monday through Saturday). The shift does not include a meal break and there is only one actual 15-min break after the first 3 h of work; there are about 20 min of nonrepetitive work while the workers clean up at the end of the shift and replenish supplies. The morning and afternoon shifts are virtually identical in terms of workload and task distribution. The workers stay on the same morning or afternoon shift for the entire week. The tasks are rotated every day, with workers moving from the folding table to the ironing area and the steam press, and the same cycle is repeated every 3 days (i.e., nondaily rotation with the morning or afternoon shift being alternated weekly).

The software (see Figure 8.62) then highlights all the tasks performed day by day by the homogenous group assigned to *drying-packaging* over a *typical week.*

Phase	Task	Mon	Tue	Wed	Thu	Fri	Sat	Sun		1950	2100
		325.0	325.0	325.0	325.0	325.0	325.0	0.00			
		Minutes worked on each day of the week per specific task							Total min. spent on each task over one week	% versus total min. worked over a week	% week min. constant
Drying	Transfer loaded cart to dryer	1.8	0.36		1.8	0.36			4	0.22%	0.21%
	Transfer partially loaded cart to area adjacent to dryers	1.8	0.36		1.8	0.36			4	0.22%	0.21%
	Load dryer	40	25		40	25			130	6.67%	6.19%
	Transfer empty cart to dryer for unloading	1.8	0.36		1.8	0.36			4	0.22%	0.21%
	Transfer loaded cart to storage area before packaging	1.8	0.36		1.8	0.36			4	0.22%	0.21%
	Empty dryer	40	25		40	25			130	6.67%	6.19%
Steam press iron	Transfer loaded cart to steam press iron		0.96			0.96			2	0.10%	0.09%
	Transfer empty cart to drying area		0.96			0.96			2	0.10%	0.09%
	Press bed and table linens on steam press iron		112.6			112.6			225	11.55%	10.72%
	Press mattress covers on steam iron		52			52			104	5.33%	4.95%
	Press pillow cases on steam iron		52			52			104	5.33%	4.95%
	Transfer loaded cart to packaging unit	1.5	1.5		1.5	1.5			6	0.31%	0.29%
	Manually iron certain types of laundry			255.4			255.4		511	26.20%	24.32%
Packaging	Transfer loaded cart to folding table	0.48			0.48				1	0.05%	0.05%
	Transfer empty cart to drying area	0.48			0.48				1	0.05%	0.05%
	Fold flat linens	45			45				90	4.62%	4.29%
	Place clean clothes in pigeon holes	9.91			9.91				20	1.02%	0.94%
	Load cart for packaging unit	9.91			9.91				20	1.02%	0.94%
	Fold sweaters	58.5		34.8	58.5		34.8		187	9.57%	8.89%
	Fold trousers	58.5		34.8	58.5		34.8		187	9.57%	8.89%
	Transfer loaded cart to packaging unit	1.5	1.5		1.5	1.5			6	0.31%	0.29%
	Packaging unit	52	52		52	52			208	10.67%	9.90%

FIGURE 8.62 Description of working hours in a representative week for a homogeneous group: drying-packaging over a 6-h shift.

The description includes the duration in minutes of each task carried out on the various days based on information obtained from the employer and the members of the homogeneous group, and on direct observations made over several working days.

Drilling down into more detail, it can be seen that the weekly shift worked by this particular homogeneous group, as described in Figure 8.62, perform the following tasks on the first day of the week (which can be defined as the *folding table day*):

- Load and empty dryer
- Fold clothes and linens
- Package clothes

On the second day of the week (the *steam press ironing day*), the main tasks performed by the workers include:

- Load and empty dryers
- Press laundry on steam iron
- Package

On day 3 (*ironing day*) the workers:

- Manually iron
- Fold clothes

The analysis of the typical day also covers all cart moving activities including the relevant technical actions.

The software automatically calculates the following values for every typical day as follows:

- Net duration of repetitive work
- Total number of minutes over the entire week devoted to a specific task
- Proportional duration (in percentage terms) of one specific task versus all tasks over the hours worked per week
- Proportional duration of all tasks versus a constant 2100 weekly minutes (five 420-min working days equal 2100 min)

The procedure now produces the daily exposure index and weekly exposure index for the homogeneous group defined as *drying and packaging over a 6-h shift.*

The resulting exposure index is viewed in a graph that describes daily trends across the week in question, for both the right and the left limb (see Figure 8.63).

The graph shows that for the group assigned to *drying-packaging over a 6-h shift*, the cycle is 3 days long. Using the time-weighted average method, the daily exposure indexes for the right and left limb, respectively, are as follows:

- "Folding table days": 21.2 and 17.7
- "Steam press ironing days": 14.8 and 13.2
- "Manual ironing days": 18.7 and 15.7

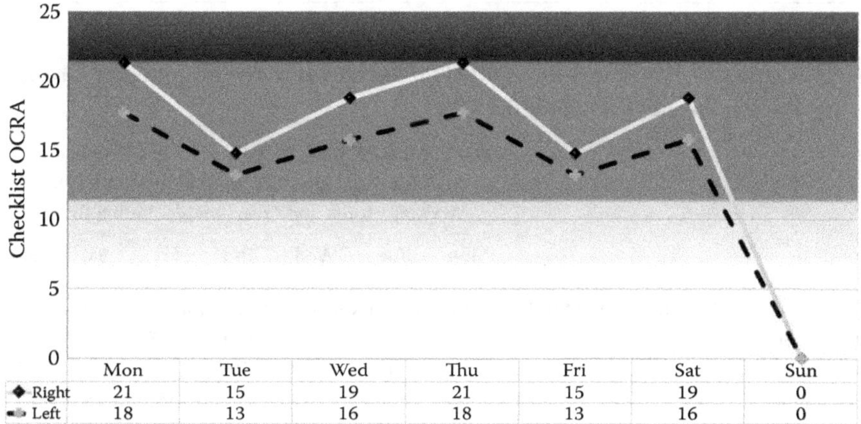

FIGURE 8.63 OCRA checklist weekly exposure index for a homogeneous group: drying-packaging over a 6-h shift.

Thus, risk is present on all three typical working days during a representative week, with taller peaks on the folding day (high—red, for the right limb). Furthermore, although the right limb is used the most and is more overloaded, the left is still exposed to definite risk on all days.

The weekly exposure index for the right and left limbs, respectively, is therefore as follows:

- With the classic time-weighted average formula: 18.2 and 15.6
- With the multitask complex formula: 21.6 and 18.1 (over a constant 5-day week)

As to the score defining weekly risk, based on current understanding we may argue that the level will be somewhere between the results obtained with the time-weighted average and the multitask complex.

8.5.6 Homogeneous Group: "Drying and Packaging over a 5-h Shift" (Tables 8.46 and 8.47; Figure 8.64)

The graph in Figure 8.64 shows that by decreasing the duration of the shift by 1 h, the daily exposure index of the homogeneous group also decreases as follows for the right and left limb, respectively:

- "Folding table days": 17.1 and 14.7
- "Steam press ironing days": 12.2 and 11.2
- "Manual ironing days": 15.9 and 12.6

TABLE 8.46
Distribution of Shifts in a Representative Week for a Homogeneous Group: Drying-Packaging over a 5-H Shift

	Distribution of Shifts in the Week for a Homogeneous Group (Total Duration of Shift in Minutes, for Working Hours and Overtime, Including Breaks)						
	Mon	Tue	Wed	Thu	Fri	Sat	Sun
Organizational Study of 1 Week, in a Representative Month							
Shift duration (min)	300	300	300	300	300	300	0
Minutes of nonrepetitive tasks in the shift	10	10	10	10	10	10	0
Total duration of breaks	15	15	15	15	15	15	0
Net duration of repetitive tasks in the shift	275	275	275	275	275	275	0
Recovery multiplier	1.20	1.20	1.20	1.20	1.20	1.20	

The right limb is no longer exposed to peaks of biomechanical overload (at purple level) and on the steam press ironing day the peak exposure level is slightly red; the left limb on that day is just short of yellow (very slight risk). The *weekly exposure index* for the right and left limbs, respectively, is therefore as follows:

- With the classic time-weighted average formula: 15.3 and 12.82
- With the multitask complex formula: 17.9 and 15 (over a constant 5-day week).

It can thus be stated that the members of the homogeneous group assigned to *drying-packaging over a 5-h shift* are exposed to an average weekly risk of biomechanical overload of the upper limbs.

8.5.7 Homogeneous Group: "Drying and Packaging over a 3-h Shift" (Tables 8.48 and 8.49; Figure 8.65)

The daily exposure index for the homogeneous group assigned to *drying-packaging over a 3-h shift* is even lower; for the right and left side, respectively, the levels are as follows:

- "Folding table days": 12.2 and 9.9
- "Steam press ironing days": 9.2 and 8.2
- "Manual ironing days": 11.3 and 9.1

The right upper limb is exposed to "slightly red" biomechanical overload on most days and the exposure is very slight (yellow) on the steam press ironing day; the left side is exposed to very slight risk (yellow) on all days.

The weekly exposure index for the right and left limbs, respectively, is therefore as follows:

TABLE 8.47
Description of Working Hours in a Representative Week for a Homogeneous Group: Drying-Packaging over a 5-H Shift

Task	Mon	Tue	Wed	Thu	Fri	Sat
Drying						
Transfer loaded cart to dryers	1.65	0.62		1.65	0.62	
Load dryer	35	23.6		35	23.6	
Transfer partially loaded cart to area adjacent to dryers	1.65	0.62		1.65	0.62	
Transfer empty cart to dryers for unloading	1.65	0.62		1.65	0.62	
Empty dryer	35	23.6		35	23.6	
Transfer loaded cart to storage area before packaging	1.65	0.62		1.65	0.62	
Ironing						
Transfer loaded cart steam press iron		0.86			0.86	
Press sheets and table cloths on steam press iron and fold manually		90			90	
Press mattress covers on steam press iron and fold manually		45			45	
Press pillow cases on steam press iron and fold manually		45			45	
Transfer empty cart to drying area		0.86			0.86	
Transfer loaded cart to packaging unit	1.4	1.4		1.4	1.4	
Manually iron certain types of laundry			110			110
Packaging						
Transfer loaded cart to folding table	0.37			0.37		
Fold flat linens	50			50		
Fold sweaters	32.5		82.5	32.5		82.5
Fold trousers	32.5		82.5	32.5		82.5
Transfer empty cart to drying area	0.37			0.37		
Place clean clothes in pigeon holes	17.45			17.45		
Load cart for packaging unit	17.45			17.45		
Transfer loaded cart to packaging unit	1.4	1.4		1.4	1.4	
Packaging unit	45	42		45	42	

- With the classic time-weighted average formula: 10.9 and 9.1
- With the multitask complex formula: 12.6 and 10.4 (over a constant 5-day week)

As the weekly exposure score will be somewhere between the results obtained with the two formulas, it can be stated that the homogeneous group defined as *drying-packaging over a 3-h shift* is exposed to slight weekly risk of biomechanical overload on the right side and borderline or very slight risk on the left side.

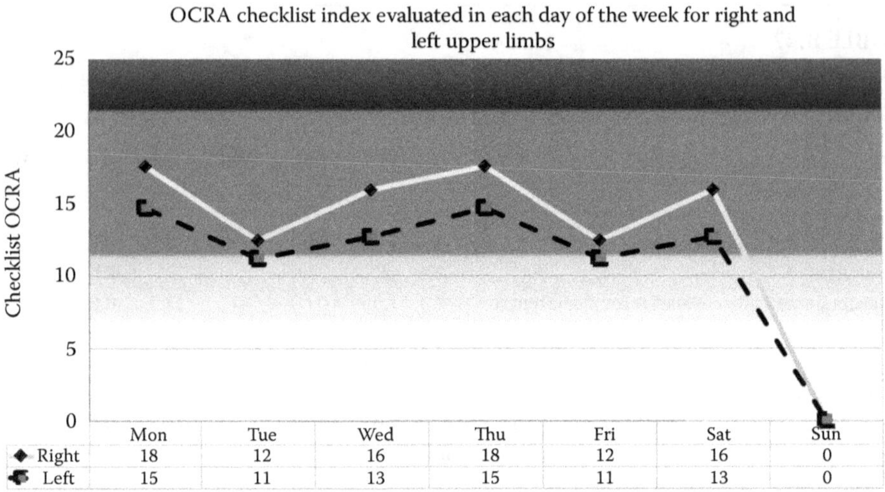

FIGURE 8.64 OCRA checklist weekly exposure index for a homogeneous group: Drying-packaging over a 5-h shift.

TABLE 8.48
Distribution of Shifts in a Representative Week for a Homogeneous Group: Drying-Packaging over a 3-H Shift

	Distribution of Shifts in the Week for a Homogeneous Group (Total Duration of Shift in Minutes, for Working Hours and Overtime, Including Breaks)						
	Mon	**Tue**	**Wed**	**Thu**	**Fri**	**Sat**	**Sun**
Organizational Study of 1 Week, in a Representative Month							
Shift duration (min)	180	180	180	180	180	180	0
Minutes of nonrepetitive tasks in the shift							0
Total duration of breaks							0
Net duration of repetitive tasks in the shift	180	180	180	180	180	180	0
Recovery multiplier	1.12	1.12	1.12	1.12	1.12	1.12	

8.5.8 Two Homogeneous Groups: Sewing and Sorting

These two homogeneous groups work 6-h shifts (7 a.m.–1 p.m. and 1 p.m.–7 p.m.); they have no meal break and stop for only one 15-min break 3 h after starting the shift. The workload and task distribution are the same every day; morning and afternoon shifts are virtually identical and workers stay on either the morning or the afternoon shift for an entire week. Task rotation is daily, and workers alternate weekly between the morning and the afternoon shift.

Organizational data relative to the tasks and task durations, and exposure indexes for the sewing and sorting groups are listed.

TABLE 8.49
Description of Working Hours in a Representative Week for a Homogeneous Group: Drying-Packaging over a 3-H Shift

Task	Mon	Tue	Wed	Thu	Fri	Sat
Drying						
Transfer loaded cart to dryer	0.9	0.25		0.9	0.25	
Load dryer	15	18		15	18	
Transfer partially loaded cart to area adjacent to dryers	0.9	0.25		0.9	0.25	
Transfer empty cart to dryer for unloading	0.85	0.25		0.85	0.25	
Empty dryer	15	18		15	18	
Transfer loaded cart to storage area before packaging	0.85	0.25		0.85	0.25	
Ironing						
Transfer loaded cart steam press iron		0.5			0.5	
Press sheets and table cloths on steam press iron and fold manually		53			53	
Press mattress covers on steam press iron and fold manually		26			26	
Press pillow cases on steam press iron and fold manually		26			26	
Transfer empty cart to drying area		0.5			0.5	
Transfer loaded cart to packaging unit	0.6	1		0.6	1	
Manually iron certain types of laundry			90			90
Packaging						
Transfer loaded cart to folding table	0.15			0.15		
Fold flat linens	43			43		
Fold sweaters	21		45	21		45
Fold trousers	21		45	21		45
Transfer empty cart to drying area	0.15			0.15		
Place clean clothes in pigeon holes	15			15		
Load cart for packaging unit	15			15		
Transfer loaded cart to packaging unit	0.6	1		0.6	1	
Packaging unit	30	35		30	35	

8.5.8.1 Homogeneous Group: Sewing (Tables 8.50 and 8.51 and Figure 8.66)

There is just one worker in the sewing group; besides stitching name tags onto clothes belonging to the residents of nursing homes and making minor repairs on a sewing machine, the worker also performs a few nonrepetitive tasks such as cleaning up at the end of the shift, supplying materials, and printing name tags.

The daily exposure index is 20.9 for the right upper limb (medium–high risk, borderline purple) and 14.1 for the left, which is thus exposed to medium risk in terms

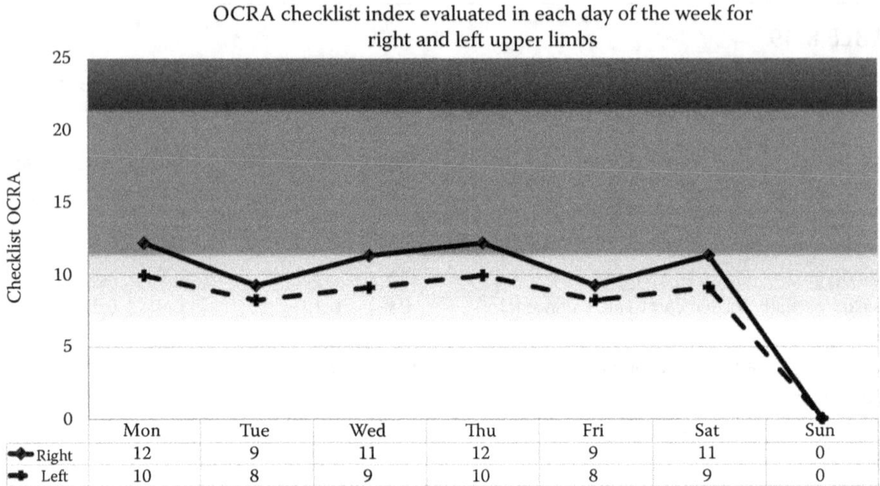

FIGURE 8.65 OCRA checklist weekly exposure index for a homogeneous group: Drying-packaging over a 3-h shift.

TABLE 8.50
Distribution of Shifts in a Representative Week for a Homogeneous Group: Sewing

	Distribution of Shifts in the Week for a Homogeneous Group (Total Duration of Shift in Minutes, for Working Hours and Overtime, Including Breaks)						
	Mon	**Tue**	**Wed**	**Thu**	**Fri**	**Sat**	**Sun**
Organizational Study of 1 Week, in a Representative Month							
Shift duration (min)	360	360	360	360	360	360	0
Minutes of nonrepetitive tasks in the shift	30	30	30	30	30	30	0
Total duration of breaks	15	15	15	15	15	15	0
Net duration of repetitive tasks in the shift	315	315	315	315	315	315	0
Recovery multiplier	1.4	1.4	1.4	1.4	1.4	1.4	

TABLE 8.51
Description of Working Hours in a Representative Week for a Homogeneous Group: Sewing

Task Duration over 1 Week (min)	Mon	Tue	Wed	Thu	Fri	Sat
Stitch name tags	157.5	157.5	157.5	157.5	157.5	157.5
Minor repairs (replace zips, buttons, etc.)	157.5	157.5	157.5	157.5	157.5	157.5

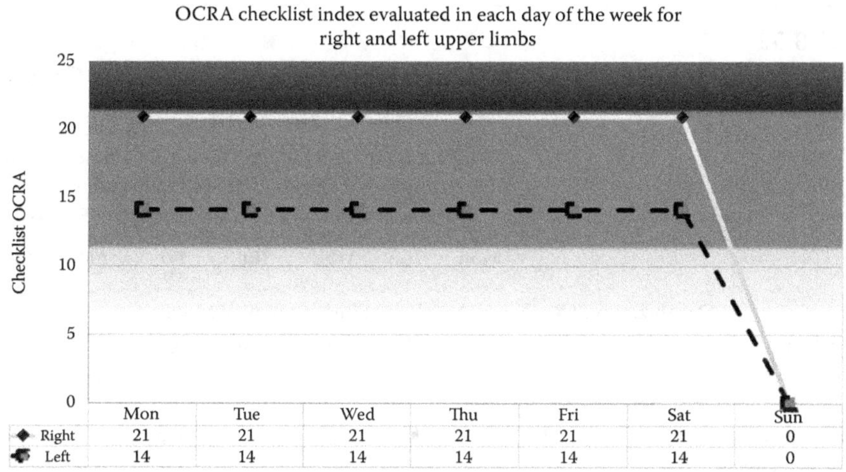

FIGURE 8.66 OCRA checklist weekly exposure index for a homogeneous group: Sewing.

of biomechanical overload. When every day is the same with respect to the number and type of tasks, the daily exposure index will also be identical throughout the entire week, which can be seen in the corresponding graph (see Figure 8.66). *The weekly exposure index* for the right and left limbs, respectively, is therefore as follows:

- With the classic time-weighted average formula: 20.9 and 14.1
- With the multitask complex formula: 23 and 15 (over a constant 5-day week)

Therefore, the right side will definitely be exposed to medium–high risk and the left to medium risk.

8.5.8.2 Homogeneous Group: Sorting (Tables 8.52 and 8.53 and Figure 8.67)

The daily exposure index is 26.4 for the right upper limb and 23.17 for the left: This is the homogeneous group with the worst exposure index and the highest risk for both upper limbs.

The intrinsic exposure indexes for each individual task performed by this group suggest that the tasks are at extremely high risk, especially due to the frequency with which the shoulder is placed in an awkward posture, and to the force required to carry out certain tasks (such as emptying bags or unloading washing machines). Since every day is the same with respect to the number and type of tasks, the daily exposure index will also be identical throughout the entire week, which can be seen in the corresponding graph (see Figure 8.67).

The weekly exposure index for the right and left limbs, respectively, is therefore as follows:

- With the classic time-weighted average formula: 26.4 and 23.17
- With the multitask complex formula: 29 and 25 (over a constant 5-day week)

TABLE 8.52
Distribution of Shifts in a Representative Week for a Homogeneous Group: Sorting

	Distribution of Shifts in the Week for a Homogeneous Group (Total Duration of Shift in Minutes, for Working Hours and Overtime, Including Breaks)						
	Mon	Tue	Wed	Thu	Fri	Sat	Sun
Organizational Study of 1 Week, in a Representative Month							
Shift duration (min)	360	360	360	360	360	360	0
Minutes of no repetitive tasks in a shift	20	20	20	20	20	20	0
Total duration of breaks	15	15	15	15	15	15	0
Net duration of repetitive task in the shift	325	325	325	325	325	325	0
Recovery multiplier	1.4	1.4	1.4	1.4	1.4	1.4	

TABLE 8.53
Description of Working Hours in a Representative Week for a Homogeneous Group: Sorting

Task Duration Over 1 Week (min)	Mon	Tue	Wed	Thu	Fri	Sat
Prepare to Sort						
Empty bags of soiled linen into wheeled hamper	40	40	40	40	40	40
Transfer soiled linen hamper to sorting area	2	2	2	2	2	2
Transfer empty hamper to soiled linen storage area	2	2	2	2	2	2
Sorting						
Manually sort linens from hamper to trolley	157	157	157	157	157	157
Transfer loaded cart to washing machines	2	2	2	2	2	2
Transfer empty cart to sorting area	2	2	2	2	2	2
Cleaning						
Manually load washing machine	56	56	56	56	56	56
Transfer empty cart to washing machine for loading	2	2	2	2	2	2
Manually empty washing machine to cart	50	50	50	50	50	50
Transfer loaded cart to drying area	2	2	2	2	2	2

Therefore, both the right and the left sides will definitely be exposed to very high risk.

8.5.9 Discussion and Conclusions

Given the type of work performed in industrial laundries, the growing number of insurance claims relating to work-related musculoskeletal disorders of the upper limbs, and the weekly exposure index for employees in this sector (red), it goes

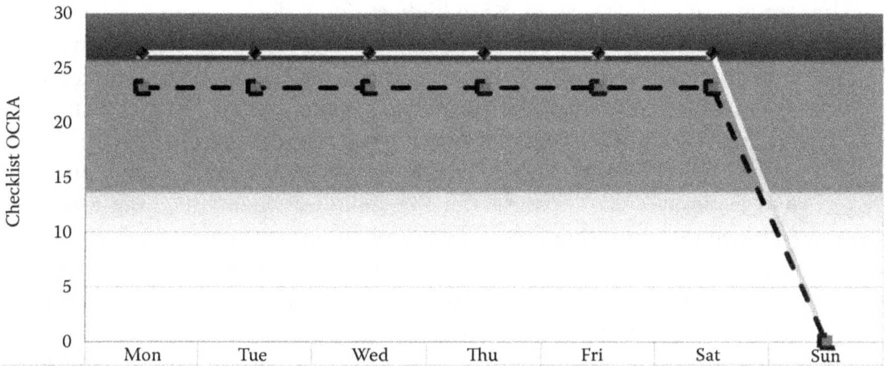

	Mon	Tue	Wed	Thu	Fri	Sat	Sun
Right	26	26	26	26	26	26	0
Left	23	23	23	23	23	23	0
Right	26	26	26	26	26	26	0
Left	23	23	23	23	23	23	0

FIGURE 8.67 OCRA checklist weekly exposure index for a homogeneous group: sorting.

without saying that there is a definite risk of biomechanical overload for the upper limbs ranging from slight to average to high, depending on the homogeneous group.

By using the OCRA checklist (multitask complex method, weekly cycle), it is not only possible to measure the level of such risk fairly accurately, but also to gather vital information for the purposes of managing risk (i.e., remedial actions, rotations, etc.) and injury (i.e., returning to the workplace, etc.).

A basic analysis of the intrinsic risk scores associated with each task will in fact highlight the main risk factors causing biomechanical overload of the upper limbs. These factors include:

- High-frequency actions
- Awkward postures for the shoulder, arm, and hand for prolonged periods during the shift
- Little or no recovery time among the various homogeneous groups in the analysis

It would therefore be advisable to put in place a preliminary program such as the one recommended here, composed of various steps that can be adopted separately, to reduce exposure to the risk of biomechanical overload of the upper limbs for laundry workers:

- Introduction of additional 8–10-min breaks for shifts with no recovery periods.
    - Staff rotation from high-risk areas to lower risk areas (e.g., the sorting area has the highest number of at-risk tasks; workers should therefore rotate to lower risk tasks such as drying-packaging).

- Implementation of staff assigned to highest risk tasks.
  - Adjustment of workbench height to avoid awkward postures of the shoulder.
  - For the "unload washing machine and dryer" tasks, if economically sustainable, adopt radical primary prevention measures, such as replacing traditional machines with modern models featuring a tilting drum, and providing tools to enable frequency of action to be reduced and avoid awkward postures for the shoulder when stretching too far to reach and grasp laundry.
  - Introduce more carts with adjustable-height basket for loading/unloading washers and dryers so as to reduce the number of "reaching" actions that entail awkward postures of the shoulder.
  - Modify "customer-laundry" relations: Ask customers to sort clothes by type and color and place them in different colored bags so as to minimize the need for sorting, a tiring activity that overloads the upper limbs.
  - Provide all employees with adequate information and training, for example, to enhance awareness of risks and reduce the number of unnecessary and hazardous actions.

These are just a few of the many recommendations that can be given to an employer based on a preliminary in-depth interpretation of an assessment designed to measure the risk of biomechanical overload of the upper limbs with the OCRA method.

If the study were extended to other more or less automated industrial laundries with a view to identifying all the various tasks involved (bearing in mind that if the same task is performed in a different manner it counts as a new task), it might be possible to produce a sample classification of all tasks, to be preassessed using the OCRA checklist so as to obtain the relevant intrinsic risk scores. Once all this data is entered into a specific basic software program for laundries (Excel would not suffice), even nonexperts would be able to conduct a risk analysis because all they would need to do is enter the organizational information regarding the homogeneous groups (i.e., weekly shifts, tasks, and task durations) and the risk indexes would be generated automatically.

8.6 BUILDING AND CONSTRUCTION: BIOMECHANICAL OVERLOAD ANALYSIS IN SPECIFIC OCCUPATIONS

8.6.1 Introduction

The building and construction industry is an extremely complex sector and is often viewed as one of the drivers of the real economy.

Today, the industry accounts for

- 11% of the gross domestic product (GDP) of the European Union, with approximately 17 million workers.
- 10% of Italy's GDP, with approximately 2 million workers, including 65% in FT employment.

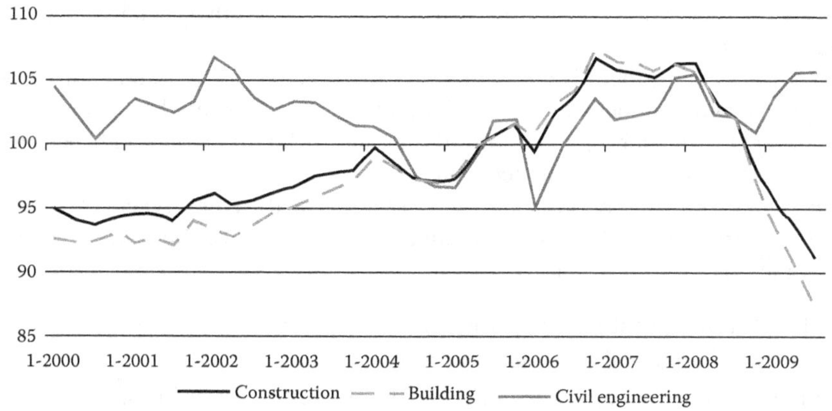

FIGURE 8.68 Production indexes for the building and construction industry, 2000–2009 (seasonally adjusted). (From Eurostat, 2009.)

Construction activities in the EU-27 provided employment to an estimated 14.8 million persons while generating an estimated €562 billion in value added (9.3% of the nonfinancial sector's total value added). Most construction enterprises serve a local market and, consequently, the construction sector is characterized by a high number of small enterprises, and relatively few large ones. Micro and small businesses (with less than 50 employees) together employed 72.1% of the EU-27 construction sector workforce in 2006. Figure 8.68 shows industrial production index trends per type of construction between 2000 and 2009.

The building and construction industry is actually composed of many segments. Each segment calls for specific skills. The following are just some of the jobs performed in the construction industry:

- Demolition
- Excavation, earthmoving
- Foundations
- Bricklaying, partitioning, and paneling, including rendering
- Painting
- Waterproofing and insulation (e.g., soundproofing)
- Flooring and walls
- Reinforced concreting, including foundations
- Steelwork
- Plasterboard structures
- Iron/steel/aluminum structures
- Glazing
- Plumbing
- Electrical installations
- Heating, ventilation, and air-conditioning (HVAC)
- Marble working
- Carpentry

Businesses and workers in the construction sector generally specialize in just one of these activities.

Across the board, there is a close correlation between quality and continuity of employment and fragmentation, and duration of employment. Most construction workers move from job to job and building site to building site, with periods of unemployment and inactivity in between.

The manufacturing process encompasses many different types of workers. *Unskilled workers* may perform any number of tasks, from hauling sacks of cement to cleaning up the building site. *Skilled workers* are assigned to specific tasks.

8.6.1.1 Occupational Diseases and Disorders in the Building and Construction Industry

There are countless papers in the international literature about job safety, but very few epidemiological studies on work-related musculoskeletal disorders and their prevention in the building and construction industry.

This lack of information may be due to the complexity of this sector, which covers countless different and often quite unique activities; the provisional nature of building sites is a good example, along with the complex organization of construction work. Moreover, a range of different businesses and workers can often be found on the same building site, with different skills; there are subcontractors, hired hands, casual workers, and often also workers hired off the books.

8.6.1.2 Assessing the Risk of Biomechanical Overload of the Upper Limbs in a Small Construction Company: Description of Tasks and Phases

Here, we summarize the results of a risk assessment with respect to biomechanical overload of the upper limbs among a representative sample of comparable construction companies. These small businesses carry out a variety of different jobs especially renovations and painting. An in-depth analysis of this assessment came up with the following activities, which can be identified as *phases*:

1. Balcony repairs
2. Install marble tiles on balcony parapets
3. Erect plasterboard structures
4. Paint outdoor walls
5. Paint indoor walls

These *phases* have been analyzed in-depth along with the tasks that they are comprised of. Detailed descriptions are provided as follows:

- *Balcony repairs*: Remove damaged concrete and/or plaster using a hammer drill; brush down steel reinforcement using stiff scrub brush and apply anticorrosion coating; repair balcony edges with cement mortar and restore to original condition.
- *Install new marble tiles*: Clean surface using scraper and brush (dry); apply cement mortar using a notched trowel and spatula. Apply mortar then correctly position and lay marble tiles.

- *Erect plasterboard structures* (walls, partitions and ceilings): Inspect and outline the position of the structures then erect the load-bearing structure using metal profiles. Apply plasterboard sheets to the load-bearing structure. Cut the plasterboard sheets to size.
- *Paint outdoor walls*: First prep surfaces by filling holes and patching cracks with spackle. Apply tape around door and window frames to protect from paint bleeds. Prepare paint, mix materials in sacks and add water to achieve the desired consistency. Before painting, certain surfaces require the application of a stabilizing primer using a roller. Once the paint is ready, apply it to the surface using a roller. Use brush to paint corners. Remove protective tape.
- *Paint interior walls*: Apply washable or enamel paint to surfaces. First prep walls by manually repairing cracks or holes with spackle. To paint old or very rough walls, first apply spackle by hand then use power sander to smooth over. Before painting, certain surfaces require the application of a stabilizing primer using a roller. Commercially bought paint then needs preparing in suitable buckets (stirred and, if necessary, diluted with water to bring it to the right consistency for painting). Once the paint is ready, apply it to the surface using a roller. Use brush to paint corners. After painting remove protective tape.

8.6.1.3 Classification of Tasks by Phase and Calculation of Intrinsic Risk Indexes for Each Task

As described previously, five phases have been identified and each one is composed of a certain number of tasks. Each task (see Table 8.54) was filmed and a sample minute was identified as a representative cycle based on which intrinsic risk was calculated as if the same task were performed for the whole duration of the shift (see Chapter 6).

The estimated intrinsic indexes indicate the following risks for the right upper limb:

- Ten high risk tasks (purple)
- Fourteen medium or low risk tasks (red)
- Two borderline risks (yellow)

The outcomes for the left upper limb are as follows:

- Four high risk tasks (purple)
- Nine medium or low risk tasks (red)
- Eleven borderline risks (yellow)
- Two risk-free tasks (green)

8.6.2 Identification of Homogeneous Groups and Calculation of Multitask Exposure

As per the methodology described in Chapter 6, homogeneous groups were formed based on risk, that is, groups of workers performing the same tasks for the same length of time.

TABLE 8.54
Phases and Tasks: Identification and Analysis

Repair Balcony Edges	**Paint Exterior Surfaces**
Remove damaged concrete/plaster—tray	Prepare paint
Remove damaged concrete/plaster—hammer drill	Apply paint with roller
Brush steel bars with grinder	Apply paint with brush and roller
Apply anticorrosion coating	**Paint Interior Surfaces**
Wet surfaces	Sandpaper corners manually
Prepare cement mortar	Sand walls using power sander
Apply cement mortar	Prepare spackle by hand
Replace Marble Tiles	Spackle ceiling irregularities
Prepare surface for laying	Spackle wall irregularities
Apply cement mortar and lay tiles	Paint ceiling
Apply external tile grout	Paint walls
Plasterboard Structures	
Install ceiling profiles	
Lay wall panels	
Install ceiling panels	
Prepare spackling	
Apply spackling	

Based on the organizational analysis that was performed, three homogeneous groups emerged:

- The *first homogeneous group* (see Figure 8.69) consists primarily of builders:
  - Repair balcony edges
  - Replace marble tiles
- The *second homogeneous group* (see Figure 8.70) consists primarily of painters:
  - Exterior wall painting
  - Interior wall painting
- The *third homogeneous group* (see Figure 8.71) consists of workers who combine building, painting, and plastering tasks:
  - Repair balcony edges
  - Erect plasterboard partitions
  - Paint exterior surfaces

FIGURE 8.69 Principal building tasks.

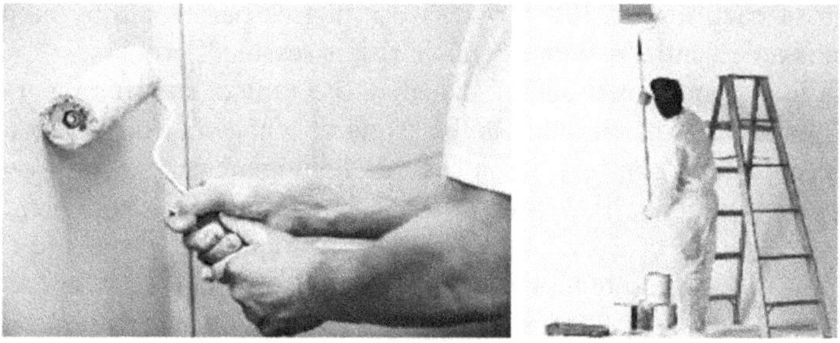

FIGURE 8.70 Principal painting tasks.

FIGURE 8.71 Principal plastering tasks.

Table 8.55 shows the shifts, breaks, and duration of nonrepetitive tasks for each group in the *representative observation period* (day, week, and month). All of the groups had the same working hours and the same duration and distribution of breaks. The net duration of repetitive work varied, depending on the days, from 250 to 380 min, and the same applied to the duration of nonrepetitive work.

The most complex phase, requiring the most attention, involves studying task distribution over the week and/or month (in which case, the work cycle is best represented by the month) and calculating task duration in minutes per shift for each homogeneous group. Although expert supervisors will again find it challenging to indicate

TABLE 8.55
Representative Average Working Hours

Shift duration	480 min
1 or 2 breaks	Variable duration
1 meal break duration	60–80 min
Duration of nonrepetitive tasks	100–200 min
Repetitive tasks	250–380 min

the duration of each task in the various work phases performed by homogeneous groups, this is an essential step in a reliable risk assessment process.

In the building and construction industry, the unit of measurement that most effectively quantifies work duration is the "square footage of work." Having chosen a modal/representative contract that a specific homogeneous group has worked on during the year (1 month), discussions took place with the employers regarding the average square footage of work completed over 1 month with respect to each phase performed by that specific homogeneous group. Also discussed with the employer was the average number of representative days actually worked over a month, the result being 2 m^2 per person, per day. Then, once the net duration of the repetitive work in the shift was known, and each task was weighted in percentage terms based on its relevance in the phase, the minutes worked on each task were calculated and distributed over each day of the 4-week period. Table 8.56 shows an example of how the duration of a specific task is measured in minutes based on the actual square footage of work completed. If no other method is available, this is one possible approach toward estimating work duration based on the available data and the use of common sense. In fact, in the interview, the employer eventually stated that it would be possible, and in fact very useful, to know beforehand what the estimated average duration of specific tasks would be in relation to a certain unit of measurement, such as 10 m^2, for example: with this data it would be possible to more effectively calculate the timing and costs of a contract, besides obviously obtaining information about potential risks.

Figures 8.72 through 8.74 illustrate the distribution in minutes of the tasks performed by three homogeneous groups over the 4 weeks of a representative month.

Considering the monthly distribution of tasks in phases, the phases performed in the month by the first group (actual builders) were: repair balconies for about 80% of the time, and replace marble tiles for about 20%. The monthly phases performed by the second group (painters) were: paint exterior surfaces for about 85% of the time, and paint interior surfaces for about 15%. The monthly phases performed by the third group (mixed) were: repair balconies for about 20% of the time, paint exterior

TABLE 8.56
Estimated Task Duration Based on Square Footage Actually Painted

Phase: Paint Exterior Surfaces	
Net duration in minutes of repetitive task in the shift	300 min
Average square footage completed over the month	4000 m^2 by the 10 members of the homogeneous group
Average square footage per worker/shift	40 m^2
Tasks Present in the Phase: Duration (in % and minutes)	
1-prepare paint	10% = 60 min
2-apply paint with roller	30% = 90 min
3-apply paint with brush and roller	60% = 180 min

Repetitive task			Week 1					Week 2					Week 3					Week 4					Total min per month
			Mon	Tue	Wed	Thu	Fri	Mon	Tue	Wed	Thu	Fri	Mon	Tue	Wed	Thu	Fri	Mon	Tue	Wed	Thu	Fri	Min
Phase: Repair balcony edges																							
1	Remove damaged concrete/plaster—tray	20%	120	120				120	120				120	120									720
2	Remove damaged concrete/plaster—hammer drill	20%	120	120				120	120				120	120									720
3	Brush steel bars with grinder	10%			120					120					120								360
4	Apply anticorrosion coating	10%			120					120					120								360
5	Wet surfaces	5%				20	20				20	20				20	20	20	20				160
6	Prepare cement mortar	12%				60	60				60	60				60	60	60	60				480
7	Apply cement mortar	12%				60	60				60	60				60	60	60	60				480
8	Smooth with spatula and sponge	12%				60	60				60	60				60	60	60	60				480
Phase: Replace marble tiles																							
9	Prepare surface for laying	5%																		15	15	15	45
10	Apply cement mortar and lay tiles	85%																		250	250	250	750
11	Apply external tile grout	10%																		30	30	30	90

FIGURE 8.72 First homogeneous group: Builders. Repetitive tasks and task duration in minutes over 1 month.

Task	Week 1					Week 2					Week 3					Week 4					Total min per month
	Mon	Tue	Wed	Thu	Fri	Mon	Tue	Wed	Thu	Fri	Mon	Tue	Wed	Thu	Fri	Mon	Tue	Wed	Thu	Fri	min.
Phase: paint esteriore surface																					
17 Prepare paint	30	30	30	30	30	30	30	30	30	30	30	30	30	30	30	30	30				510
18 Apply paint with roller	90	90	90	90	90	90	90	90	90	90	90	90	90	90	90	90	90				1530
19 Apply paint with brush and roller	180	180	180	180	180	180	180	180	180	180	180	180	180	180	180	180	180				3060
Phase: Paint interior surfaces																					
20 Sandpaper corners manually																		60	60		120
21 Sand walls using power sander																		90	90		180
22 Prepare spackle by hand																		15	15		30
23 Spackle ceiling irregularities																		60	60		120
24 Spackle wall irregularities																		60	60		120
25 Paint ceiling																		30	30	30	90
26 Paint walls																				300	300

FIGURE 8.73 Second homogeneous group: Painters. Repetitive tasks and task duration in minutes over 1 month.

Phase		Task		Week 1					Week 2					Week 3					Week 4					Min in a month
				Mon	Tue	Wed	Thu	Fri	Mon	Tue	Wed	Thu	Fri	Mon	Tue	Wed	Thu	Fri	Mon	Tue	Wed	Thu	Fri	Min.
Repair balcony edges	1	Remove damaged concrete/plaster—tray	20%	120	120																			240
	2	Remov damaged concrete/plaster—hammer drill	20%	120	120																			240
	3	Brush steel bars with grinder	10%			120																		120
	4	Apply anticorrosion coating	10%			120																		120
	5	Wet surfaces	5%				20	20																40
	6	Prepare cement mortar	12%				60	60																120
	7	Apply cement mortar	12%				60	60																120
	8	Smooth with spatula and sponge	12%				60	60																120
Replace marble tiles	9	Prepare surface for laying	5%																					0
	10	Apply cement mortar and lay tiles	85%																					0
	11	Apply external tile grout	10%																					0
Plasterboard structures	12	Install ceiling profiles	20%																90		90	90		240
	13	Lay wall tiles	25%																90		90	90		270
	14	Install ceiling panels	25%																90		90	90		270
	15	Prepare spackling	5%																	30			30	60
	16	Apply spackling	25%																	150			150	300
Paint-exterior surfaces	17	Prepare paint	10%						30	30	30	30	30	30	30									210
	18	Apply paint with roller	30%						90	90	90	90	90	90	90									630
	19	Apply paint with brush and roller	60%						180	180	180	180	180	180	180									1260
Paint interior surfaces	20	Sandpaper corners manually	10%													60	60							120
	21	Sand walls using power sander	20%													90	90							180
	22	Prepare spackle by hand	3%													15	15							30
	23	Spackle ceiling irregularities	10%													60	60							120
	24	Spackle wall irregularities	13%													60	60							120
	25	Paint ceiling	12%													30	30	30						90
	26	Paint walls	32%															300						300

FIGURE 8.74 Third homogeneous group: Plasterers. Repetitive tasks and task duration in minutes over 1 month.

surfaces for about 45%, paint interior surfaces for about 15%, and erect plasterboard structures for about 20% of the time.

8.6.3 Calculation of Final Risk Exposure Indexes Using Two Models: Time-Weighted Average and Multitask Complex

Figures 8.75 through 8.77 report daily exposure indexes for the right upper limb only and monthly exposure indexes for both the right and left upper limbs for all groups. Indexes have been computed by both the time-weighted average and the multitask complex formula.

8.6.4 Conclusions

In this section, we have provided an example of how biomechanical overload risk can be assessed for small building renovation contractors. We do not claim that our overview is in any way exhaustive or representative; however, it has helped to highlight the following:

- Assessment issues typical of this industry relating to the assessment of this specific risk.
- The techniques required to thoroughly deal with this risk.

The risk analysis process undertaken using the weekly/monthly multitask OCRA checklist method (see Chapter 6, Section 6.4) has enabled us to identify the level of risk for biomechanical overload of the upper limbs in a variety of jobs.

The three groups of workers were found to have an exposure index of red/purple for the right side and yellow/red for the left side.

The risk factors for the various tasks with the greatest impact on the calculation are the following: frequency, peak force, and awkward postures for the shoulder.

In order to reduce the risk of biomechanical overload for the upper limbs, a preliminary corrective action plan should begin by redistributing breaks, modifying certain work tools requiring the use of considerable force, and redesigning certain work methods so as to improve the worker's posture and reduce the frequency of action.

It would also be advisable to implement a training and education program for workers and put in place an active health surveillance plan. This preliminary analysis of biomechanical overload risk for the upper limbs in the construction industry has shown that the main difficulties lie in obtaining organizational data, not only due to the large number of tasks involved, but also to their variable monthly distribution (i.e., who does what and for how long?). To obtain a reliable risk assessment that accurately reflects exposure levels, it is essential to start with an accurate preliminary organizational analysis, which must be carried out together with whoever actually organizes the work. While the analysis must go into detail regarding the most common work procedures, it must also be succinct and exclude tasks that are actually only performed sporadically.

Readers should not expect to carry out quick and effortless assessments using miraculously short questionnaires: It is far better to start by assuming that workers

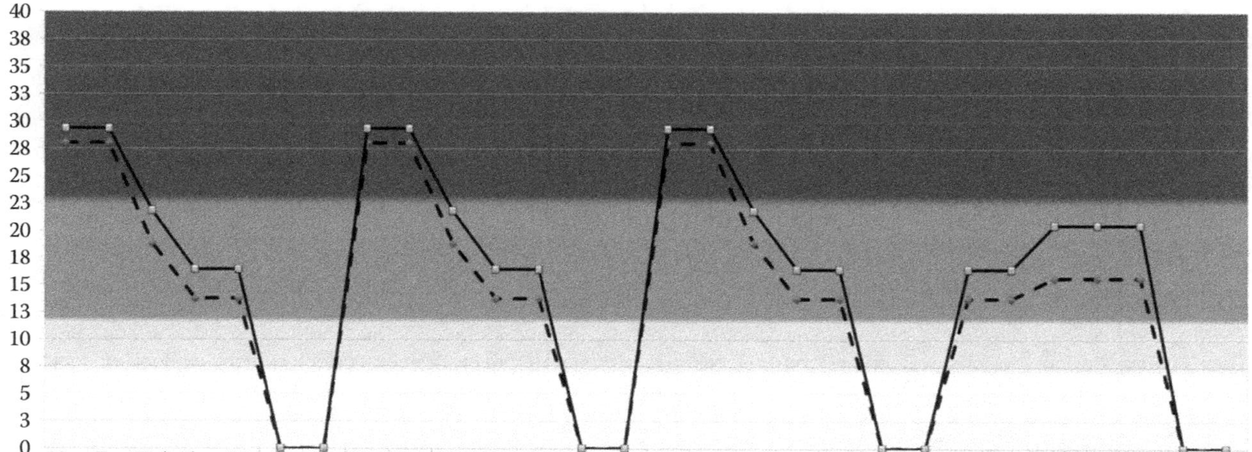

	Mon	Tue	Wed	Thu	Fri	Sat	Sun	Mon	Tue	Wed	Thu	Fri	Sat	Sun	Mon	Tue	Wed	Thu	Fri	Sat	Sun	Mon	Tue	Wed	Thu	Fri	Sat	Sun
Weighted average	28	28	19	14	14	0	0	28	28	19	14	14	0	0	28	28	19	14	14	0	0	14	14	16	16	16	0	0
Multitask	29	29	22	16	16	0	0	29	29	22	16	16	0	0	29	29	22	16	16	0	0	16	16	21	21	21	0	0

	Right	Left
Week 1	21.0	1.6
Week 2	21.0	17.6
Week 3	21.0	17.6
Week 4	15.6	9.4

Weighted average

Right	Left
19.0	15.0

	Right	Left
Week 1	24.0	21.8
Week 2	24.0	21.8
Week 3	24.0	21.8
Week 4	18.4	10.1

Multitask complex

Right	Left
21.6	18.6

FIGURE 8.75 Risk-assessment results (OCRA checklist scores) for every day, week, and month for the first homogeneous group: builders.

	Mon	Tue	Wed	Thu	Fri	Sat	Sun	Mon	Tue	Wed	Thu	Fri	Sat	Sun	Mon	Tue	Wed	Thu	Fri	Sat	Sun	Mon	Tue	Wed	Thu	Fri	Sat	Sun
Weighted average	10	10	10	10	10	0	0	10	10	10	10	10	0	0	10	10	10	10	10	0	0	10	10	24	24	24	0	0
Multitask	13	13	13	13	13	0	0	13	13	13	13	13	0	0	13	13	13	13	13	0	0	13	13	33	33	33	0	0

	Right	Left
Week 1	10.3	13,0
Week 2	10.3	13.0
Week 3	10.3	13.0
Week 4	19.2	15.1

Weighted average

Right	Left
13.2	14.4

	Right	Left
Week 1	11.5	14.2
Week 2	11.5	14,2
Week 3	11.5	14.2
Week 4	22.2	15.7

Multitask complex

Right	Left
14.2	14.5

FIGURE 8.76 Risk-assessment results (OCRA checklist scores) for every day, week, and month for the second homogeneous group: painters.

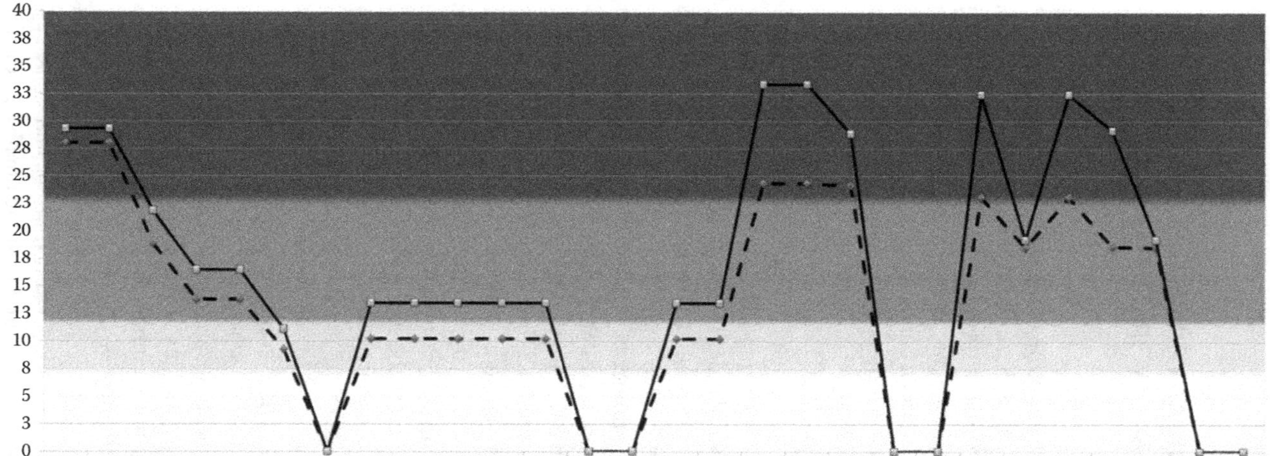

	Mon	Tue	Wed	Thu	Fri	Sat	Sun	Mon	Tue	Wed	Thu	Fri	Sat	Sun	Mon	Tue	Wed	Thu	Fri	Sat	Sun	Mon	Tue	Wed	Thu	Fri	Sat	Sun
Weighted average	28	28	19	14	14	9	0	10	10	10	10	10	0	0	10	10	24	24	24	0	0	23	19	23	19	19	0	0
Multitask	29	29	22	16	16	11	0	13	13	13	13	13	0	0	13	13	33	33	29	0	0	32	19	32	29	19	0	0

	Right	Left
Week 1	22.5	17.9
Week 2	10.3	13.0
Week 3	19.2	15,1
Week 4	20.3	7.8

Weighted average

Right	Left
17.9	13.6

	Right	Left
Week 1	26.4	23.3
Week 2	12.1	14.2
Week 3	26.6	21.5
Week 4	25.1	8.1

Multitask complex

Right	Left
19.3	14.9

FIGURE 8.77 Risk-assessment results (OCRA checklist scores) for every day, week, and month for the third homogeneous group: Builders/painters/plasterers.

in the construction industry are by default exposed to the risk of upper limb biomechanical overload, and then go on to add specific and necessarily complex risk assessments.

8.7 ANALYSIS OF BIOMECHANICAL OVERLOAD AMONG HOSPITAL AND HEALTH-CARE WORKERS

8.7.1 Foreword

In the Italian and international scientific literature, health-care workers involved in caring for noncooperative patients are among those most prone to acute and chronic musculoskeletal disorders. The lumbar spine is the most frequently affected, especially as a result of manual patient handling. This is consistent with the data from countless investigations showing that the manual handling of noncooperative patients is associated with substantial overload of the lumbar spine, often far exceeding limits deemed to be "physiological" (Marras, 2008).

According to the international literature (WHO, 2011) there are around 19,300,000 nursing staff working in health-care facilities around the globe; about 85% are women.

The World Health Organization (WHO) has reported on the dramatic shortage of nursing staff in developing countries. Worldwide, there is a shortage of at least 4.2 million nurses across 57 countries. The problem is most acutely felt in Africa and Asia, but the lack of nurses is an issue in Italy as well.

The WHO publishes a ranking of the number of nurses per thousand inhabitants, and Italy—with 5.44 nurses—is in 40th place after Turkmenistan, Kyrgyzstan and Kazakhstan. Gabon is in 50th place with 5.04 nurses. On the upside, however, while Italy's nurses are less than half those of the UK, they do provide high standards of health care. And yet they are paid less. Though working more than their British counterparts, Italian nurses earn lower wages (up to €800 less per month). Heavy workloads, high levels of responsibility, major risk of error, and suboptimal working conditions go hand in hand with wages below the European average. In short, Italy lacks nurses. Estimates suggest that the country needs 40,000 more.

Figure 8.78 shows the number of nurses in the main operational areas.

According to reliable statistics, Italy has 334,918 nurses, 48,884 rehabilitation workers, and 45,364 medical technicians. The average age is 44.6 for nurses, 46.2 for medical technicians, and 46.8 for rehabilitation workers.

It is worth emphasizing that at present in hospitals and health-care facilities for the elderly, most manual patient handling activities are performed not only by nurses but also by health-care workers who are not members of a professional association, and whose numbers are not reported nationally.

The literature provides ample evidence of the fact that biomechanical overload, which is repeated mechanical strain on tissues above critical levels, can cause degenerative joint disorders and diseases, not only of the lumbar spine but also of the upper limbs, especially the shoulder.

Moreover, in advanced health-care systems, in particular in Europe, hospitals are intervention centers for "acute" patients, which has entailed major organizational changes to several health-care sectors.

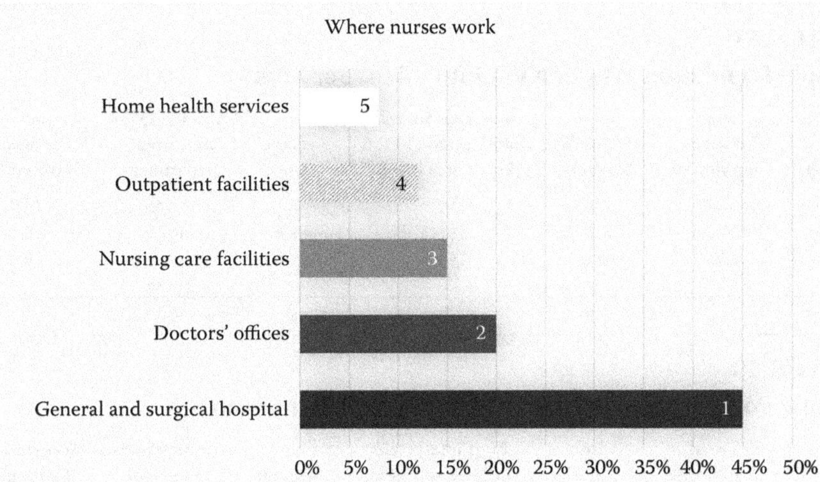

FIGURE 8.78 Distribution of nursing staff in public and private health-care facilities in Italy (2010).

However, the work organization is still "task based" in Italian hospital in-patient wards, where it remains the traditional model.

8.7.2 Epidemiological Data Concerning UL-WMSDs in the Health-Care Sector

Epidemiological data concerning work-related diseases of the spine among nurses can be found in the literature; however, there are very few studies on upper limbs diseases. The results of some preliminary research undertaken in Italy will be presented in this chapter.

The epidemiological data is drawn from health surveillance programs conducted in hospitals located in the Italian region of Liguria (2007–2008), which confirmed the excessively high number of workers with upper limb disorders and diseases (compared with a group of subjects not exposed to any specific risk). The investigation looked at 2544 workers including in-patient ward staff (78%) and outpatient services (22%).

Tables 8.57 and 8.58 show the age and gender of the sample population working in outpatient services and in-patient wards, respectively.

The results of health surveillance programs focusing on the prevalence of work-related pathologies of the upper limbs (UL-WMSDs) are shown in Figures 8.79 and 8.80 for outpatient services, and Figures 8.81 and 8.82 for in-patient ward staff.

The prevalence of affected staff is clearly very high: approximately three times higher than a reference population (Occhipinti and Colombini, 2007) not exposed to biomechanical overload of the upper limbs.

A more in-depth analysis of prevalence data broken down even further (see Figures 8.80 and 8.82) indicates that the most commonly affected body part is the shoulder, especially among women.

TABLE 8.57
Sample Population Working in Out-Patient Services

Gender	Total Subjects Interviewed	Total Exposed Subjects	% of Total Subjects Interviewed vs Total Exposed Subjects	Average Age	Average Years in Current Department	Average Years Performing Current Job
Male	127	228	65.5%	46.2	12.5	19.0
Female	417	870	52.5%	43.7	8.7	18.2
Total	544	1098	55.1%	44.3	9.6	18.4

TABLE 8.58
Sample Population Working in Hospital In-Patient Wards

Gender	Total Subjects Interviewed	Total Exposed Subjects	% of Total Subjects Interviewed vs Total Exposed Subjects	Average Age	Average Years in Current Department	Average Years Performing Current Job
Male	393	670	58.7%	42.9	8.5	15.4
Female	1601	2526	63.4%	42.2	8.2	15.0
Total	544	1098	55.1%	44.3	9.6	18.4

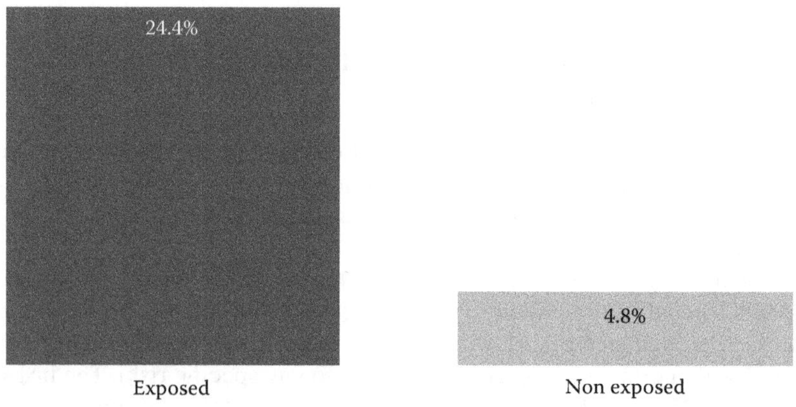

FIGURE 8.79 Prevalence (%) of subjects with UL-WMSDs in hospital out-patient services: overall data.

These results suggest that workers are exposed to biomechanical overload and that although complex, it is also necessary to evaluate risk for the upper limbs. Assessments of manual load-handling risk (for the lumbar spine) often overlook the upper limbs, as if simply carrying out the assessment sufficed to analyze all biomechanical overload situations even affecting other body parts. A risk assessment is the starting point for research and subsequently introducing preventive measures to put a stop to what could eventually become a more complex process for managing "unfit" workers.

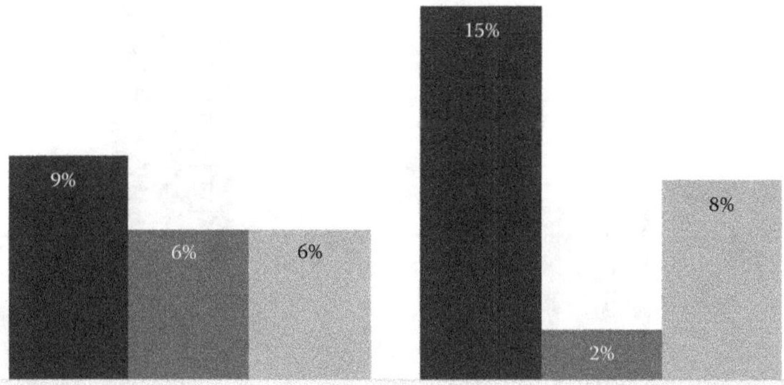

	Male	Female
■ Shoulder	9%	15%
■ Elbow	6%	2%
■ Wrist/hand	6%	8%

FIGURE 8.80 Prevalence (%) of subjects with UL-WMSDs, broken down by joint, in hospital out-patient services.

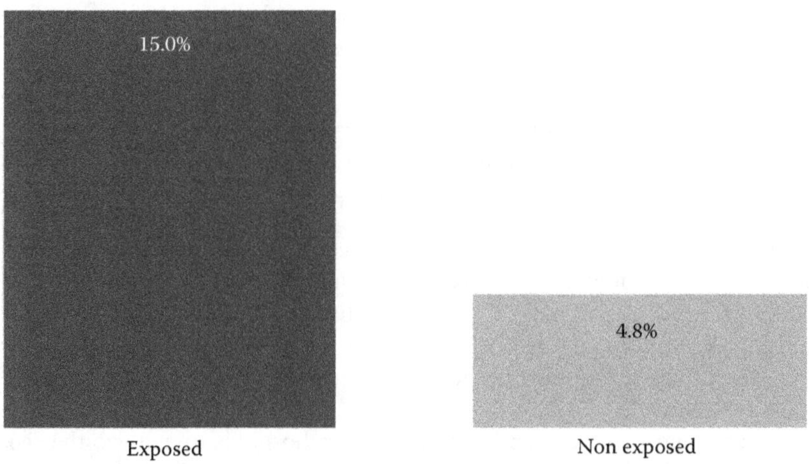

FIGURE 8.81 Prevalence (%) of subjects with UL-WMSDs in hospital in-patient wards: Overall data.

For several years now, in addition to well-known and extensively documented spinal problems, hospitals have increasingly had to deal with the problem of employees with upper limb disorders, which precludes assigning them tasks such as manual patient handling (cf. the high prevalence of musculoskeletal pathologies affecting the shoulder). This further confirms the increasing significance of the problem, as also reported in the international literature (Alexopoulos et al., 2003; Lorusso et al., 2007;

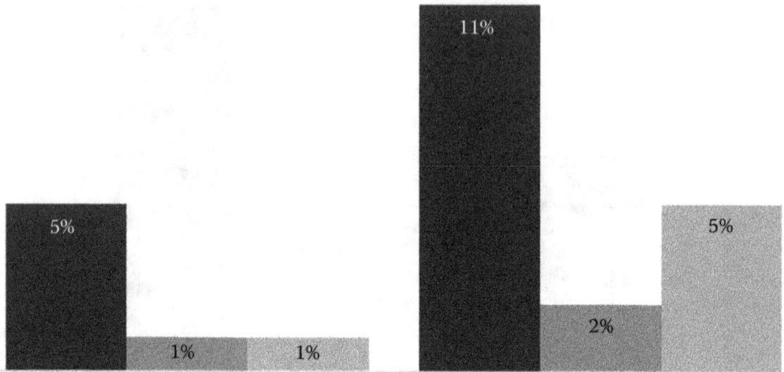

	Male	Female
■ Shoulder	5%	11%
■ Elbow	1%	2%
■ Wrist-hand	1%	5%

FIGURE 8.82 Prevalence (%) of subjects with UL-WMSDs, broken down by joint, in hospital in-patient wards.

Myers et al., 2002) and the consequent need to develop suitable tools for assessing the risk of biomechanical overload of the upper limbs.

8.7.3 Introduction to Emerging Issues in the Assessment of Risk Caused by Biomechanical Overload of the Upper Limbs in the Health-Care Sector

In order to correctly assess the risk of biomechanical overload of the upper limbs (i.e., the purpose of this study), it is worth emphasizing that hospital in-patient wards feature countless variables that make it extremely difficult to accurately and impartially quantify risk. Moreover, it is difficult to film workers as they go about their tasks due to lack of space and also because it is necessary to ask for the patient's authorization to be filmed.

Based on preliminary studies conducted in various hospitals to analyze the risk of biomechanical overload of the upper limbs, it also became apparent that different approaches need to be used depending on the department:

- *Endoscopy unit*: Here the classic OCRA method can be used (see Chapters 4 and 5). In one early study conducted in this area, (Battevi, 2009), the risk of biomechanical overload of the upper limbs was largely proportional to the number of examinations performed on a typical day.
- *Ultrasound lab technician*: These activities can be depicted as weekly or monthly task rotations (Chapter 6, Section 6.4).
- *Physiotherapy, ER, and surgical blocks*: Further studies will be required to more accurately define how the various tasks are rotated and to come up with a specific analytical technique.

- *In-patient ward patient care staff*: Accounting for about 70% of all exposed workers, this was obviously the main focus of our attention. The methods used to analyze this risk are outlined in Section 8.7.4.

8.7.4 Introductory Studies and Emerging Issues among In-Patient Ward Staff

Risk assessments present quite a number of emerging issues in this particular sector. It is, in fact, extremely difficult to quantitatively and qualitatively identify the tasks carried out by staff on a daily basis due to the following factors:

- The sheer number of tasks performed
- Variability in the way the tasks are performed
- Variability in the number of repetitions of the same task performed in the shift
- Different homogeneous groups of exposed workers who often switch tasks
- Different shifts (three shifts/month—two shifts/month—daily)
- Variability in the types of patients cared for and lifted (i.e., different body weight)
- Variability in the degree of cooperation of patients
- Variability in the environment in which the tasks are performed

Given the extreme variability and wide range of tasks involved, the first step in assessing risk exposure among operators was to accurately classify every aspect of their work. The aim of the analysis is to identify tasks and break them down into the correct *macrophases* and *phases*, as shown in Figure 8.83.

Table 8.59 provides an analytical list of the main macrophases based on organizational studies conducted in in-patient general medicine and geriatric wards. The same exercise should be conducted in different departments, such as accident and emergency, gynecology and obstetrics, and cardiology.

The main macrophase that takes between 2 and 3 h/shift is generally called the *beds and bed making* phase and includes the following phases:

- Preparation
- Make/change beds

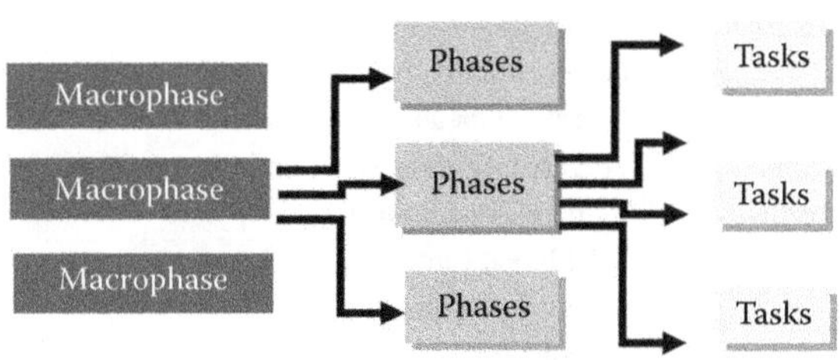

FIGURE 8.83 Classification of tasks into macrophases and phases.

TABLE 8.59
Principal Macrophases, Phases, and Tasks Identified in Hospital In-Patient Wards (Legend: OP = Operator; NC = Noncooperative; D = Disabled; PC = Partially Cooperative)

Macrophases, Phases, and Tasks in Hospital Wards		
Macrophase	**Phase**	**Task**
Beds and bed making	Preparation	Initial preparatory phase (pushing-pulling carts for D patient
		Don gloves for bedside activities
	Change bed	Completely change the bed (on height-adjustable bed) of nondisabled patient
		Make up the bed of nondisabled patient (without changing sheets)
		Completely change the bed (on height-adjustable bed) of D patient
		Make up the bed (on height-adjustable bed) of D patient (without changing sheets)
		Raise side rails on ergonomic bed of D patient
		Lower side rails on ergonomic bed of D patient
	Patient bed-bathing	Bathe the back of D patient
		Bathe the hands and face of bedridden D patient
		Change diaper
		Bathe private area of bedridden patient
		Administer enema to D patient (bedridden + NC) patient
		Dress disabled patient (bedridden and noncooperative)
	Medication	Administer simple medication to bedridden patient (bandages and bedsores)
	Pushing-pulling	Push medicine trolley
		Push medical records trolley
		Push medication trolley
		Push food trolley
		Push soiled linen trolley
		Fill and seal soiled linen bag
Beds and bed making	Patient handling	Pull D patient up in bed
		Pull up patient into sitting position—OP
		Turn bedridden patient over in electric bed to prevent bedsores
		Pull NC bedridden patient up in bed to prevent bedsores
Meals	Meal distribution	Distribute trays
		Feed bedridden patient
		Distribute bread
Therapy	Prepare and administer therapy	Prepare test tubes for exams
		Prepare and administer therapy
		Take blood sample
		Take urine sample
		Replace drip

- Patient bed-bathing and medication
- Pushing-pulling
- Patient handling

From a general point of view, the logical path of risk analysis used included the following preliminary phases:

- Interview with the head-nurse in order to identify and quantify the main variables (utilizing a protocol designed to standardize interviews)
- On-site inspection to check the accuracy of the statements made during the interview and to measure the *representative average duration of the various tasks*

Tables 8.60 through 8.62 shows the macrophases, phases, and tasks carried out in in-patient geriatric wards (in Italy defined as *Residenze Sanitarie Assistenziali*, or nursing homes), where it can be seen that there is a larger number of tasks associated with the cleaning and disinfecting of furnishings (known as *hotel duty* macrophases).

In order to better understand how tasks were identified and classified within each individual macrophase and phase, the procedure is illustrated using an example where the actual tasks are listed for each phase.

Table 8.63 shows how the *beds and bed making* macrophase encompasses six phases.

In the *bed making* phase (Figure 8.84) it was necessary to define the tasks in order to identify all the various ancillary subtasks, as well as the different procedures and techniques used.

8.7.5 Step-By-Step Presentation of the Risk Analysis Model for Biomechanical Overload of the Upper Limbs

We shall now illustrate the steps involved in acquiring the data for estimating exposure to biomechanical overload among operators caring for patients in hospital wards.

8.7.5.1 Step 1: Identification of Homogeneous Groups

The definition provided in Chapter 6, Section. 6.3.5 will be used here too. When a group of workers is assigned the same tasks with the same duration, it is possible to define them as a homogeneous group in terms of risk exposure.

For ward staff, homogeneous groups can only initially be defined based on their professional profile (i.e., nurses or nurse's aides, etc.) or the type of shifts they work.

At times, a homogeneous group may be composed of one person, if no other workers perform the same qualitative and quantitative tasks.

8.7.5.2 Step 2: Distribution of Shifts Assigned to a Specific Homogeneous Group

The second step consists of reconstructing the exact duration of the three shifts worked by a single homogeneous group, and the way the shifts are rotated over the

TABLE 8.60
Principal Macrophases, Phases, and Tasks Identified in In-Patient Geriatric Ward (Legend: OP= Operator; NC = Noncooperative; D = Disabled; PC = Partially Cooperative): Part 1

Macrophases, Phases, and Tasks in Geriatric Ward (RSA)—Part 1

Tasks	
Completely change height-adjustable bed without patient	Beds and bed making
Completely change non-height-adjustable bed without patient	
Make up bed (without changing mattress cover or sheets), without patient	
Completely change non-height-adjustable bed with D patient	
Completely change height-adjustable bed with <50 kg D patient + reposition	
Completely change height-adjustable bed with 50–70 kg D patient + reposition	
Completely change height-adjustable bed with >70 kg D patient + reposition	
Make up bed (non-height-adjustable) of D patient without repositioning	
Make up bed (height-adjustable) of D patient without repositioning	
Raise bedhead with 50 kg patient (bedridden and NC) and crank up new bed—knees bent	
Raise bedhead with 50–70 kg patient and crank up new bed—knees bent	
Raise bedhead with >70 kg patient and crank up new bed—knees bent	
Lower bedhead with hand crank—knees bent	
Lower bedhead with hand crank—standing	
Raise bedhead with old crank handle (poor maintenance), 50 kg patient	
Raise bedhead with old crank handle (poor maintenance), 50–70 kg patient	
Raise bedhead with old crank handle (poor maintenance), >70 kg patient	
Raise bedhead manually with D patient <50 kg	
Initial preparation phase (push-pull trolley—gloves, bathing materials, diaper, etc.) for all D patient beds	
Lower bedhead manually	
Slide in side rails with clip for PC patients	
Slide in side rails with clip for NC patients	
Lower side rails on electric bed	
Turn <50 kg NC patient over in bed for bathing—incorrect	
Turn 50–70 kg NC patient over in bed for bathing—incorrect	
Turn >70 kg NC patient over in bed for bathing—incorrect	
Bathe back of D patient	
Oral hygiene for D patient	
Hands and face of completely D bedridden patient	
Change diaper and bathe private parts of bedridden patient (without turning)	
Change diaper on <50 kg PC patient in manual wheelchair	
Change diaper on 50–70 kg PC patient in manual wheelchair	
Change diaper on >70 kg PC patient in manual wheelchair	
Change diaper on patient in wheelchair with hoist	
Administer enema to bedridden patient	

TABLE 8.61
Principal Macrophases, Phases, and Tasks Identified in In-Patient Geriatric Ward (Legend: OP= Operator; NC = Noncooperative; D = Disabled; PC = Partially Cooperative): Part 2

Macrophases, Phases, and Tasks in Geriatric Ward (RSA)—Part 2	
Completely bathe D patient in medical bathtub	Bathing
Completely bathe D patient in shower gurney	
Completely bathe D patient in medical shower	
Bathe PC patient in bathroom	
Dress D patient (bedridden + NC), <50 kg	Patient handling
Dress D patient (bedridden + NC), 50–70 kg	
Dress D patient (bedridden + NC) >70 kg	
Fill, close, and move soiled linen bag	
Completely undress and dress D patient, <50 kg	Bathing
Completely undress and dress D patient, 50–70 kg	
Completely undress and dress D patient, >70 kg	
Undress and basic hygiene	
Transfer to bathtub	
Personal hygiene	
Dry off and dress	
Put on elastic stockings	Patient handling
Care/medication	Therapy
Prepare and administer drugs to nondisabled patients	
Prepare and administer drugs to disabled patients	
Blood samples	
Replace drip	
IM injection, nondisabled patient	
IM injection, disabled patient	
Lift NC <50 kg patient manually from bed to wheelchair, cross-arm lift	Patient handling
Lift NC 50–70 kg patient manually from bed to wheelchair, cross-arm lift	
Lift >70 kg NC patient manually from bed t wheelchair, cross-arm lift	
Lift <50 kg NC patient from bed to wheelchair, under-arm lift	
Lift 50–70 kg NC patient from bed to wheelchair, under-arm lift	
Lift >70 kg NC patient manually from bed to wheelchair, under-arm lift	
Lift <50 kg NC patient from bed to wheelchair manually with change of posture	
Lift <70 kg NC patient from bed to wheelchair manually with change of posture	
Lift 50–70 kg NC patient from bed to wheelchair manually with change of posture	
Lift <70 kg NC patient from bed to wheelchair with electric hoist and change of posture	
Pull < 50 NC patient up in bed, cross-arm lift	
Pull 50–70 kg NC patient up in bed, cross-arm lift	
Pull >70 NC patient up in bed, cross-arm lift	
Lift <50 kg NC patient manually from wheelchair to bed, under-arm lift	
Lift 50–70 kg NC patient from wheelchair to bed, under-arm lift	
Lift >70 kg NC patient manually from wheelchair to bed, under-arm lift	

(*Continued*)

TABLE 8.61 (CONTINUED)
Principal Macrophases, Phases, and Tasks Identified in In-Patient Geriatric Ward (Legend: OP= Operator; NC = Noncooperative; D = Disabled; PC = Partially Cooperative): Part 2

Macrophases, Phases, and Tasks in Geriatric Ward (RSA)—Part 2
Pull NC patient up in bed with slide sheets (position sheets)
Pull NC patient up in bed with slide sheeds (no force)
Pull < 50 kg NC patient up in bed with draw sheet
Pull 50–70 kg NC patient up in bed with draw sheet
Pull >70 kg NC patient up in bed with draw sheet
Inadequate removal of sheet with D patient
Adequate removal of sheet with D patient
Move <70 kg PC patient from sitting to supine
Move >70 kg PC patient from sitting to supine
Move NC patient from stretcher to bed using slide board, 3 operators
Move NC patient from bed to stretcher with hoist
Manually turn 50 kg bedridden patient over in bed to prevent bedsores
Manually turn 50–70 kg bedridden patient over in bed to prevent bedsores
Manually turn >70 kg bedridden patient over in bed to prevent bedsores
Turn bedridden patient over in bed to prevent bedsores, using slide sheets

4 weeks of the month. It should be noted that in providing this organizational data, reference should be made to the *modal* shift distribution, or in other words, the most commonly worked shifts.

The shifts illustrated in Figure 8.85 are very common in hospital wards, where, in week three, staff work a double shift (morning + night).

Before 2004, in accordance with NHS regulations, ward staff in Italy generally rotated over three shifts, with a 36-h working week.

In 2004, the number of hours worked per week began to increase to compensate for chronic staff shortages.

Therefore, in wards with staff shortages, as shown in Figure 8.86, workers started working shifts even on days when they were supposed to be resting.

8.7.5.3 Step 3: Working Hours on Every Day of the Week and Distribution of Breaks

Since the study involves a weekly/monthly cycle, it is necessary to define the number and duration of breaks on every day of the month (per shift every day) as well as the number of nonrepetitive tasks.

Table 8.64 shows an example of the new organizational data for just 2 weeks of the month (instead of four, for lack of space).

Table 8.65 summarizes the duration of the weekly hours worked in hospital wards with staff shortages (as already shown in Figure 8.86).

In the first two weeks in particular, the percentage of hours worked in excess of the normal working week is quite significant.

TABLE 8.62
Principal Macrophases, Phases, and Tasks Identified in Geriatric Ward (Legend: OP = Operator; NC = Noncooperative; D = Disabled; PC = Partially Cooperative): Part 3

Macrophases, Phases, and Tasks in Geriatric Ward (RSA) Part 3	
Raise patient into sitting position to reposition pillows, and so on, 1 OP/<50 kg patient	Beds and bed making
Raise patient into sitting position to reposition pillows,and so on, 1 OP/50–70 kg patient	
Raise patient into sitting position to reposition pillows,and so on, 1 OP/>70 kg patient	
Stretcher to bed with draw sleet and slide sheets	Patient handling
Pull NC <50 kg bedridden patient up in bed to prevent bedsores, with draw sheet	
Pull NC 50–70 kg bedridden patient up in bed to prevent bedsores, with draw sheet	
Pull NC >70 kg bedridden patient up in bed to prevent bedsores, with draw sheet	
Bed to shower stretcher with incorrect draw sheet (<50 kg)	Bathing
Bed to shower stretcher with incorrect draw sheet (50–70 kg)	
Bed to shower stretcher with incorrect draw sheet (>70 kg)	
Bed tp shower stretcher with correct draw sheet	
Push bed with patient	Pushing/pulling
Push stretcher without patient	
Pull/push stretcher with patient	
Reposition NC patient in wheelchair, with sheets	Patient handling
Reposition NC <50 kg patient in wheelchair manually, cross-arm lift	
Reposition NC 50–70 kg patient in wheelchair, cross-arm lift	
Reposition NC >70 kg patient in wheelchair, cross-arm lift	
Raise PC patient from supine to standing with belt	
Move heavy PC patient from bed to wheelchair, 1 OP, with hoist	
Move <70 KG PC patient from bed to wheelchair, 1 OP, with hoist	
Set tables	Housekeeping duties
Distribute meals	
Distribute meal trays	
Distribute meals—teas, and so on.	
Feed patient in bed	
Feed bedridden patient	
Feed NC patient in wheelchair	
Sanitize tables	
Sanitize bedside cabinet	
Sanitize hoists	
Deliver water to rooms	
Prepare and slice fruit	
Put material away in kitchen	
Wash dishes	
Sanitize workbenches	
Prepare cart for setting tables	
Prepare pantry cart	
Prepare food cart	
Clean cart	
Clean bedside cabinet	
Clean bed	
Replace mattress	

TABLE 8.63
Phases Making Up the *Beds and Bed Making* Macrophase in Hospital In-Patient Wards

	Macrophase: Beds and Bed Making
Phase 1	Preparation
Phase 2	Change bed
Phase 3	Patient bed-bathing
Phase 4	Medication
Phase 5	Pushing-pulling operations
Phase 6	Manual handling of disabled patient

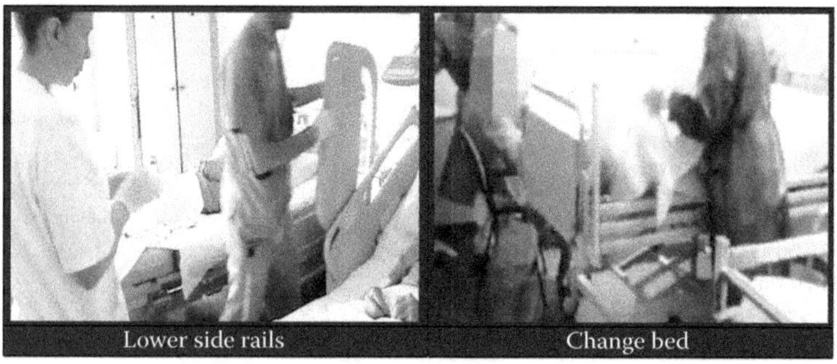

FIGURE 8.84 Change bed of partially cooperative patient: lower and raise side rails and change bed.

8.7.5.4 Step 4: Calculation of Net Duration of Repetitive Work Using Objective Variables

It is not easy to calculate the time actually spent performing repetitive tasks. Once the tasks performed during the shift by a certain homogeneous group were identified (i.e., the qualitative data), only vague and unreliable replies were given regarding the duration of these tasks (quantitative data). Therefore, other more certain variables were singled out in order to estimate task duration.

8.7.5.4.1 Organizational Data: Description of Staff and Shifts

The process will begin by defining the following:

- Number of workers belonging to a specific homogeneous group
- Breakdown of nurses or nurse's aides staff into pairs and number of pairs per shift

Table 8.66 shows that 11 staff members belong to a homogenous group. However, since some workers will be resting between shifts, there are only ever eight workers

Week 1	Indicate hours worked over 3 shifts	Mon	Tue	Wed	Thu	Fri	Sat	Sun
1st shift	7a.m.–2p.m.	×					×	
2nd shift	2p.m.–9p.m.		×					×
3rd shift	9p.m.–7a.m.			×				
Week 2	**Indicate hours worked over 3 shifts**	**Mon**	**Tue**	**Wed**	**Thu**	**Fri**	**Sat**	**Sun**
1st shift	7a.m.–2p.m.				×			
2nd shift	2p.m.–9p.m.					×		
3rd shift	9p.m.–7a.m.	×					×	
Week 3	**Indicate hours worked over 3 shifts**	**Mon**	**Tue**	**Wed**	**Thu**	**Fri**	**Sat**	**Sun**
1st shift	7a.m.–2p.m.		×				×	
2nd shift	2p.m.–9p.m.							×
3rd shift	9p.m.–7a.m.		×					
Week 4	**Indicate hours worked over 3 shifts**	**Mon**	**Tue**	**Wed**	**Thu**	**Fri**	**Sat**	**Sun**
1st shift	7a.m.–2p.m.				×			
2nd shift	2p.m.–9p.m.					×		
3rd shift	9p.m.–7a.m.	×					×	

FIGURE 8.85 Shifts distributed over a month in a hospital ward.

on three shifts over a 24-h period: four in the first shift, two in the second, and two in the third.

Staff generally work in pairs in the first shift.

8.7.5.4.2 Classification of Qualitative and Quantitative Variables Required to Objectively Measure Workload

Workloads can be measured in an objective manner by quantitatively and qualitatively defining the following variables:

- Number of beds (broken down by electrically height-adjustable, manually height-adjustable, or non-height-adjustable)
- Number of patients broken down by weight, bed type, and disability, such as
  - Cooperative (NA)
  - Totally noncooperative (NC)
  - Totally noncooperative and bedridden (NC + bedridden)
  - Partially cooperative (PC)

Week 1	Indicate hours worked over 3 shifts	Mon	Tue	Wed	Thu	Fri	Sat	Sun
1st shift	7a.m.–2p.m.	×					×	
2nd shift	2p.m.–9p.m.		×			×		×
3rd shift	9p.m.–7a.m.			×				
Week 2	Indicate hours worked over 3 shifts	Mon	Tue	Wed	Thu	Fri	Sat	Sun
1st shift	7a.m.–2p.m.				×			
2nd shift	2p.m.–9p.m.			×		×		
3rd shift	9p.m.–7a.m.	×					×	
Week 3	Indicate hours worked over 3 shifts	Mon	Tue	Wed	Thu	Fri	Sat	Sun
1st shift	7a.m.–2p.m.		×				×	
2nd shift	2p.m.–9p.m.					×		×
3rd shift	9p.m.–7a.m.		×					
Week 4	Indicate hours worked over 3 shifts	Mon	Tue	Wed	Thu	Fri	Sat	Sun
1st shift	7a.m.–2p.m.				×			
2nd shift	2p.m.–9p.m.					×		
3rd shift	9p.m.–7a.m.	×					×	

FIGURE 8.86 Shifts distributed over a month with staff shortage.

Returning to the example described in Table 8.66, which entailed identifying staff, the analysis then identifies the average number of patients generally present.

It is then possible to calculate *the number of patient units handled by each operator over a specific shift* (*patient unit/operator/shift*) either manually or automatically, if the appropriate software is used, based on the number of patients (broken down by type of bed and patient). For example, as shown in Table 8.67, with six nonbedridden patients in the first shift, each operator will handle three patients (in fact, in the first shift there are two pairs of operators), while in the second and third shifts they will handle six, since there is just one pair of operators.

Reference is made here to patients present in the ward with respect to a *modal day representative of the whole year.*

8.7.5.4.3 Determination of the Actual Average Duration of Each Repetitive Task Using Modal Units of Measurements

The following phases must be monitored to estimate the execution time of the various tasks:

TABLE 8.64
Description of Work Shifts and Break Distribution over Two Representative Weeks (Weeks 2 and 3)

Work Days and Shifts Per Day	Mon	Tue	Wed	Thu	Fri	Sat	Sun	Mon	Tue	Wed	Thu	Fri	Sat	Sun
1st Shift				X					X				X	
2nd Shift			X		X							X		X
3rd Shift	X					X			X					
	Week 1							Week 3						
	Mon	Tue	Wed	Thu	Fri	Sat	Sun	Mon	Tue	Wed	Thu	Fri	Sat	Sun
Shift duration (min)	600		420	420	420	600			1020			420	420	420
No. of contractual breaks (meal break not included)	1		1	1	1	1			2			1	1	1
No. of actual breaks lasting more than 7 consecutive minutes (meal break not included)	2		1	2	1	2			4			1	2	1
Duration of all breaks (meal break not included)	30		20	30	20	30			60			20	30	20

TABLE 8.65
Summary of Hours Worked per Week: Example of Situation in Hospital Wards with Staff Shortage

	Week 1	Week 2	Week 3	Week 4
Hours actually worked per week	45	41	38	34
Normal weekly hours as per contract	36	36	36	36
% Difference	+25%	+14%	+6%	–6%

TABLE 8.66
Breakdown of Staff per Shift in a Given Ward (Nurses or Nurse's Aides)

	No. of Pairs per Shift	No. of Workers per Shift	Total No. of Workers per Homogeneous Group	No. of Workers on Duty over 24 h
No. of pairs—1st shift	2	4	11	8
No. of pairs—2nd shift	1	2		
No. of pairs—3rd shift	1	2		

- Discussion with the head-nurse in the various hospital units regarding the *mean representative time required to perform a single task on a single patient* (based on personal experience).
- Observation and measurement of the execution time of individual tasks, per patient unit, by filming workers. Major difficulties were encountered here, due both to the small size of the rooms and also to privacy issues relating to the presence of patients.
- Measurement of the execution time of individual tasks, per patient unit, over various shifts, using a stopwatch, and definition of modal execution time.

Once the *modal time per task/patient unit* has been calculated, the analytical, logical process shown in Figure 8.87 can be used to estimate the actual duration of a certain task during the shift: dividing the number of patients requiring the task by the number of pairs of operators will produce the actual number of patients per operator. Thus, if the modal duration of the task and the number of repetitions (of the same task per patient) are known, it is possible to work out the net execution time of the task during the shift.

Table 8.68 offers an example of the final calculation of the net duration of repetitive work for each day of the week: Here, the difference between the calculated net duration (i.e., the sum of the duration of all the tasks performed using the method proposed in Figure 8.87) and the estimated net duration (*shift duration–break duration–nonrepetitive tasks = estimated net repetitive task duration*) is also reported.

TABLE 8.67
Number and Description of Patients Cared for by One Pair of Operators (Modal Data Representative of the Ward in Our Example

Bed Type	Total	List of Quantifiable Elements	No. of Patients per Bed Type	No. of Patients Handled per Shift by One Pair of Operators		
				Shift 1 (Two Pairs of Operators on Duty)	Shift 2 (One Pair of Operators on Duty)	Shift 3 (One Pair Operators on Duty)
Electric height-adjustable beds with manual bedhead	28	Number of cooperative (NA) patients	6	3	6	6
		No. of NC patients weighing less than 50 kg	1	0.5	1	1
		No. of NC patients weighing 50–70 kg	4	2	4	4
		No. of NC patients weighing more than 70 kg	3	1.5	3	3
		No. of PC patients weighing less than 70 kg	7	3.5	7	7
		No. of PC patients weighing more than 70 kg	4	2	4	4
		No. of bedridden NC patients weighing less than 50 kg	2	1	2	2
		No. of bedridden NC patients weighing 50–70 kg	1	0.5	1	1
		No. of bedridden NC patients weighing more than 70 kg		0	0	0
		Total	28			
Patients requiring medication			6	1.5	3	3
Patients requiring feeding			8	2	4	4
No. of carts for beds and bed making			2	0.5	1	1

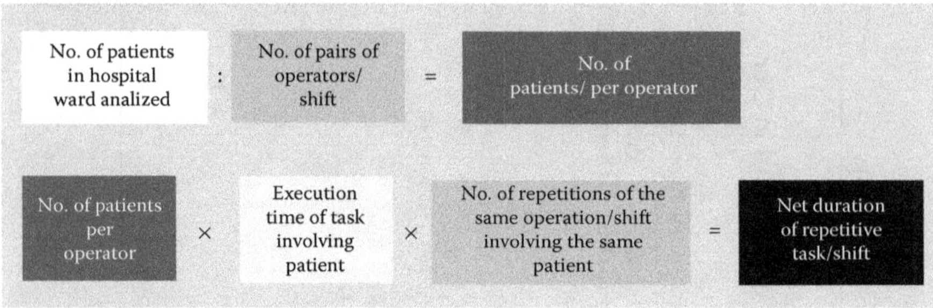

FIGURE 8.87 Analytical rationale for determining the mean representative execution time of a task during the shift.

TABLE 8.68
Final Calculation of the Net Duration of Biomechanical Overload of the Upper Limbs for Each Day of the Week

		Organizational Data for a Homogeneous Group						
		Mon	**Tue**	**Wed**	**Thu**	**Fri**	**Sat**	**Sun**
Organizational Study of Week 1 in a Representative Month								
Shift	**Shift Duration in Minutes**	420	420	600	0	420	420	420
Nonrepetitive task	Total duration (min) of nonrepetitive tasks in a shift	30	15	30	0	30	30	15
	% Duration of nonrepetitive tasks in a shift	7%	4%	5%		7%	7%	4%
Breaks	Total duration of breaks in the shift (min)	30	20	30	0	20	30	20
Estimated vs. calculated duration	Estimated net duration of repetitive tasks	360	385	540	0	370	360	385
	Calculated net duration of repetitive tasks	215	341	286	0	341	215	341
	% Difference	67%	13%	89%		9%	67%	13%

8.7.5.5 Step 5: Description of Tasks and Task Duration

As illustrated in Chapter 6.4, all that remains now is to assign the various tasks to the homogeneous group, with an indication of the following dimensions:

- Number of repetitions of the specific task per shift
- Preassessed modal duration of the individual task
- Precalculated number of patients cared for per shift and per pair of operators
- Intrinsic OCRA checklist score for each task (see Step 6)

Table 8.69 shows an example of tasks performed by a homogeneous group: to ensure adequate legibility, the example is a fairly simple one with a relatively small number of tasks.

TABLE 8.69
Example of Tasks Performed by a Homogeneous Group over Three Shifts

Tasks	No. of Repetitions of the Same Task in a Shift			No. of Patients Treated in a Shift by a Pair of Nurse's Aides			Preassigned Time Unit Duration = Seconds per Patient and per Task
	Shift 1	Shift 2	Shift 3	Shift 1	Shift 2	Shift 3	
Completely change height-adjustable bed of nondisabled patient	1			3	6	6	170
Completely change height-adjustable bed of D patient	1			11	22	22	202
Make up bed of D patient (without changing sheets)	1	2	3	11	22	22	90
Raise bedhead manually with <50 kg D patient	4	4	4	1,5	3	3	10
Raise bedhead manually with 50–70 kg patient	4	4	4	6	12	12	10
Raise bedhead manually with >70 kg patient	4	4	4	3,5	7	7	10
Lower bedhead manually	4	4	4	11	22	22	8
Turn <50 kg NC patient over in bed for bathing, INCORRECT	5	3	6	1,5	3	3	4
Turn 50–70 kg NC patient over in bed for bathing, INCORRECT	5	3	6	2,5	5	5	4
Turn >70 kg NC patient over in bed for bathing, INCORRECT	5	3	6	1,5	3	3	4
Bathe the back of D patient	1			5,5	11	11	90
Oral hygiene for D patient	1			1,5	3	3	120
Bathe hands and face of completely bedridden D patient	1			1,5	3	3	141
Change diaper and bathe private parts of bedridden patient (without turning)	2	2	3	5,5	11	11	180
Dress <50 kg D patient (bedridden + NC)	1	1		1,5	3	3	85
Dress 50–70 kg D patient (bedridden + NC)	1	1		2,5	5	5	85
Dress >70 kg D patient (bedridden + NC)	1	1		1,5	3	3	85
Fill, close, and move soiled linen bag	1	1	1	1	1	1	39
Simple medication for bedridden patient	2	2	2	1,5	3	3	109
Lift >70 kg NC patient from bed to wheelchair manually with CHANGE OF POSTURE	2	2		1,5	3	3	190
Lift 50–70 kg NC patient from bed to wheelchair manually with CHANGE OF POSTURE	2	2		2	4	4	190

(Continued)

TABLE 8.69 (CONTINUED)
Example of Tasks Performed by a Homogeneous Group over Three Shifts

Tasks	No. of Repetitions of the Same Task in a Shift			No. of Patients Treated in a Shift by a Pair of Nurse's Aides			Preassigned Time Unit Duration = Seconds per Patient and per Task
	Shift 1	Shift 2	Shift 3	Shift 1	Shift 2	Shift 3	
Pull <50 kg NC patient up in bed with draw sheet	3	3	4	1,5	3	3	40
Pull 50–70 kg NC patient up in bed with draw sheet	3	3	4	2,5	5	5	40
Pull >70 kg NC patient up in bed with draw sheet	3	3	4	1,5	3	3	40
Move <70 kg PC patient from bed to wheelchair	2	2		3,5	7	7	55
Move >70 kg PC patient from bed to wheelchair	2	2		2	4	4	55
Manually turn 50 kg bedridden patient over in bed to prevent bedsores	3	3	6	1	2	2	13
Manually turn 50–70 kg bedridden patient over in bed to prevent bedsores	3	3	6	0,5	1	1	13
Manually turn >70 kg bedridden patient over in bed to prevent bedsores	3	3	6	0	0	0	13
Raise patient into sitting position to reposition pillows and so on, 1 OP/<50 kg patient	3	3	3	1,5	3	3	8
Raise patient into sitting position to reposition pillows and so on, 1 OP/50–70 kg patient	3	3	3	6	12	12	8
Raise patient into sitting position to reposition pillows and so on, 1 OP/>70 kg patient	3	3	3	3,5	7	7	8
Pull >50 kg NC bedridden patient up in bed to prevent bedsores, with draw sheet	3	3	2	1	2	2	40
Pull 50–70 kg NC bedridden patient up in bed to prevent bedsores, with draw sheet	3	3	2	0,5	1	1	40
Pull >70 kg NC bedridden patient up in bed to prevent bedsores, with draw sheet	3	3	2	0	0	0	40
Lift <50 kg NC patient from bed to wheelchair manually with change of posture	2	2		0,5	1	1	190
Distribute trays	1	1		14	28	28	7
Feed bedridden patient	1	1		2	4	3	600

8.7.5.6 Step 6: Calculation of Intrinsic Risk Indexes for Each Task

As in all multitask assessments, the risk associated with all of the identified tasks is calculated using the OCRA checklist; again, it is worth noting that if different methods are used to perform the same operations, it should be counted as a separate task. The intrinsic risk index was calculated for all of the listed tasks, estimated as if the task had been carried out for at least 420–440 min in the shift, with one meal break and two 10-min breaks (see Chapter 6, Section 6.4).

In the health-care sector, one of the risk factors associated with patient handling tasks is the use of strength; therefore, the Borg scale was used to measure perceived exertion in relation to patient weight. Figure 8.88 shows two examples of tasks involving pulling up the patient in bed, here defined as Task A, with a disabled patient weighing 50–70 kg, and Task B, with a disabled patient weighing over 70 kg. The results refer to the right upper limb.

8.7.5.7 Step 7: Calculation of Final Risk Index

The first approach is based on the time-weighted average model. The second is the multitask complex, a method that takes the worst task score (calculated with respect to its actual duration) as the starting exposure level for all the other tasks (always with respect to their duration). Both calculation methods were used to assess multitask exposure risk with monthly rotations in the health-care sector.

8.7.6 Presentation of Final Assessments for a Representative Italian Sample Population

Some results relating to the in-patient ward described in the previous tables follow.

Figure 8.89 shows the risk indexes obtained with the OCRA checklist (using the time-weighted average formula) for each day of a representative month. The

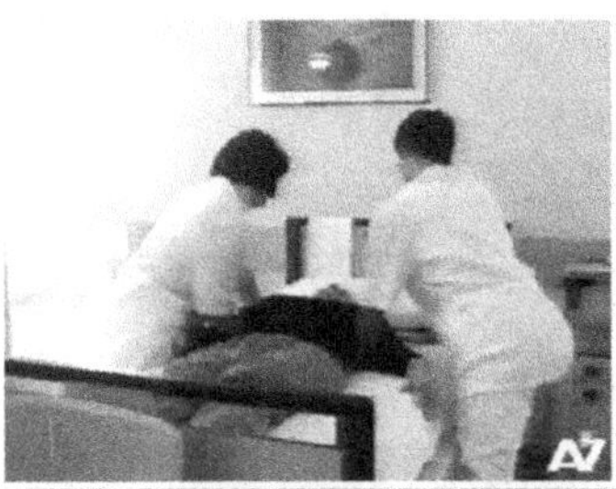

Task	Recovery multiplier	Recovery score	Frequency	Force	Side	Shoulder	Elbow	Wrist	Hand	Stereotype	Total posture	Additional	OCRA checklist
A	1.33	4	0	10	Right			1.5			1.5		15.3
B	1.33	4	0	26	Right			1.5			1.5		36.6

FIGURE 8.88 OCRA checklist intrinsic risk scores for pulling patients of different weights up in bed.

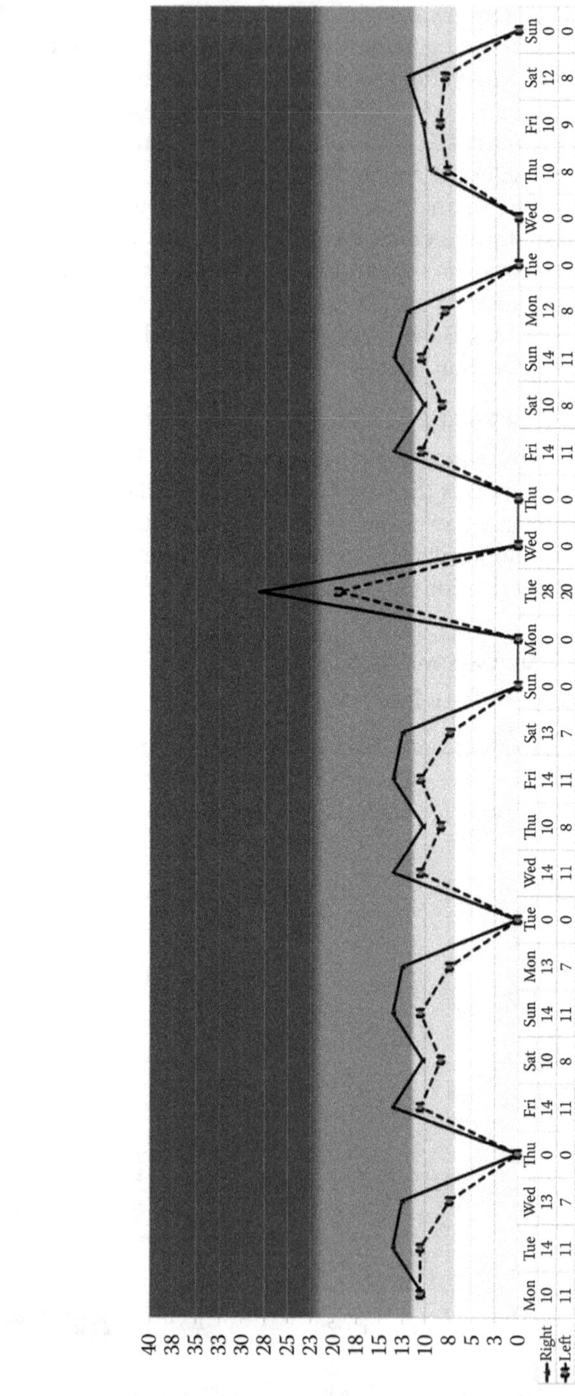

FIGURE 8.89 Risk-assessment results obtained with the OCRA checklist (time-weighted averages) over a representative month, for each day of the week and for the right and left upper limbs.

	Right	Left
Week 1	12.3	9.3
Week 2	10.4	7.2
Week 3	10.8	8.1
Week 4	8.1	6.0

	Right	Left
Week 1	12.8	9.7
Week 2	10.8	7.4
Week 3	11.2	8.2
Week 4	8.4	6.1

Right	Left
9.8	7.2

Weighted average for one month

Right	Left
10.2	7.3

Multitask complex for one month

FIGURE 8.90 Result of risk assessment: summary of risk per week and for the entire month using the time-weighted average and the multitask complex formula.

risk level "peaks" in the third week are due to the schedule entailing two shifts over 24 h.

Risk levels peak on Tuesday, which is when workers do a double shift lasting a total of over 1000 min (see Table 8.64).

In Figure 8.90, the results for each of the 4 weeks in the month and for the month as a whole using both of the calculation methods described above. There is, of course, a slight difference between the scores obtained using the time-weighted average formula versus the multitask complex. This is because the multitask complex result is affected by tasks having quite different intrinsic risk levels, ranging from none to very high risk, but also because peak force is used, and on certain days the degree of overload is extremely high.

The risk of biomechanical overload of the upper limbs is borderline red/yellow in this first situation. Although certain tasks have a high intrinsic risk score, the repetitive tasks in the example shown here have a relatively short net duration (see Figure 8.91).

8.7.7 Conclusions and Practical Recommendations

As in other sectors, assessing risk associated with biomechanical overload in healthcare workers is anything but simple. Acquiring the necessary information is made all the more challenging by factors such as the large number of tasks involved, difficulties in obtaining information, the wide variety of techniques and methods used, possible complications in filming workers, and so on. However, as stated in the introduction to this chapter, since the number of potentially work-related musculoskeletal disorders is quite large, it is essential to assess exposure risk levels, and to do so as objectively as possible, and to analyze and then redesign how the highest risk tasks are performed and/or introduce appropriate aids in order to minimize risk.

In this chapter, we have illustrated how to go about doing this by calculating net exposure duration for each repetitive task based on the *organizational variables* described previously.

The methodology requires the OCRA checklist score to be calculated for each task and for both upper limbs (i.e., an estimation of the intrinsic task values). Tasks

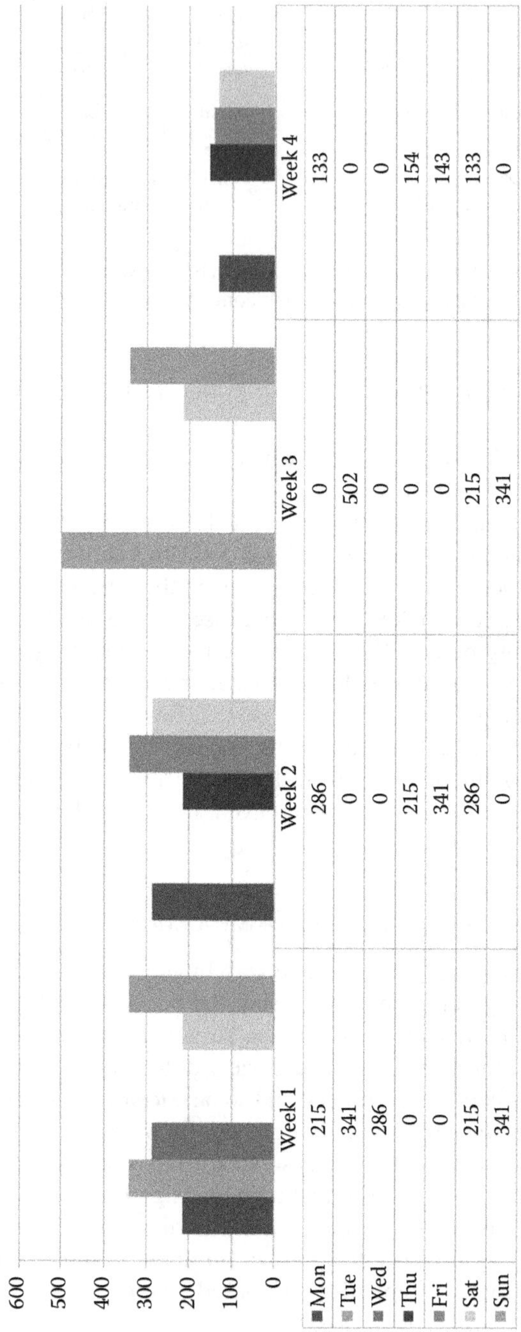

	Week 1	Week 2	Week 3	Week 4
Mon	215	286	0	133
Tue	341	0	502	0
Wed	286	0	0	0
Thu	0	215	0	154
Fri	0	341	0	143
Sat	215	286	215	133
Sun	341	0	341	0

FIGURE 8.91 Net duration of repetitive tasks on each day of a representative month.

must therefore be correctly analyzed, but first, and more importantly, they must be identified with the utmost precision.

We are often asked whether "repetitive movements" even exist in health-care settings.

If one broadens the concept of repetitive tasks to include biomechanical overload of the upper limbs, then they certainly do. Hospitals feature *patient cycles* consisting of countless tasks that are in fact constantly repeated, such as changing bed linen, pulling patients up in bed, transferring patients to wheelchairs, and so on. On their own, none of these individual tasks can be defined as repetitive, but as a whole they may definitely give rise to biomechanical overload conditions.

In terms of job structure, there are no differences between hospital workers and cleaners or kitchen workers.

At this point in time, the two final risk scores (i.e., time-weighted average and multitask complex) represent a hypothetical "exposure range" for the upper limbs reflecting both the extent of the various risk factors in relation to the actual task duration and the presence of actual risk peaks, which are here generated particularly by the use of peak force (such as when manually handling disabled patients) and by extremely long working hours.

What still remains to be done? A great deal, of course, but the first step ought to be to review and integrate the criteria for identifying the *modal duration of each task per patient unit* in all hospital wards.

The ultimate aim is to create a knowledge base and a tool for calculating exposure risk without undue effort in the earliest stages of assigning tasks to nursing staff. Such a tool will provide predefined *intrinsic risk indexes* and *modal duration units for each task* and *patient unit*. By entering only the organizational data concerning a given specific situation, it would then be possible to assess risk even before work is assigned to individual employees.

It may still be early days but we're making great progress!

9 The OCRA Index

9.1 GENERAL FRAMEWORK FOR THE ANALYSIS OF EXPOSURE TO REPETITIVE MOVEMENTS: DEFINITIONS, ORGANIZATIONAL ANALYSIS AND GENERAL CALCULATION MODEL

The occupational repetitive actions (OCRA) index is the most precise method for analyzing and evaluating the risk of exposure to repetitive tasks. The method was later followed by various simplified versions, including the OCRA checklists. This book is primarily focused on these simplified analytical models, but this chapter devoted to the original method will, hopefully, enable more expert users to achieve even higher levels of precision, especially when redesigning tasks or, better still, in the design stage itself. The OCRA index is a particularly useful tool for those involved in designing the content and duration of work cycles. The index will help them to monitor and manage not just productivity but also risk levels, the likelihood of occupational diseases and disorders, and therefore also costs. The OCRA method offers the opportunity to conduct an actual *cost-benefit analysis*.

Based on guidance gleaned from the most authoritative literature on the subject, as already discussed for the OCRA checklist in Chapter 4, it can be stated that in order to describe and assess work entailing potential biomechanical overload of the upper limbs, it is necessary to first identify and then quantify the following leading risk factors, which as a whole characterize occupational exposure in relation to their respective *duration within the overall repetitive work*:

- High frequency of action
- Excessive use of force
- Awkward and/or stereotyped posture, and movements of the upper limbs
- Additional risk factors
- Lack of adequate recovery periods

Table 9.1 reports the main terms and definitions used for analyzing repetitive tasks with the OCRA method.

The OCRA index is produced by the ratio of the absolute number of *actual technical actions (ATA) currently performed* in a work shift to the corresponding number of *recommended technical actions (RTA)*.

In practice:

$$\text{OCRA} = \frac{\text{ATA}}{\text{RTA}}$$

TABLE 9.1
Main Definition of Terms Used Commonly in Risk Assessments

Organizational Definitions

Job or Organized Work

An organized set of work activities, performed in a shift or work period; it can be composed of one or more tasks, and is also known as a *job*.

Task

A specific work activity aimed at obtaining a specific result (e.g., seaming an article of clothing and loading and unloading a pallet).

Tasks may be defined as:

Repetitive: *When actions of the upper limbs are characterized by cycles (regardless of duration) or the same gesture is repeated for most of the shift (i.e., more than half).*

Nonrepetitive: *When the actions performed by the upper limbs are noncyclic.*

Cycle

A sequence of technical actions performed by the upper limbs, which is repeated several times, always the same way.

Cycle Time

The total time assigned to carry out the sequence of technical actions that characterizes the cycle: this includes action, downtime, and any other parameters used to determine the cadence (i.e., pace of the assembly line).

Technical Action

An action involving the upper limbs. Not an individual joint movement but a set of movements performed by one or more body segments that together serve to complete a single operation.

Main Risk Factors Evaluated by the OCRA Method

Frequency

Number of technical actions per unit of time (number of actions per minute).

Force

Physical effort required by the worker to perform a technical action.

Awkward Postures and Movements

Nonneutral postures and movements of the main joints of the upper limbs adopted to complete a sequence of technical actions characterizing a cycle. Risk factors are determined by the presence of awkward postures and movements for a significant length of time.

Stereotypy

The repetition of the same gesture or series of gestures for most of the work period or shift.

Lack of Recovery Periods

A recovery period is the time within a shift during which the upper limbs are substantially inactive (i.e., the limbs are not performing any technical actions). Risk exists if there are no recovery periods or if they are too short or poorly distributed.

Additional Factors

These factors are not always present in repetitive tasks. If present, they increase the level of overall risk depending on their type, intensity, and duration.

The overall number of ATA performed per shift is a known quantity and can be reconstructed by means of an organizational analysis. At any rate:

$$\text{ATA} = \sum (F_j \times D_j)$$

where:

D_j = net duration (in minutes) of task j
F_j = average frequency of action per minute of task j

The following general formula is used to calculate RTA (*overall number of RTA per shift*):

$$\text{RTA} = \sum_{J=1}^{n} \left[\text{CF} \times (\text{Fo}_{Mj} \times \text{Po}_{Mj} \times \text{Re}_{Mj} \times \text{Ad}_{Mj}) \times D_j \right] \times (\text{Rc}_M \times \text{Du}_M)$$

where:

n = number of repetitive tasks per shift
j = generic (j-th) repetitive task performed by the upper limbs
CF = RTA frequency (30) per minute under reference conditions
Fo_{Mj}; Po_{Mj}; Re_{Mj}; Ad_{Mj} = multipliers, chosen in relation to the behavior of the force, posture, repetitiveness (stereotypy), and additional risk factors embedded in each j-th task under examination
D_j = duration (in minutes) of each jth repetitive task
Rc_M = multiplier for the "lack of recovery period" risk factor (one only for jobs with 1 or more repetitive tasks)
Du_M = multiplier that takes into account the net duration of repetitive tasks (one only for jobs with 1 or more repetitive tasks)

Once the score for each risk factor has been calculated, the relevant multiplier is then identified for each individual risk factor.

If there are no problems, the multiplier will be 1 and the recommended number of actions will not change. As the level of risk increases, the multiplier will proportionally approach 0, thus reducing the number of RTA.

The lower the number of RTA, the higher the final risk index.

9.2 ANALYSIS OF ORGANIZED WORK: SHIFT CONTENT, DISTRIBUTION, AND DURATION OF BREAKS, CALCULATION OF THE NET DURATION OF REPETITIVE WORK

The first step in analyzing organized work in order to determine whether a worker is exposed to the risk of biomechanical overload of the upper limbs, consists of identifying the following aspects of the shift:

- Working hours
- Task(s) performed per shift
- Presence of scheduled breaks or other stoppages of the worker's activities
- Presence of significant waiting times or idle time within the work cycle

9.2.1 Working Hours

The description of a work day includes the way shifts are distributed (Chapter 2, Figure 2.10): A one-shift schedule (i.e., one daily shift generally lasting 8 h or 480 min) with an unpaid meal break in the middle that is not included in the working hours), or a schedule with several shifts (two, three, or even four).

In order to conduct a detailed analysis of the duration of exposure to repetitive work, it is useful to evidence not only the official start and end of the shift, but also the actual working hours. From time to time, workers may begin working 5–10 min after clocking on, for instance when they have to don work clothes, or, more commonly, workers finish working sometimes many minutes before closing time.

Figure 9.1 reports the first part (with Example A) of the data sheet provided for calculating the risk index; this same example will be used for all the risk factors making up the index.

9.2.2 Identification of Tasks Present in the Shift

The first step in assessing exposure to biomechanical overload of the upper limbs in the workplace among individual workers or a homogeneous group of workers is to identify the presence of *repetitive tasks*, namely, tasks characterized by

- Work cycles featuring technical actions that involve the upper limbs.
- Tasks that entail repeating the same gesture for more than half the period or shift.

Workers may perform one or more repetitive tasks during the shift: Each task must be singled out and described in terms of its *net duration (in minutes) within the shift.*

Similarly, all *nonrepetitive tasks* must be singled out and quantified in terms of their duration in minutes within the shift. Such tasks must be manual but may be occasional, such as replenishing supplies, preparing, cleaning, or transporting. These

Shift description			
x	X = daily shift		
	Official start	6:00 am	
	Actual start	6:10 am	7:40
	Actual finish	1:50 pm	
	Official finish	2:00 pm	Effective shift duration in hours

FIGURE 9.1 Example A: description of shift duration.

tasks are not classified as repetitive but are not counted as either breaks or recovery periods.

If certain specific tasks, such as replenishing supplies or cleaning (when their intrinsic features classify them as repetitive) are performed for over 60 min, they must be considered as additional repetitive tasks and assessed using the OCRA index.

There are other tasks (e.g., *visual inspections*) that do not require the use of the upper limbs. These tasks can be regarded as *recovery periods for the upper limbs* and must be carefully quantified in terms of their frequency and duration.

9.2.3 Presence of Breaks and/or Stoppages

It is essential to study not only the total duration of *contractual* breaks—that is, meal and bathroom breaks—and other stoppages, but also the actual duration of individual stoppages and their distribution within the shift.

If breaks and/or stoppages are not scheduled, then it is important to report at least the modal behavior of workers during the shift. This data can be obtained by direct observation or by interviewing a significant sample of workers.

Rest and/or physiological factors, and any other elements that increase the cycle time must be highlighted, but will be counted as recovery periods only if the break and/or stoppage last at least 8–10 consecutive minutes.

Figure 9.2 shows the distribution and duration of all official and unofficial breaks and stoppages during the shift (Example A).

The total duration of the recovery periods must be subtracted from the total duration of the shift in order to obtain the net duration of repetitive work.

9.2.4 Presence of Waiting or Idle Time

Machine waiting times (or idle time) should in general not be counted, but included in the *total cycle time* (as defined in the OCRA index) or *pace* (as defined in the

Distribution of breaks				
Nr.	Start	Finish	Minutes	Notes
1	9:30	9:46	0:16	
2			0:00	
3			0:00	
4			0:00	
5			0:00	
6			0:00	
7			0:00	
Total			0:16	

Meal break			
Official start		11:50	
	Actual start	11:45	
	Actual finish	12:30	0:45
Official finish		12:20	

FIGURE 9.2 Example A: description and duration of breaks.

industry by technical staff), unless idle time recurs cyclically and lasts at least 10 consecutive seconds per minute of repetitive work (minimum ratio of work to recovery period 5:1). In this case the idle time may satisfy the need for recovery periods within the cycle (see Section 9.2.6 on recovery periods).

9.2.5 Calculation of Net Duration of Repetitive Work and Corresponding Duration Multiplier (Du_M)

Once the shift content has been estimated it is possible to calculate the *net duration of repetitive work* (Figure 9.3). The figure shows a fictitious example (Example A) describing: Duration of the shift, distribution of scheduled breaks and/or stoppages, presence and duration of meal breaks, presence of activities that qualify as recovery periods, and the presence and duration of nonrepetitive manual tasks.

The overall duration of tasks involving repetitive movements of the upper limbs represents a significant aspect of overall risk exposure.

The model for calculating the risk index is based on scenarios in which the worker performs repetitive manual tasks for a significant proportion of the shift (approx. 7–8 h). However, since situations may differ quite considerably from this *typical* scenario, such as regular overtime, part-time work, performing repetitive manual tasks for only a part of the shift, a specific multiplier has been added that takes deviations from the more routine exposure conditions into account.

Table 9.2 provides the parameters for dealing with the duration multiplier (Du_M): It should be noted that the time in minutes indicated in the table is the sum of the time spent during the shift performing all upper limb repetitive tasks.

Initially, the multiplier scores (Du_M) were chosen on the basis of the literature, although the recommendations were purely empirical (CEN, 1993; Moore and Garg, 1995); later, a special function was used to interpolate *base* scores provided for 1, 2, 4, and 8 h, and for shorter or longer duration scenarios.

Shift description		Minutes	Minutes
Shift duration	Official	480	460
	Actual	460	
Breaks or stoppages	Official	10	16
	Actual	16	
Meal break	Official	30	45
	Actual	45	
Time counted as recovery period	Due to recovery within the cycle	0	0
	Due to other causes	0	
Nonrepetitive tasks	Official	20	20
	Actual	20	
Net duration of repetitive tasks			379

FIGURE 9.3 Example A: calculation of net duration of repetitive tasks.

TABLE 9.2

Elements for Determining the Overall Duration Multiplier (in Minutes) for Repetitive Tasks per Shift (DuM)

Minutes Performing Repetitive Tasks/Shift	≤120	121–180	181–240	241–300	301–360	361–420	421–479	480–540	541–600	601–660	661–720	>720
Duration multiplier for total net duration of repetitive tasks/shift (DU_M)	2	1.7	1.5	1.3	1.2	1.1	1	0.83	0.66	0.5	0.35	0.25

9.2.6 Calculating Presence and Distribution of Recovery Periods and the Corresponding Recovery Period Multiplier (Rc_M)

A recovery period can be defined as a period in which one or more muscle-tendon units that are generally involved in performing working tasks are instead largely inactive.

The following may be regarded as recovery periods:

1. Official and unofficial breaks, including meal breaks (whether or not they are part of the work schedule and thus are paid).
2. Periods of time spent performing tasks during which muscles that are generally involved in performing working tasks are instead largely inactive (e.g., visual inspections or tasks performed using one limb at a time).
3. Periods within the cycle during which muscles that are otherwise active are completely at rest. In order for periods such as these (inspections, waiting, idle time, etc.) to be counted as significant, they must be regular and last at least 10 consecutive seconds per minutes.

Consequently, an analysis of recovery periods must begin by verifying whether they are present or absent, and if present, their duration and distribution first within the cycle, and then, at the macro level, over the entire shift.

The description and/or assessment of recovery periods should be based on:

1. Description of the actual sequence of repetitive tasks, nonrepetitive tasks and breaks during the shift.
2. Frequency and duration of recovery periods within the cycle and the shift, whether official or unofficial, as taken by most workers (*modal behavior*).

There is no agreement in the literature as to how to assess recovery periods: Bystrom (1991) made a valuable contribution in suggesting models for designing the ideal ratio of work to rest for intermittent static muscle actions (this being a duration of about 3–5 s).

However, there is a dire shortage of science-based guidance concerning recovery periods for dynamic repetitive movements (as in most work environments).

A useful *rule of thumb* comes from the Australian approach toward the prevention of repetitive strain injuries (RSI). A draft issued by the Australian Health and Safety Commission (1988), states that work involving repetitive movements lasting more than 60 min without recovery periods is unacceptable. A general rule is provided recommending a ratio of work (with repetitive movements) to recovery period of 5:1. While 4:1 might be acceptable, 10:1 would not. A similar recommendation also appears in documents released in the United States by the ACGIH (2002), indicating breaks of about 10 min every hour for repetitive manual work.

The critical adoption of both guidelines, which seem perfectly reasonable, based on current knowledge and other indications in the literature, will enable users to interpret the descriptive data concerning the sequence, duration and frequency of recovery periods within cycles of prevalently dynamic work.

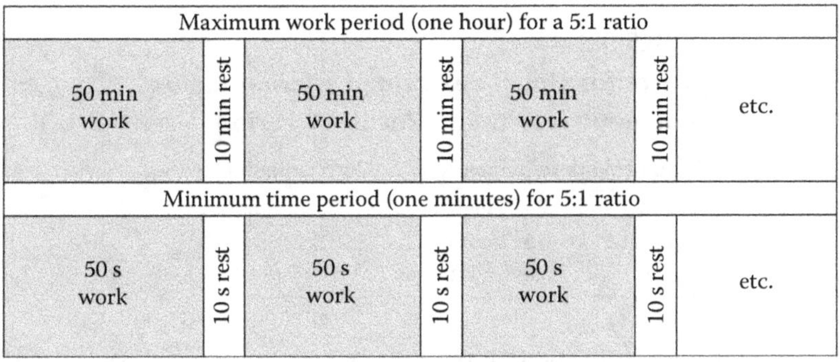

FIGURE 9.4 Maximum and minimum duration of repetitive work and recovery period in a 5:1 ratio.

In the case of repetitive work, most of the tasks entailing technical actions will consist of *movements* (rather than static actions). Based on the aforementioned recommendations, workers performing repetitive tasks should *ideally* have a recovery period every 60 min, with the ratio of work to rest being 5:1. Hence, 50 min of work followed by 10 min of recovery (Figure 9.4).

On the basis of this optimal distribution, criteria can be put together in order to use *scores* for assessing and thus classifying risk due to the absence or poor distribution of recovery periods. Importantly, this procedure does not call for the adoption of this exact allocation of work versus recovery period in all jobs featuring repetitive tasks. Adopting it simply provides a benchmark for obtaining a risk score for this factor.

To begin with, the assessment does require an accurate and correct examination of the work itself, as indicated earlier.

A straightforward analytical procedure has been formulated to determine scores, based on an observation of each individual hour making up the shift, and on verifying whether each hour includes repetitive tasks or not, and if so, if there are adequate recovery periods.

It is worth emphasizing that for the hours immediately prior to a meal break (lasting no less than 30 min, otherwise it is counted as a regular break) or the end of a shift the recovery period will be determined by these two events (i.e., meal break and end of shift).

Depending on whether or not there are adequate recovery periods within each hour of repetitive work, the hour will be ranked as *risk free* or *at risk* (due to lack of recovery periods). The OCRA method determines this factor based on the total number of hours at risk (generally between 0 and 6). More specifically, if the ratio of work to recovery period within each hour of repetitive work is between 5:1 and 6:1, the hour is considered to be risk free (score 0); if the ratio is between 7:1 and 11:1 the score will be 0.5.

If the ratio of work to recovery is above 11:1, the score will be 1 because the ratio is regarded as unacceptable (Table 9.3).

A few examples are shown here of recovery periods with different distribution patterns and the relevant scores.

Table 9.4 shows how recovery periods are distributed over a single 8 h shift with one lunch break and two 10-min breaks, one in the morning and one in the afternoon.

TABLE 9.3
Scores for Hours Featuring Different Ratios of Repetitive Work to Recovery Period

Ratio of Work to Recovery	Corresponding Score
From 5:1 to 6:1 (8–10 min)	Score = 0
From 7:1 to 11:1 (5–7 min)	Score = 0.5
Over 11:1 (less than 5 min)	Score = 1

In this case, the score for the risk posed by a lack of recovery period will be 4. This score indicates how many hours per shift are without adequate recovery.

If an 8 h shift with one lunch break had no other breaks at all then the maximum score would be 6; this is because the hour followed by a lunch break and the last hour in the shift are followed by adequate recovery periods and thus classified as *risk free*.

Table 9.5 shows how recovery periods are distributed over another 8 h shift, also with one meal break and two 10-min breaks, one in the morning and one in the afternoon. In this case, however, the two 10-min breaks do not fall at the end of the hour but within the hour.

Even if the break falls within the hour rather than at the end of the hour, the score is still 4.

The general rule, in order to simplify scoring, states that when there is a long enough break within the hour, the recovery period is regarded as adequate regardless of when it falls.

The third example is of a single 8 h shift with one lunch break and two 15-min breaks, one in the morning and one in the afternoon, both falling within the hour. While both breaks may be longer, the final score is still 4. In order to minimize risk, it is generally advisable to avoid amassing recovery periods into a small number of breaks, but rather to spread them as evenly as possible over the shift. This means avoiding breaks in the hour prior to the lunch break or in the last hour of the shift, as these hours already have adequate recovery periods. Here, changing the two 15-min breaks into three 10-min breaks (with the same total duration) will lower the risk score by one point, that is, from 4 to 3.

The fourth example is of a single 8 h shift with one lunch break and two 20-min breaks, one in the morning and one in the afternoon, both straddling 2 h. Although the breaks in this example are longer and straddle 2 h, the final score is still 4. This can be explained by the fact that breaks lasting 20 consecutive minutes are classified as having adequate recovery periods for only one of the hours that they straddle, not both. Here too, to reduce risk it is advisable to avoid amassing recovery periods into a small number of prolonged breaks: It is preferable to spread them as evenly as possible over the length of the shift, and avoid scheduling breaks in the hour before the lunch break or the last hour of the shift as they already have adequate recovery periods. Here, changing the two 20-min breaks into four 10-min breaks (with the same total duration) lowers the risk score by two points, that is, from 4 to 2.

The fifth example shown in Table 9.6 is of another 8 h shift with one 30-min meal break included in the shift and two 10-min breaks, one in the morning immediately

TABLE 9.4
First Example of Break Distribution

	10 Min Break			Meal Break		10 Min Break		
1st Hour	**2nd Hour**	**3rd Hour**	**4th Hour**	**5th Hour**	**6th Hour**	**7th Hour**	**8th Hour**	**9th Hour**
A	A	A	A		A	A	A	A

Note: The hours featuring adequate recovery periods are shown in light gray and the 4 h without adequate recovery are in dark gray.

TABLE 9.5
Second Example: The Breaks Are the Same as in Example 1, but Fall within the Hour

		10 Min Break			Meal Break			10 Min Break		
1st Hour	2nd Hour		3rd Hour	4th Hour	5th Hour	6th Hour	7th Hour		8th Hour	9th Hour
A			A	A		A			A	A

TABLE 9.6
Fifth Example: Bathroom Breaks. One Close to the Meal Break and One At the End of the Shift

				10 Min Break	Lunch Break				10 Min Break
1st Hour	2nd Hour	3rd Hour	4th Hour		5th Hour	6th Hour	7th Hour	8th Hour	
A	A	A	A			A	A	A	

prior to the lunch break and one in the afternoon immediately prior to the end of the shift. In this case the risk score is higher: No longer 4 but 5.5, as the break just before lunch (to prolong the meal break) is in fact counted together with the meal break, and the break just before the end of the shift does not reduce the risk score as the last hour of work is already considered to be *recovered* by the end of the shift.

In actual fact there is one *half hour with no recovery period*: The OCRA index provides extreme precision in counting risk scores and assigns a score of 0.5 to this portion of work time without a recovery period. This score can be assigned to periods without adequate recovery lasting from 20 to 40 min: No score is assigned for periods of less than 20 min and if more than 40 min the score will be 1.

The sixth example looks at a task in which the recovery periods are within the cycle.

There is a 60-s cycle in which 50 s are spent performing technical actions with repetitive tasks using the upper limbs (50 technical actions in 50 s = 60 actions/min) and for 10 consecutive seconds the upper limbs are at rest (e.g., waiting for the machine to process a part). The task is performed for one 6 h shift without other breaks. In this case the ratio of work to recovery period within the cycle is already 5:1.

Despite there being no other *macrobreaks* over the 6 h spent performing the task, the situation appears to be acceptable (at least for the upper limbs) as there are constant microbreaks (for the entire duration of the shift) lasting at least 10 consecutive seconds, at a frequency of at least once per minute, in a ratio of work to rest of 5:1. It should be recalled that in order for a microbreak within the cycle to be counted as recovery period, it must last for a certain number of consecutive seconds and recur constantly throughout the duration of the task. Therefore, hours of work featuring adequate microbreaks will be counted as at risk level 0, at least with regard to the *recovery period* risk factor.

Moreover, consecutive rest periods within the cycle that are used as recovery periods need to be subtracted from the total cycle time (*pace*) in order to accurately calculate both the frequency of action only with respect to the active part of the cycle and the net duration of repetitive work.

Going on to Example 7, we see the analysis of a task where the recovery periods are within the cycle (as idle time) but are not consecutive. There is a 60-s cycle in which 48 s are spent performing technical actions with repetitive tasks using the upper limbs (50 actions) and for 12 *nonconsecutive* seconds the upper limbs are at rest. The task is performed for one 6 h shift without other breaks. Here, idle time in each cycle is not counted as recovery period; frequency of action is counted with respect to the entire cycle.

The eighth example depicted in Figure 9.5 shows that it is possible to calculate the net duration of repetitive work when there are recovery periods within the cycle, using the spreadsheet for automatically calculating the OCRA index.

If *machine downtime*, *setup time*, or other interruptions lasting at least 8–10 min occur on a virtually daily basis and can be documented (such as for production changeovers, etc.), then they can be counted as recovery periods. Obviously the operator must not perform any other type of activity in the meantime (such as replenishing supplies).

Recovery periods within the cycle		
Total duration of cycle time (s)		60.0
A-consecutive period of idle time (s)		20.0 (33.3%)
B-total active time within the cycle (s)		40.0 (66.7%)
NB: For recovery periods within the cycle, the ratio of A to B must be equal to or less than 6. 5		2.0
Time that can be counted as recovery period		126.3
Description		Minutes
Shift duration	Official	460
	Actual	
Breaks or stoppages	Official	16
	Actual	
Meal break	Official	45
	Actual	
Time counted as recovery period	Recovery within the cycle	126
	Due to other causes	
Nonrepetitive tasks:	Official	20
	Actual	
Net duration of repetivive tasks		253

FIGURE 9.5 Example of automatic calculation of net repetitive work duration when there are recovery periods within the cycle.

When breaks are distributed freely and workers can decide the number and duration of their breaks (based on a fixed total duration), it is important to determine the average or modal behavior of the group by interviewing the members or by direct observation, and then to use that data to calculate the score.

Some workplaces tend to concentrate breaks just before or just after the meal break or just before the end of the shift. Since preventing musculoskeletal disorders is probably not what is driving this behavior, such trends should be discouraged via education and training and/or by reorganizing scheduled work breaks: Cramming breaks close to the meal break or the last hour of the shift actually defeats the purpose of recovery times assigned to prevent upper limb conditions.

Risk increases when workers not only concentrate their breaks but also speed up the pace of their work, shortening the cycle time in order to extend the duration of their rest period. Therefore, when there is a major discrepancy between the theoretical cycle time and the cycle time actually observed, it is crucial to closely examine the worker's behavior to measure frequency of action as well as assess the actual duration and distribution of recovery periods. It is essential to conduct a detailed study of each worker's behavior in order to determine his or her *personal* exposure level if the worker has an upper limb disorder caused by biomechanical overload, and the *risk to injury* ratio of pathology to workplace risk exposure needs to be evaluated.

The hours without adequate recovery periods will be assigned a different score for the Rc_M multiplier: If only 1 h per shift does not have an adequate recovery period then $Rc_M = 0.90$; for 2 h per shift without recovery, $Rc_M = 0.80$, and so on.

TABLE 9.7
Elements for Determining Recovery Period Multipliers (Rc_M)

No. of Hours without Adequate Recovery	< 0.5 h	1 h	1.5 h	2 h	2.5 h	3 h	3.5 h	4 h	4.5 h	5 h	5.5 h	6 h	6.5 h	7 or More Hours
Corresponding Recovery Multiplier (RCM)	1	0.90	0.85	0.80	0.75	0.70	0.65	0.60	0.52	0.45	0.3	0.25	0.17	0.10

Table 9.7 illustrates the scores used as multipliers for various numbers of hours without adequate recovery periods during a shift. Again, half an hour without adequate recovery is assigned a score of 0.5.

9.2.7 Calculation of Total Cycle Time Duration or Pace

Figure 9.6a and b show Example A with the sections in which the repetitive task(s) must be indicated along with their specific duration. Once the actual duration of each repetitive task has been identified, the next step is to count how many parts (or sets of parts making up a cycle) need to be completed by a single worker per shift.

Manufacturers generally know the number of parts (or cycles) that are completed every hour.

To obtain the *total net cycle duration* (or *pace*), which is required for an analysis using the OCRA method, it is necessary to know the number of parts (or cycles) that the worker *actually* completes per shift.

Task	Net duration of repetitive task (min)	%	No. of parts or cycles	Official	812
Load boxes and lids of refrigerator compressors	379	100%		Actual	
			Net duration of total cycle time (s)		28.0
Efficiency (in %)	100%		Observed cycle time (s)		27.0
				Difference (%)	4%

(a)

Task A:	Net duration of repetitive task (min)	%	No. of parts or cycles	Official	406
Load boxes and lids of refrigerator compressors	190	50%		Actual	
			Net duration of total cycle time (s)		28.0
Efficiency (in%)	100%		Observed cycle time (s)		27.0
				Difference (%)	4%
Task B:	**Net duration of repetitive task (min)**	**%**	**No. of parts or cycles**	**Official**	**406**
Assembly	190	50%		Actual	
			Net duration of total cycle time (s)		28
Efficiency (in%)	100%		Observed cycle time (s)		27
				Difference (%)	4%

(b)

FIGURE 9.6 (a) Example A: Definition and duration of repetitive task per shift, and number of parts and/or processing cycles. (b) Presence of one or more repetitive tasks per shift: Definition and duration of actual and proportional repetitive tasks and number of parts and/or cycle to be processed per task.

A breakdown of the *net duration of repetitive work* by the *number of parts (cycles) processed per shift* will produce the *total net cycle duration* in seconds (Figure 9.6a and b): this figure is fundamental for subsequently calculating frequency of action.

$$\text{Total net cycle time duration}\left(\text{s}\right) = \frac{\text{Net duration of repetitive task}}{\text{No. of parts }\left(\text{or no. of cycles}\right)}$$

The number of parts per shift is the number processed by an individual worker, not by the entire line. If the worker is replaced during assembly line work by a "*multipurpose* substitute worker" during breaks, the number of parts/shift to be counted will be the number actually completed by the operator, less those completed by the substitute.

Before going on to consider *frequency of action*, it is necessary to compare the duration of the *calculated total net cycle time* or *pace*, and the duration of the *observed total net cycle time*, the latter being measured by timing one or more operators in real life.

If the difference between these two cycle times is excessive (+/–5%) then the analyst must reconsider the actual content of the shift, in terms of duration of breaks, duration of nonrepetitive tasks, and so on, so as to accurately reconstruct the worker's behavior in the shift. The difference (5%), in an 8 h shift for example (480 min) translates into a period of about 20 min in which what the worker is really doing is unknown (20 min being a tolerable period albeit not short).

If the analysis of the worker's behavior during the shift has been properly conducted, the *calculated total net cycle time* will be the same as the *observed total net cycle time* (+/–5%). The aim of the organizational analysis proposed here is exclusively to assess the risk of biomechanical overload of the upper limbs, which often increases due to recurrent improper worker behavior due to insufficient information about prevention.

Workers in fact often tend to speed up their work in order to finish processing the necessary number of parts as quickly as possible in order to get ready to go home or prolong the duration of their breaks, and so on. The organizational analysis proposed here will also be useful in evidencing such behavior, weighting it in terms of risk exposure, and either taking corrective action, or disseminating best practices.

If productivity is incentivized, it will be necessary to estimate the total net cycle time (or *pace*) as a function of estimated performance (the greater the number of parts per shift, the lower the cycle time). In some cases, it may also be important to calculate *pace* as a function of *estimated performance* (positive or negative) for both individual workers and groups of workers.

9.2.8 How to Decide Which Cycle to Analyze

Let us say a task consists of packing objects into a large box, closing it, and stacking it on a pallet in 43 s: In this case, given how quickly the box is packed, it is advisable to consider the total time required to pack and stack the box as one cycle, rather than the time required to place each individual object into the box (Table 9.8).

TABLE 9.8
Pack Small Objects into a Large Box

	Take Object (1 s) Place Object in Box (1 s)	Repeat 20 Times (40 s)
Pack boxes	Close box	
ONE 43-SECOND	Pick up box	3 s
CYCLE	Position box	

Let us take another task, consisting of packing 20 Easter eggs into a large box. This time it takes 60 s to wrap each individual egg. A full box is picked up and placed on a pallet approximately every 20 min.

If the task of wrapping the object and placing it into a box consists of a complex set of technical actions (and thus has a longer duration), it is preferable to consider the wrapping of each object as a cycle and all the other operations (such as *placing the box on a pallet*) as additional actions broken down per object (in this case, 1/20). For particularly long cycle times, it is possible to consider transporting and closing the box as a *nonrepetitive* task, and thus count it as time to be subtracted from the net duration of repetitive work.

9.3 IDENTIFICATION OF TECHNICAL ACTIONS AND CALCULATION OF FREQUENCY OF ACTION

9.3.1 Identification of Technical Actions and Breakdown of the Cycle into Microphases

A proposal for measuring the frequency of *mechanical events* of the upper limbs in the cycle that is *applicable in the field* consists in analytically counting the technical actions in a cycle, and relating them to the unit of time (number of technical actions/minute = frequency of technical actions).

Therefore, in order to study *frequency of action* it is necessary to count the number of technical actions per unit of time: Then, each action must be checked for *awkward postures*, the level of *force*, and whether or not there are any *additional risk factors.*

It is virtually mandatory to film the workers and watch the recordings in slow motion for an accurate description and assessment of the various technical actions.

For technically complex tasks, the actions need to be described together with staff who are experts at that particular job.

While not identical, these technical actions closely resemble the predetermined MTM-2 and MTM-UAS time analyzing system.

Chapter 4 delves deeply into the OCRA checklist and provides examples and definitions of the most common technical actions, including in a specific annex at the bottom of the chapter.

Long and complex cycles are best analyzed by breaking the cycle down into micro-phases (Chapter 2, Section 2.7).

Operating phase (operation or elements)	% duration of phase vs total duration of cycle time	Start	Finish	Phase duration
Operation 1-load lids for refrigerator compressors	37%	0.00	10.06	10.06
Operation 2-load compressor boxes	56%	10.06	25.33	15.27
Operation 3-press start button	8%	25.33	27.46	2.13

FIGURE 9.7 Example A: breakdown of cycle in micro-phases.

Operating phase (operation or elements)	% duration phase vs total duration of cycle time	Phase duration	Load boxes and lids of refrigerator compressors	Technical action duration (s)	Total number of technical actions per cycle
	37%	10.06			
Operation 1-load refrigerator compressor lids			Take 6 lids from assembly line	1.60	6.00
			Rotate 5 lids	1.40	5.00
			Lay 5 lids on left upper limb	1.68	5.00
			Turn 6th lid (the one held in the right hand)	0.34	1.00
			Position 6th lid on machine	0.34	1.00
			Take 4 lids with the left hand	1.34	4.00
			Turn 4 lids	1.34	4.00
			Place 4 lids on machine	1.34	4.00
					Total =30
	56%	15.27	Take 3 boxes	1	3.00
Operation 2-load compressor boxes			Turn 3 boxes	2.18	3.00
			Lay 3 boxes on left upper limb	2.18	3.00
			Take 3 boxes	2.18	3.00
			Turn 3 boxes twice	4.36	6.00
			Position 3 boxes on the assembly line	1	3.00
					Total =21
Operation 3-press start bution	8%	2.13	With the left hand (the right waits)		
					Total =51

FIGURE 9.8 Example A: identification of technical actions for right upper limb within micro-phases.

Figure 9.7 shows the breakdown of the cycle and task in Example A into phases. Figure 9.8 then identifies the technical actions contained in each micro-phase.

9.3.2 Calculation of Frequency of Action and Total Actions per Shift (ATA)

The analysis described in the previous sections will define the following data:

- Net duration of repetitive task
- Number of cycles in repetitive task (or number of parts processed per shift)
- Net duration of each cycle

Once the technical actions in the cycle have been described, the next step is to evaluate:

- *Frequency of action* per unit of time: number of actions per minute.
- Total number of actions per task(s) and consequently per shift (ATA).

The result is the *net average frequency* adopted during a given period to perform each individual task.

The analysis of technical actions and their frequency also covers a special category known as *static actions* (see Chapter 4). These are actions during which the upper limb muscles remain contracted (*isometric contraction*), with the same level of force and posture, for more than four consecutive seconds (ISO, 2000).

As described in Chapter 4, static actions are present when one limb is actively holding an object or tool for a prolonged period while the other limb (generally the dominant one) is performing other tasks.

In other words, the action of *holding or gripping* will lie between the two actions of *taking* and *positioning* if it lasts for at least 5 consecutive seconds. The limb holding the object is engaged in a more or less prolonged static action.

For the purposes of scoring technical actions, every second spent performing the static action of *sustaining a grip* counts as 0.75 actions per second (essentially, 60 s of static activity will be equivalent to 45 dynamic technical actions).

The following formula is used to calculate frequency of action per minute:

$$\text{No. of technical actions per minute} = \frac{\text{No. of technical actions per cycle} \times 60}{\text{Total net cycle time duration}}$$

In Example A there are 51 actions and the total cycle time is 28 s, therefore the frequency of action per minute for the right side is 109.3

$$\text{Frequency} = \frac{51 \text{ actions} \times 60 \text{ s}}{28 \text{ s (total cycle time duration)}} = 109 \text{ actions / min}$$

TABLE 9.9
Identification of Technical Actions and Calculation of Frequency of Action with Two Tasks (A and B) per Shift

Task A; Total No. of Technical Actions per Cycle	Right		Left
Total	8		6
Calculated total net duration of cycle time		9 s	
Frequency of technical actions per cycle	53.3		40
Task B; Total No. of Technical Actions per Cycle			
Total	6.37		4
Calculated total net duration of cycle time		6 s	
Frequency of technical actions per cycle	63.7		40

The next step involves estimating the overall number of technical actions performed throughout the entire duration of the repetitive task(s). This is calculated by multiplying the frequency of action per minute by the duration (in minutes) of the repetitive task.

$$\begin{aligned}\text{ATA} &= \text{Total number of actions in a repetitive task per shift}\\ &= 109.3 \text{ actions/min} \times 379 \text{ min of total repetitive task duration}\\ &= 41{,}424 \text{ actions/shift}\end{aligned}$$

Table 9.9 describes an example of a shift during which two tasks are performed. The technical actions were counted for both tasks and both limbs, and the frequency of actions was assessed.

Table 9.10 lists the total actions performed per shift for task A versus task B for the right and left upper limb.

The figures were obtained by multiplying the duration of each task by its frequency of action per minute, initially determining the partial reference number corresponding to the actions involved in performing that particular task (per shift). The sum of the partial reference numbers will produce the total ATA performed per shift (and per limb) over several repetitive tasks.

9.3.3 Calculation of Frequency When Tasks Include Static Actions

Let us say that a cycle involves assembling components onto three different objects held steady with the left hand. The left hand picks up one object at a time and holds

TABLE 9.10
Calculation of Total Technical Actions per Repetitive Task and per Shift for Tasks A and B (Right and Left Arm)

	Right		Left	
	Task A	**Task B**	**Task A**	**Task B**
Net duration of each repetitive task per shift (min.)	226	96	226	96
Frequency of action (no. of actions/min.)	53.3	63.7	40	40
Total actions	12.046	6.115	9.040	3.840
	ATA (task A+B), right side			18.161
	ATA (task A+B), left side			12.880

it for 12 s before replacing it and picking up the second and third object; there are therefore three static actions with a total duration of 36 s.

To count the number of actions performed per cycle, the time spent performing static actions is converted into *fictitious technical actions* (36×0.75=27 fictitious actions), and added to all the other technical actions (i.e., pick up three times+replace three times=6) making for a total of 33 technical actions.

There are certain operations that involve both static and dynamic (such as holding a knife and cutting meat, or holding a screwdriver and pressing start buttons). In order to count the technical actions and therefore the frequency of action, the duration of all static actions (to be converted into *fictitious technical actions*) must be counted separately along with the simultaneous dynamic actions: The scenario to be chosen is the one determining the highest frequency of action per minute.

The example of meat cutting (Figure 9.9) shows a knife being gripped continuously for 25 s (frequency equivalent to 18.7 *fictitious technical actions*) while at the same time making 10 cuts: the representative frequency upon which the rest of the analysis will be based is the higher one, in this case corresponding to the static action of *gripping the knife* (18.7 fictitious actions, corresponding to a 25 s long grip) to be added to the initial action of *picking up the knife* and the final action of *positioning the knife*. The 10 cutting actions are in fact fewer than the 18.7 fictitious actions of holding the knife.

9.3.4 Calculating the Duration of Technical Actions

Knowing the *frequency of action* falls far short of comprehending the organizational aspects and *mechanical* content of a task.

There may be scenarios where the mechanical activity performed by an upper limb lasts less than the total time available (the total cycle time, or *pace*) thus

Load boxes	Technical action duration (s)	Total no. of dynamic technical actions	Description of static posture: enter the word hold when the worker sustains a static posture (with of without other technical actions)	Static action: duration >= 4 s	Summary: No. of technical actions dynamic and static (fictitious)
Pick up knife	0.50	1.00			1
Cut 30 times	25	10	Hold	25	18.7
Replace knife	0.50	1.00			1
		12			20.7

FIGURE 9.9 Calculation of scenario where static actions exceed dynamic actions.

saturating it to various degrees. This does not apply to very high frequencies of action in excess of 60–70 actions per minute.

When analyzing the risk of biomechanical overload of the upper limbs it is therefore important to study a particular type of *saturation* generated by the ratio, in percentage terms, of the actual duration of the activity (i.e., mechanical actions) performed by one limb and the total cycle time (otherwise known as the *pace*).

In this case, the term *limb saturation* can be used to distinguish it from other concepts and definitions of saturation more commonly used in organizational analyses.

Operationally, *limb saturation* must be studied separately for each limb, determining the total actual duration of the mechanical actions performed by the upper limb within a cycle (or during a certain observation period); it is also necessary to calculate the duration in percentage terms with respect to the total duration of the cycle (or *pace*) or of the specific observation period.

In practice:

$$\%\ \text{Upper limb saturation} = \frac{\text{Total duration of technical actions} \times 100}{\text{Total cycle time duration}}$$

We will see later that, for the purposes of conducting assessments, similar concepts can also be used to examine and evaluate the duration in percentage terms of awkward postures and stereotypy.

Limb saturation conditions approaching 100% will not for the time being be associated with a specific multiplier, but they must still be described, especially when *designing* new tasks. In this case, extremely high saturation levels (98%–100%) ought to be avoided.

Going back to Example A in Figure 9.8, it can be seen that besides the number of technical actions, the respective duration of each individual action or group of identical actions, is also indicated. From this it will be possible to obtain the following, first per phase and then per cycle:

- *Duration of the activity performed by the limb*, based on the sum total duration of all technical actions (dynamic and static).
- *Total active time* or *cycle time of the workstation*, including the sum total time devoted to executing the required phases or elements, and including both those performed by the upper limbs and all the other steps required to complete the cycle.

Once this data has been obtained, it is possible to estimate two different types of *saturation*:

- *% saturation of the limb* (i.e., *duration of limb activity* × 100/*total cycle time* or *pace*).
- *% saturation of the cycle* (i.e., *total active time* or *cycle time of the workstation* × 100/pace or *total cycle time).* This is a mandatory constraint when designing or redesigning work organization also with a view to minimizing upper limb risk exposure.

Figure 9.10 also refers to Example A and shows the results of the calculation of active time and saturation for both upper limb and task.

9.3.5 Methods for Analyzing Frequency of Action in Long Work Cycles

In some cases, workers perform *critical* tasks with long cycles lasting up to and over 10 min: generally, the work is organized in *cells* or islands, where operations are not broken down into separate tasks, but rather grouped together and assigned to an individual worker who occasionally ends up processing a certain product from start to finish. Manufacturing *cells* featuring long cycles must therefore be analyzed because although less repetitive, they may present high risk exposure levels due to other incongruent risk factors.

This type of analysis requires a certain degree of simplification since it is extremely challenging to film, analyze, and count all the technical actions performed over periods of between 10 and 30 min or more. It may be preferable to first break the work down into subgroups of tasks, which may in fact be altogether separate tasks, and then treat them accordingly as a multitask analysis. As a general rule, long tasks must be divided into phases, which makes it easier to analyze the task phase by phase.

The example is now described of assembling a sofa.

It takes approximately 30 min to assemble a certain style of sofa. The work can be broken down into four *subtasks*: fill cushions, dress up frame, insert back, final finishing. Even these *subtasks* are quite long, but the actions of which they are comprised are very homogeneous; therefore, to analyze the technical actions, it is advisable to count the actions performed over a representative period of two/three (preferably nonconsecutive) minutes for each *subtask*. The resulting frequency per minute will be taken as the *average representative frequency of the subtask.*

First microphase	
Take 6 lids from assembly line	1.60
Rotate 5 lids	1.40
Lay 5 lids on left upper limb	1.68
Turn 6th lid (the one held in the right hand)	0.34
Position 6th lid in machine	0.34
Take 4 lids with the left hand	1.34
Turn 4 lids	1.34
Place 4 lids on machine	1.34
Upper limb active time+ other active time (s)	10.1
Upper limb active time (s)	9.37
Idle time (unsaturation): s	0.00
% unsaturation right	0%
% saturation right	100%
Walk	0.7
Second microphase	
Take 3 boxes	1
Turn 3 boxes	2.18
Lay 3 boxes on left upper limb	2.18
Take 3 boxes	2.18
Turn 3 boxes twice	4.36
Position 3 boxes on the assembly line	1
Upper limb active time + other active time (s)	15.3
Upper limb active time (s)	12.91
Idle time (unsaturation): s	0.00
% unsaturation right	0%
% saturation right	100%
Walk	2.4
Third microphase	
Waits for left hand pressing button	1
Upper limb active time + other active time (s)	1.0
Upper limb active time (s)	0.00
Idle time (unsaturation): s	1.13
% unsaturation right	53%
% saturation right	47%
All tasks	
Upper limb active time + other active time (s)	26.3
Upper limb active time (s)	22.28
Time (unsaturation): s	1.13
% unsaturation right	4.12%
% saturation right	95.88%

FIGURE 9.10 Example A: calculation of active time and saturation for upper limb and task (right hand).

9.4 ANALYSIS AND ASSESSMENT OF FORCE USING THE BORG SCALE

9.4.1 Preliminary Data

Force represents the leading biomechanical effort required to perform a certain technical action (or set of actions). Force may be *external* (applied force) or *internal* (tension developed in the muscles, tendons or joints). Force may need to be used while performing work to handle or hold tools or to keep a body part in a certain position.

Force can thus be related to static or dynamic actions (contractions). The repeated use of force is reported in the literature as a risk factor for both tendons and muscles. A mutually reinforcing interaction has been documented between force and frequency of action (or repetition) in determining tendon disorders and nerve entrapment (such as carpal tunnel syndrome [CTS]).

In real-life situations it is difficult to quantify the use of force.

Some authors opt for a semiquantitative estimate of *external* force based on the weight of the object; others suggest using mechanical or electronic dynamometers. The resulting data is often used in a highly unsophisticated manner, and only occasionally is it integrated into biomechanical models for more accurately estimating the forces acting on different body parts and joints.

Surface electromyography is generally recommended for quantifying *internal* force, and when properly used, arguably represents the gold standard for studying force in work environments.

9.4.2 Application of the Borg Scale and Estimation of Physical Force (Exertion)

Every method has practical or theoretical shortcomings and is difficult to apply in the field, as force is not always able to be measured based on the weight of the object that is handled (e.g., how much force is exerted to tighten a screw using a manual screwdriver). Then again, the most appropriate measuring instruments are not always available. It has been suggested (Putz-Anderson, 1988) that such difficulties can be overcome by using a special scale (Borg CR-10 scale = category ratio of perceived exertion 10-point scale) proposed (Borg, 1998) for rating the perception of muscular force for a given body part.

When applied to a large enough group of workers, the results obtained using the Borg scale are roughly comparable with surface electromyography (Borg CR-10 scale × 10 = % MCV measured with EMG) (Grant et al., 1994).

The force perceived by the whole upper limb should be quantified for every technical action making up the cycle. For practical purposes, one can preliminary identify actions calling for minimal muscle involvement (Borg scale = 0/0.5). At a second stage, the average weighted score will then be calculated for all the actions performed in the cycle, including time fractions for every action and the relative level on the Borg CR-10 scale.

Based on past experience, a few useful practical recommendations can be offered to obtain reliable information with this method and also to overcome certain uncertainties relating to the use of *subjective* data.

Table 9.11 proposes a model for applying the Borg scale to gather information about perceived physical exertion. It is advisable to follow the operating phases in the order listed:

- First analyze the technical actions then study force: that is, how the cycle is performed.
- It may be more effective for the worker or workers to be interviewed by the technical staff that were involved in the preliminary work analysis and description of technical actions.
- The worker(s) must be asked whether there are any *technical actions* within the cycle that call for *appreciable muscle force* involving the *upper limbs*; it is important that the question be phrased like this because workers often mistake muscle exertion with the *overall fatigue* they feel at the end of the shift.
- Once the actions involving the use of force have been extrapolated, the worker(s) are asked to describe each one using the terms indicated in the Borg CR-10 scale, instead of with a number (e.g., light and moderate). Actions will then be scored from 0 to 10 and the analyst will add in the duration of each action and its time fraction with respect to the duration of the cycle.
- Since the exposure assessment is conducted primarily for preventative purposes it is important to ask the worker to explain why the actions reported as challenging required *physical force* This information is of great immediate practical interest because at times force is used to perform an action due to a technical flaw in the product, inefficient tools, faults, a poor choice of mechanical aids, or other easy to solve problems.
- The worker(s) must be consistent in defining perceived force while performing the various actions. Having external observers rate effort can lead to major errors. It is difficult for an outside observer to perceive the use of force, even when the level is quite high, in actions involving the fingertips or with the joints in certain positions (such as pressing a button or lifting/lowering a

TABLE 9.11
Subjective Assessment of Perceived Effort Using the Borg CR-10 Scale

		Borg CR-10 Scale	
0.5	Extremely light		
1	Very light	6	
2	Light	7	Very hard
3	Moderate	8	
4		9	
5	Hard	10	Extremely hard (almost maximum)

lever with the fingers, pinching, raising the arms, etc.). In fact, it is helpful for the interviewer to perform the same operation both to help the worker rate the level of force and also to experience the action first-hand.

- Once all the information has been obtained from the worker(s), the next step involves calculating the weighted average score for the entire set of actions making up the cycle. The time-weighted average effort is obtained by multiplying the score (on the Borg scale) assigned to each action by its time fraction within the cycle and then adding up the partial scores. Figure 9.11 shows the time-weighted average score (on the Borg scale) within each micro-phase and with respect to the total cycle for Example A.

Example A Load boxes and lids of refrigerator compressors	Technical action duration (s)	Borg scale CR-10 scores	Duration actions with force (s)		% time with peaks of force
First microphase					
Take 6 lids from assembly line	1.60			0.0	0.000
Rotate 5 lids	1.40	0.5	1.524	0.8	0.000
Lay 5 lids on left upper limb	1.68	0.5	1.524	0.8	0.000
Turn 6th lid (the one held in the right hand)	0.34	0.5	0.305	0.2	0.000
Position 6th lid in machine	0.34	0.5	0.305	0.2	0.000
Take 4 lids with left hand	1.34			0.0	0.000
Turn 4 lids	1.34	0.5	1.524	0.8	0.000
Place 4 lids on machine	1.34	0.5	1.524	0.8	0.000
				3.35	
	Average weighted force			0.33	
	Force multiplier			1.00	
Second microphase					
Take 3 boxes	1.00				0.000
Turn 3 boxes	2.18	3	2.18	6.5	0.000
Lay 3 boxes on left upper limb	2.18	3	2.18	6.5	0.000
Take 3 boxes	2.18				0.000
Turn 3 boxes twice	4.36	3	4.36	13.1	0.000
Position 3 boxes on the assembly line	1.00	3	1.00	3.0	0.000
				29.2	
	Average weighted force			1.9	
	Force multiplier			0.67	

All three microphases			Force	% time with peaks of force
Operation 1-load refrigerator compressor lids			3.35	
Operation 2-load refrigerator compressor boxes			29.18	
Operation 3-press start button			0.00	
			32.53	
	Average weighted force		1.16	
	Force multiplier		0.82	

FIGURE 9.11 Example A: Calculation of weighted average score for force based on force scores assigned to each action and duration. The third micro-phase does not use the hand. Scores are calculated both within each micro-phase and for the entire cycle.

- If the same task is performed by several workers (even in different shifts) it is advisable to interview as many of them as possible: The larger the number of workers interviewed the more reliable the weighted average score for physical effort. If the same work is performed by both male and female workers, it is advisable to calculate separate indexes. It is also advisable to exclude workers with existing upper limb conditions from the calculation, as well as newly hired workers (less than 1 year) and individuals with extreme anthropometric characteristics. The assessment should also overlook scores (particularly if very high) supplied by workers who fail to provide a technical explanation for their choice.
- The utmost attention should be paid to reporting any actions requiring *peak force*, that is, values equal to or above 5 on the Borg scale. For preventative purposes, these actions should be eliminated where possible or at least corrected. On the other hand, for assessment purposes, if peak force levels above 5 on the Borg scale are reported, the time fraction that they represent within the cycle time needs to be carefully examined. If their overall duration takes up at least 10% of the cycle time, the situation must be highlighted because it means that the force risk factor is exceedingly high: The corresponding multiplier for calculating the OCRA index can be as low as 0.01 and the final risk score will be untenable.

9.4.3 Force Multiplier (FoM)

It is obvious that the higher the force required to perform a series of technical actions, the lower its frequency should be.

The relationship between the frequency with which actions are performed and the average force required to perform them comes from intermediate data that later led to the drafting of the EN 1005-3 standard by a CEN technical group (1993). Based on this data, a multiplier was identified and applied to the frequency of action constant based on the *average weighted degree of force* (or perceived effort) *in the cycle*, and therefore in the task (Table 9.12), as determined using the various procedures already illustrated in this volume. Intermediate scores can be utilized.

In Example A, the respective multiplier for each force score is also shown in Figure 9.11.

To choose the multiplier, reference must be made to the *average force score weighted by the duration of the cycle*. However, if the task includes technical actions requiring a level of force substantially higher than 5 on the Borg CR-10 scale or above 50% of the maximum voluntary contraction (MVC), and lasting at least 10% of the cycle time, then the multiplier of 0.01 will be used.

It is worth bearing in mind that Table 9.12 can also be applied to results obtained using methods other than the Borg scale for measuring force.

Such other methods may include:

- *Dynamometric measurements* of the force exerted to perform the action(s).

  In this case, the result obtained for one or more sample workers is compared (in percentage terms) with a reference force (F_L) exerted to perform

TABLE 9.12
Force Scores (Obtained from the Weighed Mean Scores over the Entire Cycle) Used to Determine the Force Multiplier (Fo_M)

Force level as % MVC/F_L	**5%**	**10%**	**15%**	**20%**	**25%**	**30%**	**35%**	**40%**	**45%**	**≥50%**
Borg CR-10 scale scores	0.5	1	1.5	2	2.5	3	3.5	4	4.5	≥5
Force multiplier (FoM)	1	0.85	0.75	0.65	0.55	0.45	0.35	0.2	0.1	0.01

the same action(s). The international literature—EN 1005-3 in particular (CEN, 2002)—can also provide this data; the standard states that the reference force for a working population shall be defined as the capacity of at least 85% of its members.

- *Surface electromyography (EMG) measurements* of the muscle groups engaged in the action(s). The EMG results recorded during the activity are compared with the MVC of the same muscle groups of the workers in the study.

While their complexity would demand a more extensive analysis than is possible in this book, both methods—especially the second—can be utilized in conjunction with the OCRA index to quantify the force factor; Table 9.12 shows that both correspond very well as far as the calculation of the relative multiplier (Fo_M) is concerned.

9.5 ANALYSIS AND ASSESSMENT OF AWKWARD WORKING POSTURES AND STEREOTYPY

9.5.1 Introduction

The postures and movements of different segments of the upper limb during repetitive tasks are crucial for determining the risk of developing musculoskeletal disorders and diseases.

More specifically, there is enough consensus in the literature to argue that extreme joint postures and movements are potentially harmful, as are postures (even not extreme) that are sustained for prolonged periods of time, and excessively frequent identical movements of various joint segments (stereotypy).

An accurate description of posture and movement may also be able to predict which specific upper limb disorder might affect exposed workers when other risks are also present (frequency, force, duration).

Upper limb postures are described and assessed based on a representative cycle of each repetitive task; left and right limbs must be analyzed separately. The frequency and duration of positions and/or movements are considered for the four main anatomical segments:

- Postures and movements of the arm with respect to the shoulder (flexion, extension, abduction)
- Movements involving the elbow (arm-forearm flexion-extension, forearm pronation-supination)
- Wrist postures and movements (flexion-extension, radioulnar deviations)
- Hand postures and movements (mainly type of gripping)

To simplify the analysis of joint postural involvement, which is already complicated enough insofar as it considers four main joints and two limbs, the OCRA index assessment will refer only to *significant joint involvement.* This is classified by scores ranging from 1 to 24 and over, based on weighting not only using range of motion but also with respect to a subjective perception of joint involvement. To study this latter aspect, reference was made to *psychophysical* research in which the perceived effort linked to postures and/or movements of the main upper limb joints was scored. According to the results, significant wrist extension was, for instance, perceived as much more challenging than radial or ulnar deviations (with flexion in between); similarly, elbow supination was more challenging than pronation.

It also emerged, however, that all major scapula-humeral joint movements are perceived as very hard.

Once conditions of *significant postural involvement* have been identified and scored for each joint and limb, the next step is to measure the proportion of time (in the cycle and consequently in the task) that the position is held for (i.e., the duration), which is generally expressed using fractions such as one-third (from 25% to 50%), two-thirds (from 51% to 80%), and three-thirds (from 81% to 100%).

TABLE 9.13
Summary of Degrees in Excess of 50% of Maximum Upper Limb Joint ROM and Relative Weighted Score (for One-Third of Cycle Time) versus Subjective Perception

Joint	Movement	50% Maximum Joint ROM	Score
Shoulder	Abduction	45°–80° (25%–50%)	4
	Flexion/abduction	+80° (10%–20%)	4
	Extension	+20°	4
Elbow	Supination	+60°	4
	Pronation	+60°	2
	Flexion-extension	+60°	2
Wrist	Flexion	+45°	3
	Radial deviation	+15°	2
	Ulnar deviation	+20°	2
	Extension	+45°	4

TABLE 9.14
Summary of Weighted Scores for the Main Types of Grip (Lasting One-Third of the Cycle Time) with Respect to the Ability to Develop Force

Wide grip (4–5 cm)	Score 1
Narrow grip (1.5 cm)	Score 2
Small finger movements	Score 3
Pinch	Score 3
Palmar grip	Score 4
Hook grip	Score 4

Table 9.13 summarizes the main joints of the upper limb, degrees above around 50% of the maximum joint range of motion, and relevant scores, weighted with respect to subjective perceptions: the scores refer to a duration of one-third of the cycle time (from 25% to 50%) or, for the shoulder, 10% of the cycle time.

With regard to the hand, there is clear evidence that certain types of holds (pinch, palmar grasp, etc.) are more critical than *grips* or *power grips,* and are therefore judged as medium/high involvement. Taking this into consideration, grip involvement scores were ranked as shown in Table 9.14.

Figure 9.12 illustrates the main types of hand grips.

The various upper limb joint movements included in the analysis will be described in separate sections with respect to the risk exposure level of each joint segment.

As already reported speaking about the analysis of frequency of action, also for postures it is necessary to conduct the workstation analysis by filming the workers.

9.5.2 Description and Assessment of Awkward Postures and Movements of the Scapulohumeral Joint (Shoulder)

Highly regarded studies on the movements and postures of the shoulder (Punnett et al., 2000) have emphasized that dynamic or static actions in which the arm is

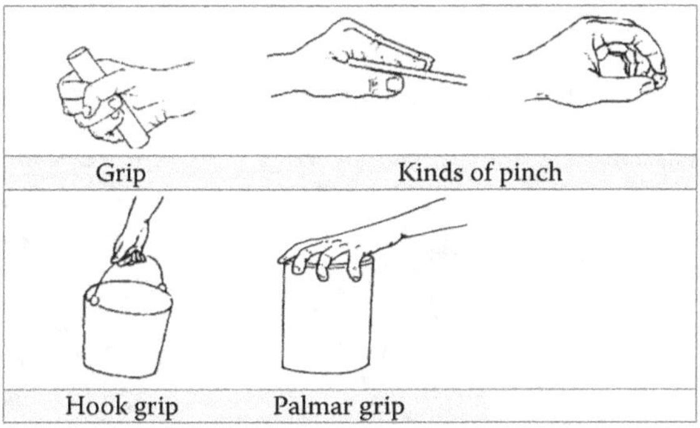

FIGURE 9.12 Main types of grip.

elevated (in flexion or abduction) to roughly shoulder height for even only 10% of the duration of a cycle/task, will place the worker at risk of disorders of this joint. The same study estimates that compared with workers not exposed to this condition (but who may also operate with the arm in intermediate positions), the risk increases 1.4-fold for every 10% increase in the time spent with the arm raised to shoulder height.

These results have been useful for putting together a more detailed analysis of shoulder postures and movements, as well as identifying postures in which the arm is elevated to shoulder height (in flexion or abduction) and *timing* them in percentage terms versus the entire duration of the cycle/task. The analysis will then serve to assign scores based on the postural involvement of the shoulder as shown in Figure 9.13.

When static actions call for raising the arm considerably above the shoulder, the scores provided in the aforementioned figure should be used with respect to the duration of the position.

Figure 9.13 also includes drawing of risk areas that can be reached when the arm is moved or kept in a sustained position in

- *Abduction* (over 45°)
- *Flexion* (over 80°)
- *Extension* (over 20°)

The main risk scores assigned to flexion and/or abduction at or above 80° are

- 4 if the movements or postures in risk area take up between 10% and 20% of the time
- 8 up to 30%
- 12 up to 40%
- 16 up to 50%
- 24 over 50%

	5%	1%	15%	20%	25%	30%	35%	40%	45%	51%	55%	60%	65%	70%	75%	81%	>84%
ABD 45°	0.25	0.5	1.3	2.4	4	4	4.4	5.2	6.5	8	8	8.3	8.9	9.7	10.7	12	12
EXT°	0.25	0.5	1.3	2.4	4	4	4.4	5.2	6.5	8	8	8.3	8.9	9.7	10.7	12	12
FL-AB 80°	2	4	6	8	10	12	14	16	19	24	24	24.6	25.5	26.8	28	28	28

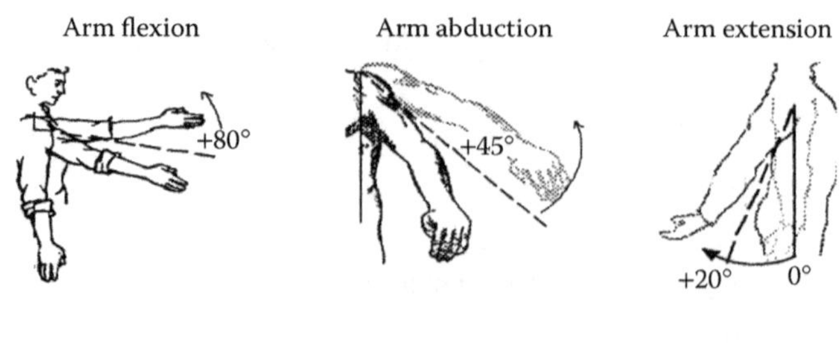

FIGURE 9.13 Assessment of awkward shoulder movements.

The main risk scores assigned to abduction between 45° and 80°, or extension above 20° are

- 4 if the movements or postures in risk area take up between one-third of the time.
- 8 if two-thirds of the time.
- 12 if three-thirds of the time. Intermediate scores are provided.

9.5.3 Description and Assessment of Elbow Movements

Figure 9.14 shows risk areas for pronation and supination movements (above 60°) and, separately, for flexion-extension movements of the elbow (above 60° in total, regardless of the starting position).

Descriptions of movements are envisaged but not *sustained positions*, for both pronation-supination and flexion-extension of the elbow. Certain positions that are sustained for a period of time (such as with the elbow flexed or the wrist pronated) are often uninfluential and occasionally even restful; only seldom do such positions pose problems (e.g., holding a tray with extreme supination of the elbow).

When movements approach the maximum pronation of the elbow, a score of 2 is recommended (for one-third of the cycle time).

Conversely, maximum supination places greater strain on the elbow (score 4 for one-third of the cycle time). Flexion-extension exceeding 60° calls for less effort (score 2 for one-third of the cycle time).

To help better understand how to assign scores based on the duration of joint movements, these simple strategies are recommended:

- Assign 1 s to each at-risk supination
- Assign 0.5 s to each at-risk pronation
- Pronation-supination is present only if objects are rotated
- Assign 0.5 s to each at-risk flexion or extension

0	5%	1%	15%	20%	25%	30%	35%	40%	45%	51%	55%	60%	65%	70%	75%	81%	>84%
FLX/EXT	0	0.3	0.7	1.3	2	2	2.2	2.5	3.1	4	4	4.1	4.3	4.7	5.3	6	6
PRON	0	0.3	0.7	1.3	2	2	2.2	2.5	3.1	4	4	4.1	4.3	4.7	5.3	6	6
SUPIN	0.25	0.5	1.3	2.4	4	4	4.4	5.2	6.5	8	8	8.3	8.9	9.7	10.7	12	12

Elbow pronation-supination

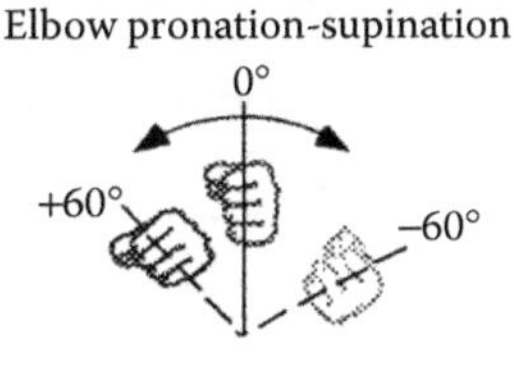

Elbow flexion-extension

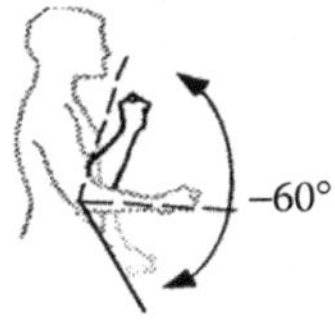

FIGURE 9.14 Assessment of awkward elbow movements.

0	5%	1%	15%	20%	25%	30%	35%	40%	45%	51%	55%	60%	65%	70%	75%	81%	>84%
Radioulnar	0	0.3	0.7	1.3	2	2	2.2	2.5	3.1	4	4	4.1	4.3	4.7	5.3	6	6
FLEX	0.2	0.4	1	1.8	3	3	3.3	3.8	4.7	6	6	6.2	6.6	7.2	8	9	9
EXT	0.25	0.5	1.3	2.4	4	4	4.4	5.2	6.5	8	8	8.3	8.9	9.7	10.7	12	12

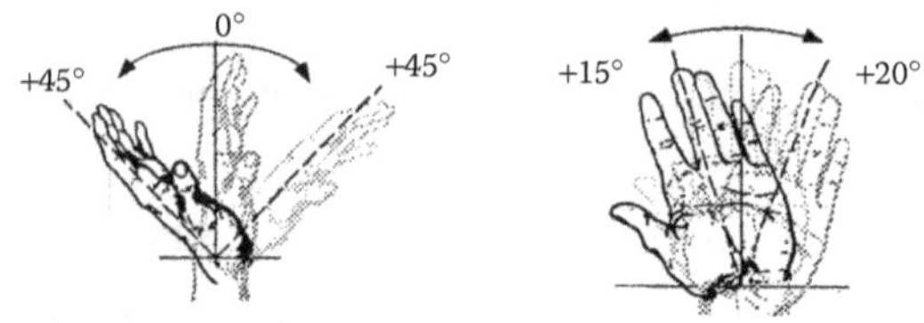

FIGURE 9.15 Assessment of awkward wrist movements.

- Elbow flexion-extension is at-risk generally speaking when an object is grasped and replaced with an intermediate distance of at least 40 cm.

9.5.4 Description and Assessment of Awkward Postures and Movements of the Wrist

In Figure 9.15, the first picture describes areas at risk for postures/movements of the wrist in flexion and extension (above 45°). The second picture describes areas at risk for radioulnar deviations (15° for radial deviations, 20° for ulnar deviations).

For extension above 45°, sustained for at least one-third of the cycle time, the score will be 4 (high risk); for flexion the score is 3, for radioulnar deviations the score is 2. For other time combinations Figure 9.15 clearly indicates the relative scores.

9.5.5 Description and Assessment of Awkward Postures and Movements of the Hand

In light of the extreme complexity and variability of hand postures and movements, for the sake of simplicity the main focus will be on describing hand-finger grasping positions.

Figure 9.16 suggests how to assign postural risk scores for all kinds of grips.

0	5%	1%	15%	20%	25%	30%	35%	40%	45%	51%	55%	60%	65%	70%	75%	81%	>84%
Wide grip	0	0.15	0.35	0.6	1	1	1.05	1.2	1.5	2	2	2.05	2.2	2.4	2.7	3	3
Narrow grip	0	0.3	0.7	1.3	2	2	2.2	2.5	3.1	4	4	4.1	4.3	4.7	5.3	6	6
Pinch	0.2	0.4	1	1.8	3	3	3.3	3.8	4.7	6	6	6.2	6.6	7.2	8	9	9
Palmar/hook	0.25	0.5	1.3	2.4	4	4	4.4	5.2	6.5	8	8	8.3	8.9	9.7	10.7	12	12

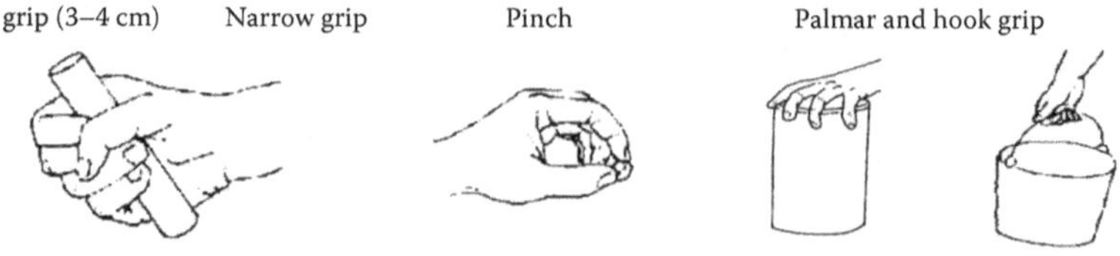

FIGURE 9.16 Assessment of awkward postures and movements of the hand and fingers.

As a general rule, when a worker uses a different grip to those described here, the score ranges from 1 to 4 (for one-third of the time) depending on the extent to which it resembles one of those shown.

Any finger movements included in a task or performed to achieve operational results and that are difficult to describe as separate movements are called *fine finger movements*.

9.5.6 Posture Multiplier (PoM)

In the models presented here, when postures and/or movements exceed 40%–50% of the joint range of motion and involve the joint or segment in question for at least one-third of the cycle time, the condition has been is defined as at risk. For the shoulder, holding the arm at shoulder height for as little as 10% of the time is regarded as potentially harmful to the joint.

Combinations exceeding this *minimal* postural effort (e.g., wide arm gestures performed for the whole cycle time) give rise to potentially greater risk.

These elements all contribute toward designing a table to help identify the PoM, according to the classification based on descriptions and assessments. In this table, *postural effort 4* corresponds to a multiplier of 0.70, which *per se* reduces the number of reference technical actions per unit of time by about 30%.

For higher levels of postural effort, the multiplier score will be increasingly unfavorable.

Table 9.15 shows the elements required to go from the descriptive effort score to the corresponding PoM. This table will be used for each segment of the upper limb and separately for each limb. The table also indicates intermediate scores, which have recently been added to automatic calculation software.

For the purposes of calculating the OCRA index, the most *penalizing* posture multiplier (Po_M) is used, corresponding to the highest postural effort from among the scores for the hand, wrist, elbow, or shoulder of each limb.

9.5.7 Analysis of the Risk Factor for *Stereotypy* or Lack of Variations in the Task

Once the task has been checked for awkward postures for the sustained periods indicated in Section 9.5.6, the next step is to check for *stereotypy* or a *lack of variations in the task*, in other words:

- Identical technical actions performed with the same posture (awkward or not awkward) taking up a good proportion of the cycle time (or duration of the repetitive task)

or

- Groups of actions, even if different and using different postures, but which are performed in a very short cycle (lasting equal to or less than 15 s)

Since there are so many possible combinations of actions, postures, and cycle times, the operational references in Table 9.16 may help the analyst check for stereotypy and assign the appropriate risk score.

TABLE 9.15
Elements for Determining Basic and Intermediate Postural Effort Multipliers (Po_M)

Score (select the highest from among the shoulder, elbow, wrist, hand)	0–3	4–7	8–11	12–15	16–19	20–23	24–27	≥28
Awkward posture and movement multiplier (Po_M)	1	0.70	0.60	0.50	0.33	0.1	0.07	0.03

0	4	5.4	6.7	8	9.4	10.7	12	13.4	14.7	16	17.4	18.7	20	21.4	22.7	24	25.4	26.7	28
1	0.7	0.6667	0.6333	0.6	0.5667	0.5333	0.5	0.4433	0.3867	0.33	0.2533	0.1767	0.1	0.09	0.08	0.07	0.0567	0.0433	0.03

TABLE 9.16
Operational References for Assigning Scores in Relation to High and Moderate Stereotypy

	High Stereotypy: Score 4
A1	Identical technical actions or groups of identical technical actions repeated for almost the entire cycle time (more than 80%). Static postures sustained for over 80% of the cycle time (e.g., prolonged gripping of a knife or screwdriver).
A2	Extremely short cycles lasting less than 8 s, featuring actions that involve the upper limbs.
	Moderate Stereotypy: Score 2
B1	Identical technical actions or groups of identical technical actions repeated for over 50% of the cycle time. Static postures sustained for over 50% of the cycle time (e.g., prolonged gripping of a knife or screwdriver).
B2	Extremely short cycles lasting less than 15 s, featuring actions that involve the upper limbs.

9.5.8 Stereotypy Multiplier (Re_M)

It has been reported in the literature that when identical actions are performed for more than 50% of the duration of the cycle (and thus the task) the situation (also defined as *stereotypy*) may pose a potential risk (i.e., high repetitiveness). Likewise, stereotypy (high repetitiveness) also occurs when technical actions featuring the frequent repetition (≥4 times a minute) of the same set of actions and postures are performed within an extremely short cycle (i.e., less than 15 min). Stereotypy has also been broken down into two levels, based on the proportion of time spent performing the same actions or, alternatively, the duration in seconds of short cycles spent performing mechanical actions.

Depending on which ones are chosen, it is thus possible to determine stereotypy multipliers (Re_M) as shown in Table 9.17, which proposes two different multipliers for two different stereotypy scenarios.

9.5.9 Assessment of Awkward Postures and Stereotypy (Example A)

Figures 9.17 through 9.19 refer to the example used in this chapter to assign posture scores for the shoulder and hand, respectively (no other joint segments are at risk) and stereotypy scores.

The recommended spreadsheet was used to calculate risk scores and their respective multipliers.

First the analysis of the results of the assessment of awkward postures and movements was applied within each phase, and the relevant risk scores were obtained for the phase, and then automatically to the entire cycle and the task.

Figure 9.17 shows how the spreadsheet is used to easily calculate the time spent in awkward postures and/or movements. When an X is entered for an awkward posture, the software automatically adds a time, obtained from the duration of the technical

TABLE 9.17
Elements for Determining the Stereotypy Multiplier (Re_M)

	Stereotypy Multiplier (RE_M): High Stereotypy
0.7	Identical technical actions or groups of identical technical actions repeated for almost the entire cycle (more than 80%).
	Static postures sustained for over 80% of the cycle time (e.g., pro-longed gripping of a knife or screwdriver).
	Extremely short cycles lasting less than 8 s, featuring actions that involve the upper limbs.
	Stereotypy Multiplier (RE_M): Moderate Stereotypy
0.85	Identical technical actions or groups of identical technical actions repeated for over 50% of the cycle time.
	Static postures sustained for over 50% of the cycle time (e.g., prolonged gripping of a knife or screwdriver).
	Extremely short cycles lasting less than 15 s, featuring actions that involve the upper limbs.

action(s) that generated it. The score can be manually *adjusted* to reflect technical action or actions generating less awkwardness with respect to their actual duration.

Figure 9.18 shows how risk scores are calculated first, followed by the posture/movement multiplier (for Example A), when there is some form of awkwardness within every joint segment. In this case, to obtain a single risk score for each joint, a sort of *weighted average score* is calculated.

Once risk scores have been obtained for each joint, the *driving* score, based on which the awkward PoM will be found, is the worst one of all the joint segments.

When calculating scores within a micro-phase, stereotypy is always absent by default (multiplier 1).

Figure 9.19 shows the final assessment of awkward posture and stereotypy risk factors. Only the three phases making up the task are indicated and for each awkward posture/movement the overall duration of the various awkward postures and the length of time they are sustained by the various joint segments are proposed as a total (rather than for an individual action or group of actions). All the elements required to calculate first the risk scores and then the corresponding multipliers are therefore at hand, and the procedure is the same one explained in Figure 9.18 for the micro-phase. It is at this stage that, to complete the analysis, it will be decided whether the stereotypy is *absent*, *moderate*, or *high*, along with the corresponding multiplier, which in this case is *moderate.*

9.6 IDENTIFICATION AND QUANTIFICATION OF ADDITIONAL RISK FACTORS

9.6.1 Description of Additional Risk Factors

Alongside the main risk factors described, the literature reports various other work-related factors that must be taken into consideration when assessing

Technical actions description for right side	Action	Shoulder: awkward postures and movements			Elbow: awkward movements			Wrist: awkward postures and movements				Hand and fingers: awkward postures and movements						Stereotypy
Load boxes and lids of refrigerator compressors	Technical action duration (s)	Flexion abduction >80°	Abduction 45÷80	Extension >+.20°	Pronation >60°	Supination >60°	Flexion abduction >60°	Flexion >.45°	Extension >45	Radial deviation >15°	Ulnar deviation >.20°	Wide grip (3–5cm)	Narrow grip (1–2 cm)	Pinch	Palmar grip	Hook grip	Small finger movements	
Take 6 lids from assembly line	2.01					× 2.0												
Rotate 5 lids	1.68				× 1.7									× 1.7				
Lay 5 lids on left upper limb	1.68									1				× 1.7				
Turn 6th lid (the one held in the right hand	0.34				× 0.3									× 0.3				
Position 6th lid in machine	0.34	× 0.3					× 0.3							× 0.3				
Take 4 lids with left hand	1.34																	
Turn 4 lids	1.34				× 1.3									× 1.3				
Place 4 lids on machine	1.34	× 1.3					× 1.3							× 1.3				
% duration of awkward postures and movements in whole phase		17%	0%	0%	33%	20%	17%	0%	0%	0%	0%	0%	0%	67%	0%	0%	0%	

FIGURE 9.17 Example A: analysis of awkward postures and movements per type, duration and joint with respect to technical actions in micro-phase 1.

Right side	Shoulder: awkward postures and movements	Elbow: awkward movements	Wrist: awkward postures and movements	Hand and fingers: awkward postures and movements	Stereotypy
Total time (seconds) spent in awkward posture/movement per joint	1.6767	3.3533; 2.012; 1.6767		6.7067	
% Time spent in awkward posture/movement per Joint	16.7%	70.0%	0.0%	66.7%	
Score for each awkward posture/movement considering the % shown on	6.00; 1.30; 1.30	4.70; 9.70; 4.70		2.20; 4.30; 6.60; 8.90; 8.90; 6.60	
% Evaluation of distribution activity for the single awkward posture/movement of each joint	100.0%	47.6%; 28.6%; 23.8%		100.0%	
Evaluation of actual scores based on actual duration of individual awkward posture/movement per joint	6.0	2.2; 2.8; 1.1		6.6	
Risk score for each joint (as a sum of all score)	Score 6.00	Score 6.13	Score 0.00	Score 6.60	
Multiplier for joint	0.67	0.67	1.00	0.67	
Choose worst multiplier between posture/movement and stereotypy	Posture multiplier (Po_M) = 0.67			Stereotypy multiplier (Re_M) = 1.00	

FIGURE 9.18 Example A: calculation of risk and awkward posture and movement scores in a single micro-phase.

Microphase duration	Microphase summary of results- right	Shoulder: awkward postures and movements			Elbow: awkward movements			Wrist: awkward postures and movements				Hand and fingers: awkward postures and movements						Stereotypy moderate = 1 High = 2
		Flexion abduction >80°	Abduction 45÷80	Extension >.20°	Pronation >60°	Supination >60°	Flexion extension >.60°	Flexion >.45°	Extension >45	Radial deviation >15°	Ulnar deviation >.20°	Wide grip (3–5 cm)	Narrow grip (1–2 cm)	Pinch	Palmar grip	Hook grip	Small finger movements	
10.1	Operation 1- load refrigerator compressor lids	1.68	0.00	0.00	3.35	2.01	1.68	0.00	0.00	0.00	0.00	0.00	0.00	6.71	0.00	0.00	0.00	1
15.3	Operation 2- load compressor boxes	0.00	0.00	0.00	6.56	4.38	2.18	0.00	0.00	0.00	0.00	0.00	0.00	0.00	10.91	0.00	0.00	
2.1	Operation 2- press start buttom	0.00	0.00	0.00	0.00	0.00	0.00	0.00	0.00	0.00	0.00	0.00	0.00	0.00	0.00	0.00	0.00	
		6%	0%	0%	35%	23%	14%	0%	0%	0%	0%	0%	0%	24%	39%	0%	0%	
		Score 2.00			Score 6.29			Score 1.00				Score 7.50						
		1.00			0.67			1.00				0.63						
		Posture multuplier (Po_M) = 0.63										Stereotypy multiplier (Re_M) = 0.85						

FIGURE 9.19 Example A: calculation of risk and awkward postures and movement scores in the task (all micro-phases).

exposure. Here, they are described as additional, not because they are less important but because each of them could be present or absent from time to time in the workplace.

Additional risk factors can be divided into two categories that correspond to two assessment models: The first includes scenarios featuring additional physical and mechanical risk factors, the second includes organizational risk factors.

Albeit not necessarily exhaustive, the list of additional physical and mechanical factors includes:

- Operating vibrating tools
- Extreme precision in positioning objects
- Localized compression of anatomical structures of the hand or forearm by tools, objects, or work areas
- Exposure to very low ambient temperatures or contact with very cold surfaces
- Using gloves that hamper the dexterity required by the task
- Parts with slippery surfaces
- Jerky or forceful movements
- Movements with contact stress or repeated shocks (e.g., hammering or striking hard surfaces, using the hand like a tool)

1. *Operating vibrating tools*; Vibrating tools place the upper limbs at risk, however, the level of vibration must be assessed using the appropriate analytical procedures also in compliance with recent EU regulations (EU Directive 44/2002). A discussion of these analytical aspects would be outside the scope of this book; therefore, it is sufficient to detect the use of vibrating tools transferring more or less high levels of vibration intensity to the hand/arm.

   Latest generation screwdrivers designed to automatically stop when the preset torque has been reached may not expose workers to any harmful vibration levels. Conversely, impact screwdrivers should be considered at risk. In any case, if the screwdriver does not have a *soft stop* feature it should always be considered at risk.
2. *High-precision work* (tolerance of approximately ±1.5 mm in positioning a part, with the eyes very close to the part); In this case the cervical, shoulders and arms muscles will be contracted due to the precision required to perform the task.
3. *Localized compression of anatomical structures of the hand or forearm by tools, objects or work areas*; such situations may give rise to traumatic tendinitis or bursitis. From the practical standpoint it is advisable to observe workers' hands (especially the palm and fingers), and check for redness and/or calluses that may suggest the presence of this risk factor.
4. *Exposure to very low ambient temperatures or contact with very cold surfaces*; ambient temperatures of 0° or less, or contact with surfaces that have a temperature of 0° or less (e.g., frozen meat and ice cream).

5. Use of *inadequate gloves* that hamper the ability to grip (e.g., wrong size, bulkiness preventing a part from being grasped, or requiring force to be used).
6. Parts with *slippery surfaces* (e.g., parts coated with lubricant or slippery foods).
7. Execution of *jerky or forceful movements* (e.g., throwing objects, tearing off adhesive tape or ripping up boxes) at a frequency of twice a minute or more.
8. Actions with *contact stress* (e.g., hammering or striking hard surfaces) at a frequency of over twice a minute; *using the hand like a tool* more than 10 times an hour.

Alongside physical or mechanical factors there are others that are indicated generally as *psycho-social*, which are associated in the literature with upper limb work-related musculoskeletal disorders (US-WMSDs). Some of these factors are clearly individual and are therefore not covered by methods designed to detect *collective* and *work-related* risk exposure for a group.

On the other hand, there are other distinctly organizational factors (such as regular overtime, incentive work, inadequate training, working at the pace set by a machine, processing rapidly moving parts), that enhance collective risk for biomechanical overload of the upper limbs (as well as other health problems) and are included here in the risk model if they are measurable.

Some of these factors (e.g., incentive work and long working hours) are dealt with in other stages of the analysis in terms of frequency of action, and overall duration of daily repetitive tasks. Others, especially working according to the preset pace of a machine, are examined with respect to two separate scenarios:

- The work is performed at the speed set by the machine but there are *buffer zones* where the pace can be increased or decreased to some extent. A *buffer* is a space between parts that gives the worker a very brief pause (e.g., long enough to sip some water). If buffers allow the worker to leave the line for at least 5 min then there is no risk associated with the machine working at a preset pace.
- The pace of the work is determined exclusively by the machine: Generally, speaking, this is when the worker is obliged to work at a strictly preset speed, especially with moving objects.

The following needs to be detected for each factor: Time spent with the factor present (with respect to the duration of the cycle and then the task), or the frequency of actions performed with the factor present (especially jerky movements and actions with contact stress), or even the quantitative level (e.g., for vibrations).

For the purposes of the assessment, except for the *vibrations* factor, which has specific exposure risk estimation procedures, the analyst will have to examine every deviation from the optimal condition (where there are no or only very minor additional risk factors), incrementally as the number of additional factors increases, along with the relative duration and/or frequency and/or level.

Here, the description of organizational factors will be restricted to checking whether or not assembly lines force workers to work at a certain pace or feature a preset speed but with buffers.

9.6.2 Assigning Scores for Additional Factors

Each additional risk factor shown here has been assigned the same risk score (score = 4 if the duration is one-third of the cycle time, 8 for two-thirds, and 12 for three-thirds). As additional risk factors cannot be quantified beforehand, it is possible to choose a score of between 2 and 4 to indicate intensity. (2—low, etc.) with respect to exposure lasting one-third, two-thirds, or three-thirds of the cycle time due to the presence of other additional risk factors judged to be hazardous for musculoskeletal disorders of the upper limbs.

For intensely vibrating tools such as jackhammers, grinders, and so on, the scores for exposure to vibration risk will be: 8 for one-third of the time, 12 for two-thirds, and 16 for three-thirds.

For organizational factors (such as pace preset by the machine), the scores are generally assigned to the entire cycle:

- Eight for pace set by the machine but with buffers
- Twelve for pace set exclusively by the machine

9.6.3 Multiplier for Additional Factors (AD_M)

There is no clear-cut evidence in the literature for developing assessment grids and hence multipliers for the presence and extent of additional risk factors. Therefore, a range of scores is suggested for assigning to the relative multiplier (Ad_M), based on the presence and extent of the various additional factors as classified in the descriptive analysis. However, it can be assumed that additional risk factors may affect the number of reference technical actions per unit of time by up to 20%.

Table 9.18 shows the elements used to assign the additional risk factor multiplier (Ad_M), based on the descriptive classifications presented.

TABLE 9.18
Elements for Determining the Additional Risk Factor Multiplier (Ad_M)

Scores for additional risk factors	0–3	4–7	8–11	12–15	≥16
Additional multiplier (AD_M)	1	0.95	0.90	0.85	0.80

9.7 MODEL FOR CALCULATING THE OCRA INDEX

The OCRA index is produced by the ratio of the actual number of ATA (derived from tasks with repetitive movements) *actually* performed in a shift to the corresponding number of specifically RTA. In other words:

$$\text{OCRA} = \frac{\text{ATA}}{\text{RTA}}$$

The total number of ATA performed in the shift is known and can be reconstructed via an organizational analysis. For Example A, Table 9.19 shows all the steps in the calculation.

These are the steps for calculating RTA (*total number of recommended actions in the shift*) (Table 9.20):

- Start from the reference frequency of action per minute of each repetitive task (frequency constant (CF) = 30 actions/min). This will be a constant for all repetitive tasks, as the other risk factors are optimal or uninfluential (i.e., force, posture, additional factors, lack of recovery time) and for a duration of repetitive work totaling 7–8 h in the shift.
- The same frequency is adjusted in relation to the presence and extent of risk factors for force, posture, repetitiveness (stereotypy), and additional factors for each task. Tables are provided for this purpose with scores obtained from the specific multiplier as a function of the extent of each of the various risk factors.
- The weighted frequency thus obtained for each task is multiplied by the number of minutes actually spent performing each task (Dj) to obtain a partial reference number of recommended actions (RPA_j) in the task (regardless of *recovery* and *duration* factors).
- The RPA scores obtained for the various tasks are added up (this step is unnecessary if it is considered that there is only one repetitive task).

TABLE 9.19
Example A: Calculation Steps for Estimating ATA (Right)

Frequency of Technical Actions per Minute	**109.3**
	x
Net duration of repetitive task/s (min)	379.0
	=
ATA (technical actions actually performed in the shift)	41,424

TABLE 9.20
Example A: Calculation Steps for Estimating RTA (Right)

CF	Frequency Constant—30 Actions per Minute	30
X		x
FoM	Force multiplier	0.79
X		x
PoM	Awkward posture/movement multiplier	0.63
X		x
ReM	Stereotype multiplier	0.85
X		x
AdM	Additional risk factor multiplier	0.9
X		x
RcM	Lack of recovery period multiplier	0.6
X		x
DuM	Duration multiplier	1.1
X		x
D	Net duration of repetitive task/s per shift	379
		=
RTA	Recommended technical actions	2857.15

- The result (RPA_{tot}), which represents the partial reference number of recommended actions in the shift, is now *adjusted* via the multipliers for
  - The presence and sequence of *recovery periods* (Rc_M) within the entire shift
  - The *total time* (in minutes) in the shift spent *performing* one or more *repetitive tasks* with the upper limbs (Du_M)
- The resulting value (RTA) represents the total number of recommended actions in the shift. This value will be a function of the various risk factors affecting the work scenario examined in this analysis. The RTA represents the denominator of the fraction expressing the synthetic exposure index (OCRA). The numerator will be the total number of actions actually performed in all the repetitive tasks examined (ATA).

In Example A, ATA is equal to 41,419.3 while RTA is equal to 2,857.15. The resulting OCRA index score is therefore 14.5 (very high risk).

For a thorough interpretation and classification of the OCRA index results, the reader should go to Chapter 4, Section 4.2.

9.8 ANALYSIS OF MULTIPLE TASK ROTATION WITH THE OCRA INDEX

Chapter 6 (Section 6.2) shows how to deal with assessments using the OCRA checklist, when there is a rotation, mostly daily, but also weekly, monthly, or yearly, among many repetitive tasks. Similar problems arise when using calculation models with the OCRA index. In this section the issue of *infrequent* task rotations will be looked at (i.e., less than once every 90 min), as well as situations in which the calculation model used in the previous section is unsuitable.

The approach described in Section 9.7 generates results that can be defined as *time-weighted average.* It is appropriate only when task rotations are very frequent, for instance, at least once every 90 min, or, especially, when individual tasks are in reality *subtasks* making up a *complex* general task (with a cycle time generally lasting several minutes). Here, *high* exposure levels could be offset by being alternated by *low* exposure levels in very quick succession.

Conversely, when repetitive tasks are rotated less frequently (e.g., once every 100 min or more), the *time-weighted average* approach might lead to underestimating the actual risk level due to the flattening of exposure peaks. It is more realistic in such scenarios to adopt an alternative approach based on the concept of the *most overloading task as the minimum.* The result will be a minimum level equivalent to the OCRA index for the most overloading task in relation to its actual duration and a maximum level equal to the OCRA index for the same task in relation (but only theoretically) to the overall duration of all the repetitive tasks examined.

A special procedure, based on the same data collected for assessments with the OCRA index described in this chapter, enables the user to precisely estimate the actual indicator within the range of scores between the hypothetical minimum and maximum levels.

The index calculated with this procedure is the *OCRA Index Multitask Complex (cOCRA).*

The procedure is based on the following formula:

$$\text{OCRA Index Multitask Complex } (\mathbf{cOCRA}) = \mathbf{ocra}_{1(\text{Dum}1)} + (\boldsymbol{\Delta}\mathbf{ocra} \times \mathbf{K})$$

where:

1,2,3,…,N = repetitive tasks listed according to their OCRA index score (1 = highest; N = lowest) calculated considering the respective actual duration multiplier (Dum_i) and the recovery period multiplier Rc_M (the same for all tasks)

$\mathbf{ocra}_1$ = OCRA index for task_1 considering Dum_1

$\mathbf{Dum}_i$ = duration multiplier according to the actual duration of the task

$\mathbf{Dum}_{tot}$ = duration multiplier for the total duration of all repetitive tasks

$\boldsymbol{\Delta}\mathbf{ocra}_1$ = highest OCRA index considering Dum_{tot} (chosen from among N tasks)—OCRA index for task_1 considering Dum_1

$$\mathbf{K} = \frac{(\mathbf{ocra}_{1max} * \mathbf{FT}_1) + (\mathbf{ocra}_{2max} * \mathbf{FT}_2) + \cdots + (\mathbf{ocra}_{Nmax} * \mathbf{FT}_N)}{(\mathbf{ocra}_{1max})}$$

ocra$_{1.N\ max}$ = OCRA index for tasks from 1 to N considering Dum_{tot}
ocra$_{i\ max}$ = highest OCRA index (for $task_i$) considering Dum_{tot}
FT$_i$ = time fraction (scores between 0 and 1) for $task_i$ with respect to the total duration of repetitive work

In other words, to calculate the cOCRA, the following steps must be followed:

- For each task, calculate the traditional OCRA index for a single task (ATA/RTA) considering the intrinsic duration (actual) of the task (as if every task were the only one performed per shift), the relative Du_{Mi} and Rc_M (the same for every task). **Ocra**$_{\mathbf{int}}$
- For each task, calculate the same OCRA index (ATA/RTA) leaving all parameters unchanged, including Rc_M, except for Du_{Mtot} which in this case will be considered in relation to the total duration of all the repetitive tasks. **Ocra**$_{\mathbf{max}}$
- Number the tasks (1, 2, 3, etc.) in decreasing order according to their $Ocra_{int}$ score; the highest $Ocra_{int}$ score will identify task 1.
- Calculate "**Δocra**$_{\mathbf{1}}$" for task 1.
- Consider the *total duration of repetitive work* given by the total duration (in minutes) of all repetitive tasks.
- Calculate the *time fraction* for task 1, 2, 3, and so on (**FT**$_i$), by dividing their respective duration (in minutes) by the total repetitive duration calculated in the previous point.
- Calculate **"K"** with the formula; in short:
  - Multiply each individual "$Ocra_{1\ max}$" by FT_1; "$Ocra_{2\ max}$" by FT_2; "$Ocra_{3max}$" per FT_3, and so on.
  - Add the resulting values.
  - Divide this amount by "**Ocra**$_{1\ max}$"
- The result (K) will be within the range of 0 and 1.
- Calculate the cOCRA using the general formula; in short:
  - Start from $Ocra_{1\ int}$ (OCRA index for the most overloading task calculated considering its actual duration).
  - Add "$Δocra_1$" (the difference between $Ocra_{max}$ and $Ocra_{1int}$) multiplied (weighted) by "K."

Example of cOCRA calculation
Let us say that there is repetitive work with a total shift duration of 322 min composed of two tasks:

A. *Weld bracket* with a duration of 226 min
B. *Punch terminal hole* with a duration of 96 min

In this example (Table 9.21), the following results were obtained in terms of frequency of action, cycle time, and number of actions actually performed per shift for each task (right limb): the following multipliers for force (Fo_M), posture (Po_M), stereotypy (Re_M), and additional risk factors were also obtained (Ad_M).

TABLE 9.21
Results Obtained in the Example in Terms of Frequency of Action, Cycle Time, and Number of Actions Actually Performed per Shift for Each Task and the Multipliers

Technical Actions

	Task A	Task B
Technical action frequency (per minute)	53.3	63.7
Total duration of cycle time (in seconds)	9	6
ATA	12,046	6,115

Multipliers

	Fo_M	Po_M	Re_M	Ad_M
Task A	0.91	0.5	0.85	1
Task B	0.75	0.5	0.7	1

Lastly, the following general multipliers must be considered for recovery periods and the total duration of repetitive work: $Rc_M = 0.6$ (4 h without adequate recovery); $Du_M = 1.2$ (for a total duration of 322 min).

The resulting OCRA index, according to the time-weighted average procedure described in the previous section, is 7.5.

According to the new method proposed here (cOCRA), the following data will now be considered:

Task A
ATA = 12,046
RTA (con $Du_{MA} = 1.5$ for 226 min) = $30 \times 0.91 \times 0.5 \times 0.85 \times 1 \times 226 \times 0.6 \times 1.5 = 2360$
RTA (with $Du_{Mtot} = 1.2$ for 322 min) = $30 \times 0.91 \times 0.5 \times 0.85 \times 1 \times 226 \times 0.6 \times 1.2 = 1888$

$$\text{Ocra A}_{\text{int}} = 5.1$$

$$\text{Ocra A}_{\text{max}} = 6.4$$

Task B
ATA = 6115
RTA (with $Du_{MB} = 2$ for 96 min) = $30 \times 0.75 \times 0.5 \times 0.7 \times 1 \times 96 \times 0.6 \times 2 = 907$
RTA (with $Du_{Mtot} = 1.2$ for 322 min) = $30 \times 0.75 \times 0.5 \times 0.7 \times 1 \times 96 \times 0.6 \times 1.2 = 544$

$$\text{Ocra B}_{\text{int}} = 6.7$$

$$\text{Ocra B}_{\text{max}} = 11.2$$

Task B has a higher Ocra_{int} and therefore this will be the starting point for applying the general formula to calculate cOCRA (task1); as well, the time fraction (FT) for task A is 0.7 (226/322 min), while the FT for task B is 0.3 (96/322 min).

Using this data and applying the formula to calculate cOCRA, the result will be

$$\text{ocra}_1 = \text{ocra index for task}_B \text{ considering Dum}_B = 6.7$$

$$\Delta\ \text{ocra}_1 = (11.2 - 6.7) = 4.5$$

$$\text{K} = ((6.4 \times 0.7) + (11.2 \times 0.3)/11.2) = 0.7$$

therefore:

$$\text{cOCRA} = 6.7 + (4.5 \times 0.7) = 9.85$$

The resulting cOCRA is 9.85; this outcome is acceptable when one of the tasks examined here is performed for its entire duration and is then followed by the next task.

Conversely, if two tasks are alternated every 90 min the result obtained using the time-weighted average approach will be more fitting (7.5 in the example shown here).

9.9 NEW DEVELOPMENTS: MICRO-PHASES AND MINI-INDEXES PER PHASE

As mentioned in Chapter 2, the organizational studies used by analysts to structure production activities and then assign tasks to individual workplaces call for the identification of *micro-phases* (also known technically as *phases* or *elements*). These undividable *bricks* are assembled, taken apart and reassembled to create the tasks assigned to each workplace along a production line to improve productivity (Figure 2.16). The procedure is called *line balancing*.

More expert analysts would like to detect risk as early as possible, that is, at the micro-phase level.

Therefore, besides recommending the use of the OCRA index, which is the easiest way to explore risk factors, by breaking each task down into micro-phases it has been possible to obtain a kind of miniOCRA index for each one.

Figure 9.20 shows the mini-indexes for the three phases into which the task in Example A was broken down.

This is the classic OCRA index calculation, but in this case:

- Each micro-phase is treated like a cycle, with its own cycle duration.
- Frequency is assessed with respect to the number of actions performed during the micro-phase and its duration; the same goes for awkward postures, force, and additional risk factors.

First microphase		
1.00	Force	0.73
0.67	Posture	0.53
0.90	Additional	0.90
78,727.6	No. of actions	23,027.8
4,752.00	No. of RTAs	2,789.9
16.57	Mini-index	8.25

Second microphase		
0.62	Force	0.70
0.57	Posture	0.50
0.90	Additional	0.90
36,06.4	No. of actions	20,746.5
2,510.0	No. of RTAs	2,09.8
14.46	Mini-index	8.27

Third microphase		
1.00	Force	1.00

Mini OCRA risk index for each microphase
(Evaluated as a constant: 420–440 min. of net duration of repetitive task, 2 breaks + 1 meal break; no stereotypy)

FIGURE 9.20 Example A: OCRA mini-indexes per phase.

- Stereotypy receives a multiplier of 1, since this factor can be assigned only to a complete task and not a micro-phase.
- *Standard* organizational data is used, which is all that is available at this preliminary stage in the risk assessment process: Task duration 440 min, two properly scheduled 10-min breaks, and one properly scheduled meal break. Essentially, the process involves calculating an intrinsic risk score using the same constants for all microphases. The duration multiplier is therefore always equal to 1, and the lack of recovery period multiplier will always be 0.6 (4 h without adequate recovery).

All the data obtained for each micro-phase is then put together again to estimate the final index. In other words, the following figures are drawn from each individual micro-phase (along with the sum total for all micro-phases): number of actions, duration, time-weighted mean force, duration of awkward postures, and duration of additional factors.

The elements introduced now to calculate the final index are

Right upper limb	Final OCRA index	Left upper limb
0.79	Force	0.75
0.63	Posture	0.53
0.85	Stereotypy	0.85
0.90	Additional	0.90
0.60	Recovery	0.60
1.10	Duration	1.10
41,412.00	Total No. of techical actions in shift	16,869.30
2,857.15	Total No. of recommended technical actions	2,281.93
14.49	OCRA index	7.39

FIGURE 9.21 Example A: final exposure index for the specific task.

- Actual organizational data relating to the net duration of repetitive work
- Actual organizational data relating to the distribution of recovery periods
- The pace of the equipment or machinery (fixed, with buffers, or variable)
- Moderate or high stereotypy

In constructing the task with various micro-phases, the analyst will be able to balance not only productivity, but also the consequent level of exposure.

Figure 9.21 shows the final index for Example A.

10 Health Surveillance

10.1 INTRODUCTION

The purpose of this chapter is to consider the aims, organization, operating aspects, and impacts of health surveillance on working populations exposed to conditions of biomechanical overload of the upper limbs. Since the nature and main objective of this analysis is risk analysis, the pathogenesis and diagnostic framing of the relevant diseases and disorders will only be touched on. Readers should consult the literature for additional details about the clinical aspects of the various health issues.

According to European and Italian legislation, employers must call on medical specialists to oversee the health of workers who, following risk assessments, are significantly exposed to health risks, including biomechanical overload of the upper limbs.

Health surveillance must be carried out by the company doctor and include, among other things:

- Preventative medical examinations to screen for conditions that might be incompatible with the job the worker has been assigned, to ensure his or her suitability for the specific tasks involved.
- Periodic medical examinations to regularly monitor the worker's health.

Both examinations require:

- Clinical and diagnostic tests targeting the specific risk (i.e., disorders due to biomechanical overload).
- Statement as to the worker's fitness to work.

10.2 AIMS OF HEALTH SURVEILLANCE

Health surveillance programs conducted for individual workers and working populations as a whole, are managed by company doctors and focus on disorders and, more importantly, diseases due to biomechanical overload, primarily for preventive purposes.

The aims, at the *individual level*, are to

- Identify workers who are *hypersusceptible* to workplace risks in order to adopt the most suitable precautions for avoiding the onset of the specific disease or disorder.
- Identify any conditions in the early preclinical stages in order to prevent them from developing into overt disease.

- Identify workers affected by overt disease, in order to adopt the most suitable protective measures and fulfill any medicolegal requirements.

At the *collective level*, the aims are to

- Contribute toward a more in-depth and precise risk evaluation, including through the use of disease and disorder occurrence data in various groups of exposed workers.
- Draft collective health reports to assess the effectiveness of existing preventive actions and plan additional ones.
- Contribute toward a better understanding of the diseases and disorders concerned, and draw comparisons with other groups of workers. Collective data is valuable for conducting comparative analyses to evidence any disease peaks among the specific group concerned.

10.3 PATHOLOGIES OF INTEREST

Health surveillance covers all known diseases and disorders of the musculoskeletal system and peripheral nerves involving the upper limbs, both occupational and nonoccupational.

However, the focus will be primarily on the work-related pathologies as those listed in the annex to the Italian Ministerial decree dated 09/04/2008 and reported in detail in Table 10.1 of this book.

In addition, diseases due to biomechanical overload are included on other official lists 2 and 3 (that must be reported to the supervisory authority for epidemiological purposes), defined as *unlikely or possibly* occupational.

List 2 includes ulnar nerve entrapment syndrome (UTS) of the elbow, tendinopathy of the distal attachment of the triceps, and Guyon's Canal syndrome (GCS). List 3 includes thoracic outlet syndrome (TOS) (excluding vascular forms) and Dupuytren's contracture (DC).

Also to be considered are disorders and diseases of the cervical spine, especially those that may have an impact on the upper limbs (e.g., cervicobrachial syndrome).

10.4 PRINCIPAL CLINICAL PICTURES AND A FEW BRIEF COMMENTS ON THE PATHOGENESIS

Clinically, musculoskeletal disorders of the upper limb, associated with biomechanical overload of the upper limbs, feature variable characteristics that are not always easy to differentiate. The causes may be changes to the periarticular soft tissues (tendinitis, tenosynovitis, and bursitis), peripheral nerves (nerve entrapment and TOS) or to certain joints (metacarpal trapezium joint arthritis). The symptoms of these conditions are well known and in the classic forms often present quite characteristically (Cailliet, 1973; Kasdan, 1991). Over recent years, ultrasound (US) techniques have greatly helped to pinpoint the location and status of these diseases and also to study the morphology and ecostructure of muscles, tendons, and nerves. In this chapter we

TABLE 10.1
List of Notifiable Occupational Diseases (Pursuant to Italian Ministerial Decree of 11 December 2009 Originally Issued on 27 April 2004)

Cause	Disease
List 1: Diseases Very Likely to be of Occupational Origin	
Microtrauma and awkward postures sustained by the upper limbs in tasks characterized by continuous pace and repeated for at least half of the shift	Biomechanical overload syndrome of the shoulder: • Supraspinatus tendinitis (or rotator cuff tendinitis) • Tendinitis of the long head of the biceps • Calcific tendinitis (Duplay's disease) • Bursitis Biomechanical overload syndrome of the elbow: • Lateral epicondylitis • Medial epicondylitis • Olecranon bursitis Biomechanical overload syndrome of the wrist-hand: • Flexor/extensor tendinitis (wrist-finger) • DQS • Trigger finger • CTS
List 2: Diseases Unlikely to be of Occupational Origin	
Microtrauma and awkward postures sustained by the upper limbs in tasks characterized by continuous pace and repeated for at least half of the shift	Biomechanical overload syndrome: • UTS of the elbow • Tendinopathy of the distal insertion of the triceps • Ulnar tunnel syndrome (UTS) (GCS)
List 3: Diseases Possibly of Occupational Origin	
Microtrauma and awkward postures sustained by the upper limbs in tasks characterized by continuous pace and repeated for at least half of the shift	Biomechanical overload syndrome: • TOS (excluding vascular forms) • DC

will be examining the genesis and clinical characteristics of the diseases and their most frequent sites of involvement.

10.4.1 Tendinopathies

Tendinopathy (both insertional and noninsertional), tenosynovitis, and stenosing tendinitis are conditions where there is inflammation along the length of the tendons or at their attachment to muscle and bone. Table 10.1 shows the location of the most common tendinopathies due to biomechanical overload. When making a differential diagnosis, it is advisable to consider systemic diseases (such as amyloidosis, seronegative arthritis, and rheumatoid arthritis) that may also involve muscles and tendons, as well as replacement/neoplastic diseases that occasionally mimic inflammatory pain (Missere, 1998).

10.4.1.1 Shoulder Tendinopathies

In shoulder tendinitis, the most frequent diagnosis is "impingement syndrome" (or "scapulohumeral periarthritis"), which may lead to calcific tendinitis or Duplay's periarthritis syndrome and adhesive pericapsulitis or "frozen shoulder." This can generally be defined as "rotator cuff tendonitis" (Figure 10.1), an inflammation of one or more of the following muscle-tendon units: supraspinatus, biceps brachii, teres minor, subscapularis, and infraspinatus. The first two are the most commonly affected.

Repeated mechanical strain may cause the tendons running through the narrow bony space between the humerus and acromion to thin and tear. Subsequently, calcification is caused by insufficient blood flow.

The gliding surfaces of bursae are also prone to inflammation (commonly subacromial bursitis and subdeltoid bursitis).

The most characteristic symptom is pain during shoulder movements, especially rotation and abduction ("painful arc syndrome," typically between 70° and 120° of abduction). Initially, the pain is intermittent and only later does it become constant. Pain may also radiate to the deltoid muscle or arm, and in later stages the pain may also be felt at rest. Adhesive pericapsulitis and advanced tendinitis cause loss of limb function.

10.4.1.2 Elbow Tendinopathies

The commonest tendinopathies affecting the elbow are lateral epicondylitis and epitrochleitis (or medial epicondylitis). Here it is the attachment that becomes inflamed.

Biomechanical overload has a rapid effect on these areas (epicondyle and epitrochlea) of the elbow (Figure 10.2). With so many muscle insertions in such a small space, mechanical stresses may give rise to microfractures of the collagen fibers connecting the tendon to the periosteum, with consequent inflammatory reactions.

If the inflammation persists, fibrosis eventually develops (epicondylosis) followed by intramural calcification.

Symptoms are initially felt in the elbow, especially when gripping an object firmly and/or moving the wrist. Later, the pain may radiate along the dorsum of the arm to the shoulder or forearm. Localized pain is sharp at palpation of the epicondyle/epitrochlea or may be triggered by passive or active tension applied to the affected

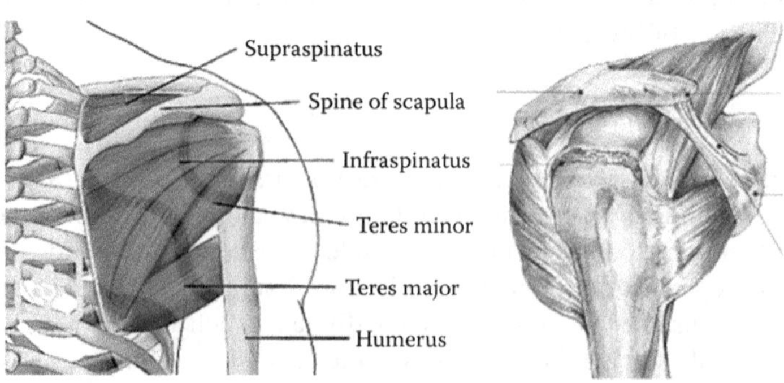

FIGURE 10.1 – Rotator cuff.

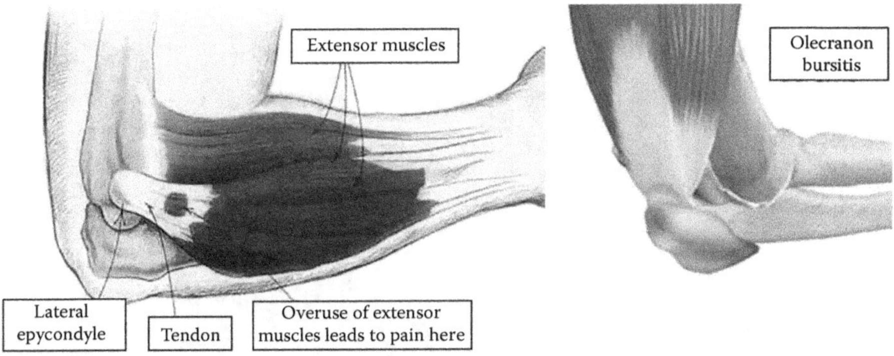

FIGURE 10.2 Lateral epicondyle and olecranon bursa.

tendons. The differential diagnosis must consider neurological conditions generally radiating from the elbow.

A less common tendinitis is enthesopathy of the distal attachment of the triceps. Elbow pain is not always localized at the olecranon and therefore may be felt more generally in the elbow area. Olecranon bursitis is an inflammatory complication in which the bursa is full of exudate; it may be associated with various inflammatory conditions (arthritis). This form of tendinitis is generally degenerative, although repeated micro or macrotraumas can lead to inflammatory changes and thus are tendinopathies that may or may not be associated with olecranon bursitis. Occasionally, biomechanical overload may be caused by activities calling for the worker to constantly lean on the olecranon, or repeated stresses exerted on the triceps muscle with contractions and releases.

10.4.1.3 Wrist/Hand Tendinopathies

The wrist and hand are often affected by chronic tendinitis, the main symptom being pain, exacerbated by movement, localized along inflamed tendons. The terms *peritendinitis*, *tenosynovitis*, and *tenovaginitis* are synonyms.

The inflammatory process is triggered when joints are overused and the synovial membrane secretes excess fluid. Inflammation is generated by the inability to meet the demand for additional joint lubrication.

There is often inflammation of the flexor tendons with compression of the median nerve resulting in carpal tunnel syndrome (CTS) (Figure 10.3).

Similar abnormalities may also stem from the presence of awkward wrist postures in extreme positions and the consequent lengthening of the tendon and nerve structures, as well as increased pressure within the carpal tunnel leading to an inflammatory reaction.

Stenosing tenosyinovitis, including De Quervain's syndrome (DQS) or "trigger finger," are primarily determined by fibrosis and adhesions between the sheath lining and adjacent tendon fibers.

In DQS, pain is localized in the anatomical snuff box and is exacerbated by ulnar and radial deviations of the hand and wrist; the pain may also radiate down the thumb or up the forearm.

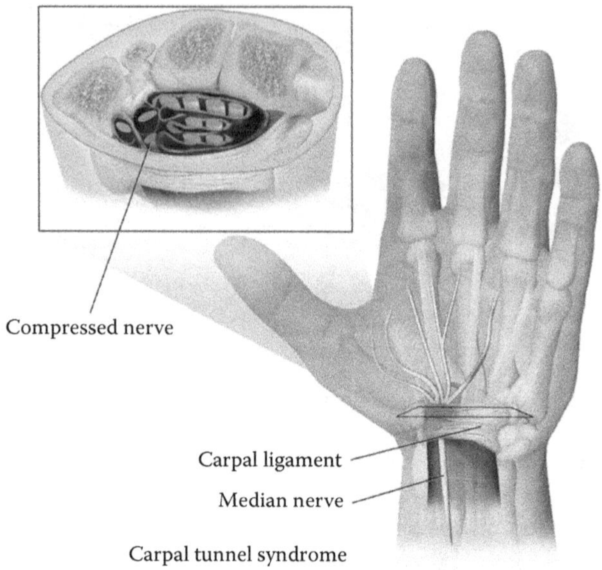

FIGURE 10.3 Carpal tunnel and median nerve.

Trigger finger (caused by tendons bunching into small lumps at the metacarpophalangeal level), does not present with the typical painful symptoms but is a condition where a characteristic snap or click is felt and the finger is locked into a flexed or extended position.

Volar or dorsal ganglion cysts of the wrist were formerly diagnosed by physical examination. Sonography of the wrist now enables such abnormalities to be broken down into synovial cysts (prevalently dorsal, affecting the joint) and ganglion cysts proper.

Other forms of tendinitis can affect the distal insertion of the extensor carpi radialis and flexor carpi radialis muscles.

Hand problems may also include trapeziometacarpal rhizoarthrosis: the main symptom is sharp pain at abduction and opposition of the thumb. Severity can also progress to joint deformity and pain at rest.

10.4.2 Entrapment or Canalicular Syndromes

Entrapment or canalicular syndromes (CS) are caused by the compression of a nerve traveling through a tunnel of tissue or along an inextensible space. Table 10.2 lists the leading canalicular disorders of the upper limb caused by biomechanical overload. Nerve trunk involvement is often the consequence of a tendon or muscle condition that causes changes to the nerve either directly (by compression) or indirectly (by increasing pressure in inextensible anatomical tunnels).

The third column of Table 10.2 shows the strong link between specific conditions and occupational activities. The weaker association with TOS is also due to the lack of universally recognized diagnostic criteria.

TABLE 10.2
Work-Related Canalicular and Related Upper Limb Conditions and Relationship between Risk and Injury

Pathology	Joint	Association
TOS	Shoulder girdle	Doubtful
Ulnar entrapment at the elbow	Elbow	Certain
CTS	Wrist	Certain
UTS (GCS)	Wrist	Certain

TOS is caused by the compression of the brachial plexus and the subclavian artery and vein in the interscalene triangle or between the first rib and the clavicle, or under the pectoralis minor muscle. The syndrome is caused by functional factors that may be associated with anatomical changes such as a cervical rib, mega-apophysis of the C7 transverse process, fibrous bands, or malunion of a clavicular fracture.

Various neurovascular signs are ascribed to TOS. Vascular impairment generally develops later and as yet there is no proof that it is related to any particular condition.

Therefore, only "neurogenic TOS" is included among the pathologies of interest. Symptoms are noncharacteristic and consist of upper limb pain possibly radiating to the neck, paresthesia and dysesthesia of the arm down to the hand, often originating from poor posture; the condition is often worse at night and most often involves the ulnar nerve. In advanced cases there is weakness and swelling of the hand. The syndrome may be associated with Raynaud's phenomenon.

The most common entrapment syndrome is *CTS*. It develops as a result of compression of the medial nerve and symptoms typically include paresthesia of the thumb, index, and middle fingers and half of the ring finger, first noticed at night and causing waking. The tingling may radiate up the forearm and arm. Eventually, the paresthesia may also occur during the day, with numbness, wrist pain radiating to the hand and forearm, weakness, and hypotrophy of the thenar eminence.

Less frequent is *GCS*, a condition due to compression of the ulnar nerve in the ulnar tunnel. There is paresthesia of the fourth and fifth fingers, and it primarily occurs at night, but may also appear during the day. The pain may radiate along the forearm and, in more advanced cases, leads to numbness and weakened finger abduction and adduction.

Ulnar entrapment syndrome at the elbow is sometimes associated with epitrochleitis. The bony profile of the olecranon is a "critical zone" through which the ulnar nerve travels to the elbow. Compression may lead to swelling at the attachment to the medial epicondyle.

Clinical evidence of entrapment syndromes is difficult to detect in the early stages. Therefore, reported symptoms are of the utmost importance. A careful clinical history can supply very useful information for the purposes of recommending diagnostic instrumental exams.

10.5 ORGANIZATION OF HEALTH SURVEILLANCE PROGRAMS

10.5.1 Activation

Essentially, there are two criteria that either separately, or together, will determine how a targeted health surveillance program should be activated with respect to a specific group of workers:

1. Existence of a potential occupational risk
2. Reports of work-related pathologies of interest

Both of these criteria will be examined in depth.

10.5.1.1 Is There a Potential Occupational Risk?

The most effective way to determine whether there is a potential occupational risk is to analyze and assess working conditions, according to the procedures described in this book. More specifically, when the occupational repetitive actions (OCRA) checklist calculation procedure is applied, health surveillance should be activated as follows:

- Starting with cases given an OCRA checklist score of between 7.6 and 11 (yellow). In this scenario, a nonaggressive health surveillance plan can be put in place on a one-off basis, consisting only of recording the workers' symptoms in last 12 months, by means of standardized questionnaires or protocols. The collective results of the exercise will then provide a basis for taking further measures, such as reassessing risk exposure, and either confirming or dropping systematic surveillance programs.
- Clearly, in cases where the assessment reaches an OCRA checklist score of >11 (red), more thorough health surveillance must be activated.
- If the analytical assessment has not yet been conducted or completed, then the outcomes of the "quick assessment" for repetitive manual tasks may be used (see Chapter 3). In this case, it is unnecessary to activate health surveillance only for situations found to be at "acceptable risk" (green).
- Alternatively, it may be decided that health surveillance needs to be activated based on the presence of at least one of the four "possible risk indicators" shown in Table 10.3.

10.5.1.2 Are There Any Cases of Work-Related Musculoskeletal Diseases or Disorders?

Regardless of the criteria used to determine the situation as illustrated in Section 10.5.1.1, it may emerge, based on systematic sources such as statistics on sick days or, more often, quite randomly following medical exams requested by workers, that in certain workplaces (manufacturing units, departments, or lines) there are one or more cases of work-related upper limb conditions (such as entrapment syndromes

TABLE 10.3
Signs of Potential Exposure to Repetitive Upper Arm Movements and Effort

1. Repetitiveness

Jobs featuring cyclic tasks entailing the execution of the same movement (or short set of movements) of the upper arms every few seconds, or the repetition of a cycle of movements more than twice a minute for at least 2 h during the shift.

2. Use of force

This is when force is used repeatedly (at least once every 5 min) for at least two consecutive hours in the shift. Indicative parameters would include:

- *Gripping an unsupported object weighing more than 2.7 kg, or applying an equivalent grip force.*
- *Pinching an unsupported object weighing more than 900 g or applying an equivalent pinch force.*
- *Using virtually peak force to operate levers, buttons and so on (e.g., tightening bolts with a wrench and tightening screws with a manual screwdriver).*

3. Awkward postures

Tasks that entail adopting or retaining extreme positions of the shoulder or wrist for periods of one continuous hour or periods of up to 2 h in the shift.

Indicative parameters would include:

- *Hands held above the head and/or arms held at shoulder height.*
- *Positions featuring an obvious deviation of the wrist.*

4. Repeated impacts

Jobs in which the hand is used as if it were a tool (e.g., using the hand as a hammer), more than 10 times an hour for a total of at least 2 h over the shift.

or tendinopathies). Clinically confirmed pathologies must be counted and used to decide whether or not to activate systematic health surveillance programs.

Clearly, in workplaces where there are only a few workers, even a single case should suffice to embark on a targeted health surveillance program (at least Level 1, i.e., based on symptoms reported by the worker).

Where the workforce is more numerous, it is necessary to appraise a larger number of cases. There are no hard and fast rules for deciding what level of action should be taken: generally speaking, specific health surveillance programs should be put in place based on an annual *incidence* of over 1.2% "new" cases of overt work-related disorders and diseases, or a *prevalence* of total cases of overt pathology that is twice as high as the unexposed working population (indicatively, >7.5% of workers with at least one clinically diagnosed work-related musculoskeletal disorder).

10.5.2 Health Surveillance Structure and Content

When the focus is biomechanical overload of the upper limbs, health surveillance programs are conducted:

- Prior to assigning workers to tasks that entail a specific potential risk.
- Periodically, as per the general principles prevailing in this field.

Preventive and, more importantly, periodical health surveillance programs for upper limb pathologies, may be organized on two levels:

- Level 1 programs are generalized, addressing all exposed workers and aiming to uncover "anamnestic cases." At this level, the medical histories of individual workers are recorded via interviews conducted by trained health-care staff.
- Level 2 programs consist in further exploring the clinical status only of workers with positive medical histories; the aim here is to detect clinically proven cases.

Since collecting comprehensive medical histories may be challenging, the utmost attention must be paid to diagnosing existing upper limb pathologies also based on instrumental exams such as US and electromyography/electroneurography (EMG/ENG). However, such testing would follow preliminary clinical screening by a company doctor.

Level 1 programs may be run using the history-taking tools described in Section 10.6.1. Worker-reported symptoms are very important for this group of pathologies, since most appear early on and, if well documented, may represent valuable indicators.

The minimum criteria for defining "anamnestic cases" are the following (Hagberg et al., 1995; De Marco et al., 1996a):

- Pain and/or paresthesia (pins and needles, numbness, tingling, a burning sensation, etc.) affecting the upper limb over the previous 12 months and lasting at least 1 week or occurring at least once a month
- Onset unrelated to acute trauma

Once medical histories have been taken, all "anamnestic cases" are further explored via clinical and instrumental diagnostic exams to single out clinically proven cases that will be handled accordingly.

If the outcome is negative, the worker will still be defined as an "anamnestic case" to be monitored more closely than other workers classified as normal.

Therefore, in any case, the company doctor will adopt a specific monitoring program for both an amnestic and clinically proven cases.

Table 10.4 summarizes the steps involved in the surveillance program described here.

TABLE 10.4
Steps in Health Surveillance Process

Health Surveillance

Level 1	Anamnestic cases (positive threshold)	⇒	Occurrence of anamnestic cases	⇒	Statistical comparisons Additional surveillance Preventive measures
Level 2	Suspicion of existing pathology	⇒	Clinical or instrumental analysis	⇒	Diagnosis Statistical comparisons Ability to work Complaint/report

10.5.3 Periodicity in Relation to Exposure Levels

Analyzing the individual and, more importantly, collective results of health surveillance programs represents a vital tool for judging the quality and adequacy of risk exposure assessments and for evaluating the effectiveness of preventive measures; the results are also vital for scheduling additional preventive measures. Evidence of increasing or excessively high levels of specific pathologies in a specific group of exposed workers signals the need to further explore or at least strengthen the strategies and initiatives adopted to prevent upper limb problems in that group.

More specifically, collective health surveillance data can be used to plan subsequent programs (*periodicity*). With regard also to exposure analysis and assessment, Table 10.5 may be a useful contribution.

Table 10.5 shows exposure levels expressed in OCRA checklist scores combined with the collective outcomes of health surveillance (even as a one-off exercise involving only history-taking). Several possible combinations are generated, each one suggesting a different testing periodicity.

Worker-requested medical exams must also be taken into consideration regardless of their periodicity.

Lastly, individual workers with pathologies and "anamnestic cases" will require the company doctor to draw up an individual follow-up plan stop.

TABLE 10.5
Scheme for Using the Results of Exposure Assessments and Health Surveillance to Determine the Periodicity of Follow-Ups

Outcome of Exposure Assessment	Periodicity Suggested by Health Surveillance		
	Disease Occurrence	Program	Other Measures
Exposure acceptable (green) or borderline (yellow)	Similar to reference data: Excess cases: *<1.5 times higher with stable trends*	No generalized health surveillance	Follow-up only affected workers
Exposure acceptable (green) or borderline (yellow)	Excess cases: *>1.5 times higher versus reference data or positive trends*	Periodicity to be decided based on further assessment	Conduct further risk assessment and decide periodicity of additional health checks
Exposure borderline (yellow) or low (light red)	Excess cases: *1.5–3 times higher versus reference data*	Every 3 years	Follow-up affected workers on a case-by-case basis
Exposure low, medium or high (red or purple)	Excess cases: *More than three times higher than reference data*	Every year/2 years based on risk level and level of excess cases reported	Follow-up affected workers on a case-by-case basis

10.6 HEALTH SURVEILLANCE TOOLS

Health surveillance programs to detect diseases and disorders caused by biomechanical overload of the upper limb must be conducted in a stepwise fashion (Table 10.4).

Level 1 programs generally involve all exposed workers to single out "anamnestic cases" by administering a standard questionnaire.

Level 2 programs consist in investigating anamnestic cases via clinical exams and instrumental testing to reach a diagnosis (proven cases).

Here, we will be presenting the reference tools for conducting health surveillance: anamnestic questionnaire, patient record, and criteria for conducting instrumental exams.

10.6.1 Anamnestic Questionnaire

In the study (and therefore surveillance) of musculoskeletal disorders and diseases of the upper limb, reported symptoms are of great value as they generally have an early onset and, if recorded correctly, may be useful adjuncts to diagnostic suspicion and in deciding whether or not to pursue further clinical and/or instrumental testing.

The decision to report a detailed and orderly description of symptoms based on recent medical history may determine a more precise diagnostic process in individual cases as well as help compare epidemiological data from different sources. Comparisons can be made if data is collected in a standard fashion and when an "anamnestic threshold" is adopted for the relevant conditions (Menoni et al., 1996).

Here, specific reference is made to a procedure for gathering anamnestic data concerning upper limb problems over the previous 12 months (Figure 10.4).

The procedure is part of a more general protocol, omitted here, documenting the collection of personal details, the occupational history of performing tasks at risk for upper limb pathologies, the past medical history of the relevant conditions and affected joints, together with the date of onset, instrumental and/or clinical tests, and whether or not treatment was followed. The general protocol also covers disorders of the cervical spine, included here for the purposes of making a differential diagnosis.

Recent medical history includes symptoms broken down by the affected joint over the previous 12 months and divided into two categories:

- *Pain*
- *Paresthesia*

Both groups of symptoms must be investigated separately and recorded in separate parts of the patient file, one for each limb.

Articular and periarticular pain must be reported separately for each joint, as well as any pain radiating approximately or distally. It should also be indicated whether the pain appears while moving the joint, lifting weights, or at rest.

Medical Questionnaire for the Study of Upper Limb Disorders

Date ____________________

Company	**Department**	**Work station and job**
Name and surname:		
Date of birth:	Age:	Gender:
Position held since:	Employed by company since:	

Disorders in the last 12 months

Shoulder ☐No Yes☐	Disorders present since (year)	Right	Left
R L 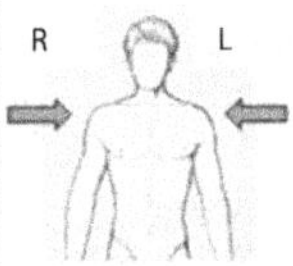 **Disorder-related treatment:** ☐Medication **Have you carried out clinical tests like:** ☐Physiotherapy ☐Orthopaedic examination/physiatrist ☐RX ☐Ultrasound/MRI	Pain during movement		
	Pain at rest		
	Positive threshold ☐Continuous pain ☐Pain for at least one week in the last 12 months ☐Pain at least once a month in the last 12 months		
	Minor disorders Sub-threshold pain		

Elbow ☐No Yes ☐	Disorders present since (year)	Right	Left
R L 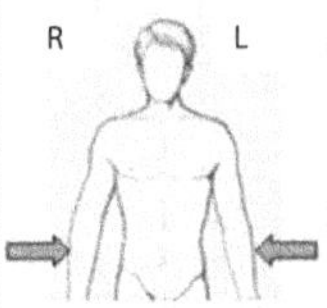 **Disorder-related treatment:** ☐Medication **Have you carried out clinical tests like:** ☐Physiotherapy ☐Orthopedic clinical examination/physiatrist ☐RX ☐Ultrasound/MRI ☐EMG (electromyography)	Pain gripping objects or lifting weights		
	Pain at rest		
	Positive threshold ☐Continuous pain ☐Pain for at least one week in the last 12 months ☐Pain at least once a month in the last 12 months		
	Minor disorders Sub-threshold pain		

Wrist/hand ☐No Yes☐	Disorders present since (year)	Right	Left
NB: indicate location of pain. **Disorder-related treatment:** ☐Medication **Have you carried out clinical tests like:** ☐Physiotherapy ☐Orthopaedic clinical examination/physiatrist ☐RX ☐Ultrasound/MRI ☐EMG (electromyography) R L 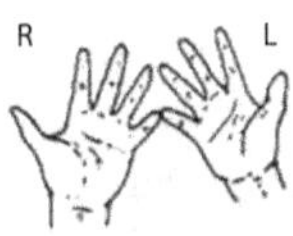	Pain when gripping		
	Pain in movement		
	Pain at rest		
	Pain in 1st finger		
	Pain in all fingers		
	Pain in palm		
	Pain on back of hand		
	Positive threshold ☐Continuous pain ☐Pain for at least one week in the last 12 months ☐Pain at least once a month in the last 12 months		
	Minor disorders Pain below the threshold		

FIGURE 10.4 (a) Anamnestic questionnaire for upper limb disorders—Part one.

For the hand, pain must be accurately localized. The following information must also be included for each joint:

- Past treatment
- Clinical tests/instrumental exams performed
- Months or years onset of condition

The second group of symptoms includes paresthesia and dysaesthesia (pins and needles, burning, numbness, tingling, prickling) and whether the symptoms occur during the day or night.

Nocturnal paresthesia is when the symptoms appear during sleep (therefore including when night shift workers are asleep during the day: paresthesia occurring while night shift workers are at work is classified as daytime). In both cases, the localization, frequency and mode of onset are reported as well as time in months since onset.

Nocturnal paresthesia ☐ No Yes ☐		Disorders present since (year)	Right	Left
Tingling, burning, numbness, pins and needles	**Disorder-related treatment:** ☐ Medication	Arm		
		Forearm		
		Hand		
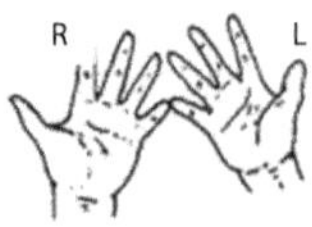	**Have you carried out clinical tests like:** ☐ Physiotherapy	Lasting less than 10 min		
		Last more than 10 min		
		During sleep		
	☐ Orthopedic clinical examination/ physiatrist	Upon awakening		
NB: indicate location	☐ RX	Positive threshold ☐ Continuous pain ☐ At least one week of pain in the last 12 months ☐ Pain at least once a month in the last 12 months		
	☐ Ultrasound/MRI	Minor disorders Pain below the threshold		
Daytime paresthesia			Right	Left
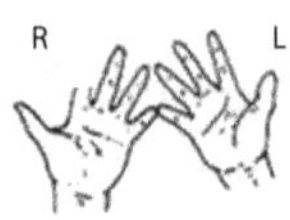	**Disorder-related treatment:** ☐ Medication	Arm		
		Forearm		
		Hand		
	Have you carried out clinical test like: ☐ Physiotherapy	Lasting less than 10 min		
		Last more than 10 min		
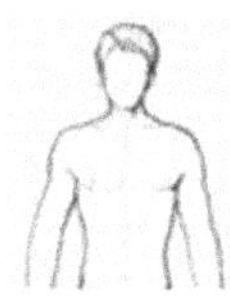	☐ Orthopedic clinical examination/ physiatrist	With arms raised		
		Resting on elbow		
		Grasping with force		
	☐ RX	Positive threshold ☐ Continuous pain ☐ At least one week of pain in the last 12 months ☐ Pain at least once a month in the last 12 months		
	☐ Ultrasound/magnetic Resonance computed	Minor disorders Pain episodes the threshold		

No. of days off work due to upper limb disorders ________			
Trauma-diagnosis (if known) ☐ ***Yes*** ☐ ***No***			
Shoulder (frozen shoulder, tendinitis, etc.)	☐ Yes	☐ No	When?
Elbow (epicondylitis, medial epicondylitis, etc.)	☐ Yes	☐ No	When?
Wrist/hand tendinitis; ganglion cyst, etc.)	☐ Yes	☐ No	When?
Wrist/hand (Carpal tunnel syndrome, Guyon)	☐ Yes	☐ No	When?
Total number of days with disorder in the last 12 months			
Section to be completed by the company doctor			
Notes			
Call employee for visit			
Ask employee to bring in clinical and instrumental test results			
Advise employee to contact the company doctor when symptoms recur			
Observations			

Signature of company doctor Date ____________

FIGURE 10.4 (CONTINUED) (b) Anamnestic questionnaire for upper limb disorders—Part two.

The progression of the condition is investigated by a standard question including

- Number of episodes of pain or paresthesia over the last 12 months.
- Duration of episodes in days (including treatment days even after subsidence of pain). It should be explained that the "duration of paresthesia episodes" refers to the consecutive number of days or nights during which paresthesia was present, even if only for part of the day or night. "Paresthesia duration" (on the contrary) refers to the duration in minutes of individual episodes.

With a detailed and structured description of symptoms as described, it is possible to identify the joint involved and the characteristics of the condition. The analysis may guide the company doctor toward developing a specific diagnostic suspicion and/or recommending instrumental tests.

The duration and frequency of pain and paresthesia allow workers to be classified as anamnestic cases when they are above the anamnestic threshold defined as: *presence of pain or paresthesia, duration at least 1 week, or at least one episode a month in the past 12 months.*

This definition of an "anamnestic case" is arguably useful in epidemiological research and studies relating risk to injury, while it does not qualify as a clinical-diagnostic parameter.

Therefore, since all the conditions considered here progress at a different rate, these criteria may provide useful guidance in managing individual clinical cases, if used cautiously and associated with other anamnestic and diagnostic data.

Based on the outcome of the patient history questionnaire, a few operational recommendations are provided to help the company doctor manage the caseload.

When studying large populations, while it may be impossible to examine every single worker within a limited time frame, it is still necessary to check for significant symptoms.

Bearing this in mind, self-administered questionnaires may be an option; data can be collected for epidemiological purposes in two different ways:

- Method 1: Guided self-administered questionnaire
- Method 2: Self-compiled questionnaire reviewed by the company doctor prior to meeting the subject

The first method involves collecting patient histories by administering the questionnaire to groups of between 15 and 20 exposed workers, with the help of trained health-care workers (nurses or nurses' aides). This method provides company doctors with relatively reliable data for an initial anamnestic screening program.

With the second method, subjects complete the questionnaire entirely on their own, and the company doctor reviews it later. The results provide a rapid "rough estimation" and thus a preliminary statistical overview of workers affected by more or less significant conditions of the upper limb; these workers will be called in for a clinical examination.

10.6.2 Clinical Examinations

Once a clinical suspicion has been raised, it may be advisable to conduct physical exams or instrumental testing, if necessary referring to a specialist (orthopedic surgeon, physiatrist, neurologist).

The clinical signs of tendinopathy are more informative than those of entrapment syndromes. A negative response to pain-eliciting stimuli may in fact rule out tendon disorders or at least suggest early-stage changes, obviously in the presence of significant symptoms.

Conversely, in the case of entrapment, a negative response does not rule out the possibility of disease, as will be discussed further on.

The discussion here will focus on interpreting the results of a systematic review of symptoms in order to guide the subsequent diagnostic process.

Moreover, several specific clinical exams will be recommended that are simple to perform and interpret, bearing in mind that they will be part of a health surveillance program rather than used a specialist setting.

The clinical tests and instrumental exams are also chosen based on their sensitivity and specificity, and positive or negative predictive value, if such data is available in the literature.

The appropriate instrumental testing and whether or not recourse should be made to specialists (physiatrist, orthopedic surgeon, urologist) will be discussed with respect to the various conditions.

10.6.2.1 Shoulder Tendinitis

Shoulder tendinitis should be suspected when pain is reported (in movement or when lifting weights) during the 12 previous months, with reference to the threshold indicated in Section 10.6.1. An association with pain at rest suggests more advanced tendinitis. No clinical exam will be required if symptoms are episodic and the subject has been pain free for over 30 days: in this case, if above the anamnestic threshold, the subject will be followed up 6 months later or invited back sooner if a flare-up occurs.

A clinical exam will be recommended if the subject reports continuous pain, current episodic pain, or pain relating to a specific trigger point. An episode is defined as current when the pain-free interval is less than 30 days (this also applies to disorders reported subsequently).

During the clinical exam, pain is evoked by palpation of specific *trigger points* (anterolateral and posterior) and during global movements of the scapular girdle.

The following global movements of the scapular girdle are performed actively by the subject sitting or standing:

1. *Flexion*: The subject lifts the painful arm forward and above the head to the vertical position (180° flexion).
2. *Abduction*: The subject performs a complete 180° abduction of the painful arm on the frontal plane. The "painful arc" is of particular importance (between 70° and 120° of active abduction), and is regarded as pathognomonic for impingement syndrome.
3. *External rotation and abduction:* The subject raises the painful arm and bends it behind the neck, touching the superior-medial angle of the opposite scapula.
4. *Internal rotation and abduction:* The subject reaches behind the back and touches the inferior aspect of the opposite scapula.

The subject is checked for *pain* reported at any time during the exams.

Tendinitis of the long head biceps may be detected by having the subject flex his or her elbow against resistance with the forearm supinated and the arm bent (80°–90°): the test is positive for pain elicited on the anterior aspect of the shoulder.

Scapulohumeral joint palpation is performed to locate a painful anterior trigger point, often indicative of impingement syndrome. With the patient sitting or standing and the upper limbs relaxed, the examiner palpates the shoulder joint anteriorly, laterally, and posteriorly.

If a "painful arc" or a sensitive anterior trigger point is detected in at least two of the tests, or if pain is reported in the long head of the biceps test, a shoulder US is recommended; in most cases, this will clarify the diagnosis by excluding or identifying the location of tendon inflammation.

If US is not an option, it may be helpful to seek specialist advice from an orthopedic surgeon or physiatrist.

Shoulder x-rays appear useful only for detecting tendon calcification (Duplay's disease) or degenerative joint disease.

10.6.2.2 Elbow Tendinitis and Bursitis

Epicondylitis or epitrochleitis (medial epicondylitis) should be suspected when pain is reported at the tendinous insertion and appears when the subject grips an object or lifts weights.

In more advanced forms, there may also be pain on movement.

No clinical exam will be required if symptoms are episodic (and the number of episodes is below the aforementioned threshold) and the subject has been pain free for over 30 days.

If above the anamnestic threshold, the subject will be followed up 6 months later or invited back sooner if a flare-up occurs.

A clinical exam will be recommended if the subject reports continuous pain, episodic pain with a pain-free interval of less than 30 days, or pain relating to a specific trigger point.

The clinical exam involves inspecting and invoking pain at palpation or while stretching epicondyle muscles. The latter test is positive if pain is reported at the epicondyle of the elbow.

Visible swelling over the medial, lateral, or dorsal aspects of the elbow, even without clinical signs, suggests the need for a specialist referral.

Clinical examination of epicondylitis involves

1. Palpation of the epicondyle
2. Palpation of the epicondyloid muscle tendons, approximately 2 cm below the epicondyle
3. Passive flexion of the wrist with elbow extended (stretching of the epicondyle muscles)

Clinical examination of epitrochleitis involves

1. Palpation of the epitrochlea
2. Palpation of the epitrochlear muscle tendons, approximately 2 cm below the epitrochlea

Clinical examination of olecranon bursitis involves

1. Palpation of the olecranon, with the elbow flexed

If pain is reported, an elbow US is recommended; in most cases, this will clarify the diagnosis by excluding or identifying the location of the inflammation.

If US is not an option, it may be helpful to seek specialist advice. Elbow x-rays appear useful only for detecting tendon calcification or degenerative joint disease.

10.6.2.3 Wrist and Hand Tendinitis and Degenerative Disease

Tendinitis or wrist and hand degenerative disorders may be suspected if pain is felt when gripping objects or moving the wrist and hand. Localized hand pain must be precisely localized.

There is no need for a clinical examination if the symptoms are episodic (and the number of episodes is below the aforementioned threshold) and the subject has been pain free for the last 30 days or more. If above the anamnestic threshold, the subject will be followed up 6 months later or invited back sooner if a flare-up occurs. A clinical exam will be recommended if the subject reports continuous pain, episodic pain with a pain-free interval of less than 30 days, or pain relating to a specific trigger point.

The inspection is performed even in the absence of pain if the subject reports discomfort when extending a finger (*trigger finger may be suspected*), discomfort when fully opening the hand (*possibly Dupuytren's contracture*), or nodules (*possible ganglion cysts*).

The wrist is checked for localized swelling (volar, dorsal, radial styloid, ulnar styloid), ganglion cysts, hypertrophic wrist muscles, contracture of the palmar aponeurosis, or trapeziometacarpal joint deformation.

If tendonitis of the extensors or flexors of the wrist and hand are suspected, the clinical exam will check for pain when bending and extending the wrist both against resistance and up to the wrist's maximum passive range of motion.

US is recommended when the wrist is swollen and movement is painful.

If the pain involves the first finger, the trapeziometacarpal joint must be palpated for arthrosis (just below the flexion wrinkles on the palmar side of the flexed wrist); in advanced cases, there is joint deformity. The Finkelstein test must also be performed: the subject is asked to flex his or her thumb and clench his or her fist over the thumb, then to actively bend the wrist toward the ulnar side and keep it slightly flexed. The test is positive for *DQS* if pain is felt along the radial edge of the wrist.

Lastly, the radial styloid is palpated and will be positive if pain is elicited.

If any of the abovementioned clinical signs are positive, the subject will be referred to an orthopedic specialist or physiatrist.

If trigger finger is suspected, the subject is asked to straighten the finger with force; the test is positive if the finger catches and then locks in position. US may be performed electively; if ultrasound US is not an option, it may be useful to refer the subject to a specialist.

10.6.2.4 Proximal Entrapment Syndromes

In the case of cervical pain radiating toward the upper limbs, with frequency and duration suggesting that the case is above the anamnestic threshold for the cervical spine (Occhipinti and Colombini, 1989), a clinical exam is recommended and should include eliciting pain by applying pressure to the apophysal spinous process, cervical intervertebral space and palpation of the relevant muscles (cervical paravertebral and upper trapezius muscles). Pain may also be reported while performing specific movements of the cervical spine (flexion, extension, right and left rotation, and leaning to the left or right). Based on a combination of anamnestic, clinical, and functional findings, the case may be classified as Grade 1, 2, or 3 clinico-functional spondyloarthropathy of the cervical spine.

If the subject reports continuous (or subcontinuous) cervicobrachial symptoms over the previous 12 months, an x-ray of the cervical spine is recommended (standard and oblique projections). If the exam reveals a prominent C7 transverse process, cervical rib, narrow foramina, severe degenerative joint disease, and Klippel–Feil syndrome, the subject will be referred to a specialist with a differential diagnosis of *TOS*.

Conversely, TOS (a canalicular disorder of the proximal upper limb) may be suspected in the presence of poorly localized daytime postural-onset paresthesia associated with nonsystematic upper limb pain or Raynaud's syndrome.

In this case the clinical exam will include the "candlestick position" and "fatigability test" to elicit symptoms of paresthesia.

The "fatigability test" involves asking the subject to raise his or her arms and open and close his or her hands for 30 s. The test is positive if the symptoms of paresthesia as reported in the patient history appear, or if cramping develops.

For the "candlestick test," the subject abducts the upper limbs 90°, keeping them in extra-rotation with elbows bent 90° and the hands up. The position must be sustained for 30 s. Again, the test will be positive if paresthesia develops.

These maneuvers investigate the neurogenic signs of TOS. If vascular signs are observed, the subject should be referred to a specialist.

If TOS is suspected, clinical tests are recommended if the subject crosses the anamnestic threshold mentioned in the introduction.

TOS is particularly difficult to diagnose given the lack of targeted instrumental methods, although recent imaging techniques, for example, electrical capacitance tomography (ECT), can provide valuable help in studying the structure of the scalene muscles, veins and arteries, and brachial plexus. TOS may often be diagnosed by exclusion.

10.6.2.5 Distal Entrapment Syndrome

The most characteristic symptom of distal entrapment syndrome is paresthesia of the upper limbs, initially during sleep. In advanced cases, there may also be pain or hypoesthesia. Entrapment syndrome should be suspected when paresthesia is continuous, or has appeared at regular intermittent intervals for at least 3 out of the previous 12 months. This latter parameter was added because entrapment syndromes initially present with cyclic symptoms.

If the subject crosses the anamnestic threshold (and is classed as an "anamnestic case"), but the paresthesia is not as previously described, the subject should be called back after 6 months for a follow-up.

If nocturnal paresthesia is reported only when the upper limbs have been in awkward postures, then there will be no need for instrumental testing.

Entrapment syndrome of the median nerve of the wrist (*CTS*) is suspected if paresthesia is reported primarily during sleep or upon awakening and affects the first three fingers and half the fourth.

Nocturnal paresthesia may be associated with daytime paresthesia, and, in more advanced cases, with hypoesthesia (i.e., when the subject complains of dropping small objects or difficulty when performing tasks such as tightening or loosening a screw).

Entrapment syndrome of the ulnar nerve of the wrist (*GCS*) will be suspected if the subject complains of nocturnal paresthesia affecting the fourth and fifth finger, with or without daytime paresthesia or hypoesthesia.

Paresthesia also involving the forearm and hand, with or without elbow or forearm pain, suggests the need to consider a differential diagnosis including *UTS at the elbow* (or "epitrochlear-olecranon groove syndrome").

In this case, the physical exam will seek to elicit paresthesia radiating from the elbow toward the fifth finger, with the examiner pressing on the epitrochlear-olecranon groove, and with the subject's elbow flexed.

In any case, instrumental investigations will be required if a CS is suspected, to confirm the suspicion, detect the affected nerve, and determine the exact location and severity of the nerve injury.

US of the wrist should be ordered if paresthesia is observed at examination and is reported as continuous or episodic for over 6 months.

If US provides evidence of median or ulnar nerve compression, then EMG should be performed to assess the severity of changes in nerve conduction. If US is not an option and the paresthesia is continuous, EMG should be performed, while in the case of episodic paresthesia it may be preferable consult a specialist before going on to instrumental exams.

For episodic paresthesia lasting less than 6 months' duration, or dating back more than 6 months but in complete remission at the time of the examination, clinical tests should be performed as described in the following lines.

If one test result is positive, US and/or EMG is necessary, whereas if test results are negative then the worker should be given a follow-up appointment.

The recommendation to begin with instrumental exams when continuous paresthesia is reported stems primarily from the poor specificity and sensitivity of the most common clinical tests used in orthopedics and neurology (Tinel's test, Phalen's test, reverse Phalen's test, compression test), which therefore provide no additional information on the presence of characteristic symptoms (Kuschner et al., 1992; Megele, 1992; Williams et al., 1992). According to the literature, the tests that best help raise the suspicion of CS are the compression test and Phalen's test (Durkan, 1991; Williams et al., 1992). The clinical exam includes only these two tests and is recommended exclusively for cases of noncharacteristic paresthesia symptoms. Phalen's test consists of asking the subject to push the dorsal surfaces of both hands together and flex the wrists to 90° for 60 s.

The test is described as positive when it causes localized paresthesia.

For the compression test, the examiner applies pressure with his or her thumb to the volar aspect of the subject's wrist over the area of the carpal tunnel for 30 s. Again, the test will be positive if localized paresthesia develops.

The decision-making process will vary according to the reported symptoms.

10.6.3 Instrumental Exams

In many cases, instrumental exams are required for a complete diagnosis.

It is not possible here to go into too much detail; therefore, a general overview will be provided of the various instrumental exams that may be helpful for diagnosing different conditions.

Tendinopathy can be diagnosed by means of imaging techniques such as US and magnetic resonance imaging MRI) (Gagey et al., 1991; Hannesschlager et al., 1989; Wang et al., 1994; Yu, 1994).

US can be very helpful to assess tendons for edema, calcification, or strain; MRI can be used when there are strong clinical suspicions, but a US has not detected abnormalities because the tendons cannot be accessed or because the lesion may not involve tendons (Beltran et al., 1990).

In the case of entrapment syndrome, especially CS, US can assess the extent and origin of the nerve compression, and can also detect the syndrome in the early stages.

With surface EMG it is possible to study the sensory and motor conduction velocity of peripheral nerves, and identify the exact location of the nerve injury.

It should be emphasized that in their clinical experience, the authors have come across numerous cases of subject median nerve compression readily identified by US but showing no EMG changes.

In a nutshell, the following recommendations can be provided regarding instrumental exams:

US is advisable if the following are suspected:

- Rotator cuff tendinitis
- Epicondylitis/epitrochleitis
- Ganglion cysts
- Digital flexor tendinitis
- DQS
- Tenosynovitis of the extensor carpi radialis
- Tenosynovitis of the flexor carpi radialis and ulnaris
- DC

US is seldom used to investigate peripheral nerve disorders and is viewed as an ancillary test for suspected CTS (but only with concomitant digital flexor tendinitis).

EMG is recommended if the following conditions are suspected:

- Brachial plexus syndrome (or TOS)
- Median nerve compression syndrome
- Ulnar nerve compression syndrome
- Radial nerve compression syndrome

Diagnostic techniques such as computed tomography (CT) and MRI are helpful particularly for making a differential diagnosis and for medicolegal and/or insurance purposes, while classic radiographs are useful only in a limited number of cases.

Laboratory tests may be useful for making a differential diagnosis, and for assessing non-work-related predisposing factors, and may include tests for connective tissue diseases, dysmetabolic syndromes, and so on.

10.6.4 Medical Records

Soft- or hard- copy medical records may be used to store the results of physical exams, instrumental exams, diagnoses, and measures adopted during health surveillance.

Figure 10.5 provides a sample medical record in line with the approach adopted here for health surveillance programs relating to diseases and disorders caused by biomechanical overload of the upper limbs.

Table 10.6 lists the degree of severity of the conditions diagnosed.

Clinical evaluation model for upper LIMB musculoskeletal disorders

Company name	**Department**	**Job description**	
Name and surname:			
Date of birth:		Age:	Gender: M F
Dominant limb: Right Left		Workstation:	
OCRA risk index Right	OCRA checklist index Right	OCRA risk index Left	OCRA checklist score Left
Length of time in position:		Number of years with company:	
Diagnostic conclusions			

Severity of work-related musculoskeletal disorders of the upper limbs (UL-WMSDs)

- ☐ Mild (1 "mild" condition)
- ☐ Moderate (1 "moderate" or 2 "mild" conditions)
- ☐ Severe (1 "severe" or 2 "moderate" conditions)

Fitness to work

- ☐ Fit
- ☐ Unfit .... Permanent .... Temporary

Restrictions and/or limitations

Occupational diseases

- ☐ Based on assessment of work-related causation Yes .... No ....
- ☐ Case reported to supervisory authority Date: ....
- ☐ Case reported to insurance system Date:

Notes

To be reexamined within ____________

Company doctor (signature) ____________ Date ____________

FIGURE 10.5 (a) Medical record: clinical assessment of upper limbs—Part one.

Physical Examination

a) Shoulder joint: maneuvers performed ☐ Yes

Shoulder disorder right for ______ months **Shoulder disorder left for ______ months**

Shoulder palpation

Front pain		Lateral pain		Posterior pain	
☐ Right	☐ Left	☐ Right	☐ Left	☐ Right	☐ Left

Mobility of shoulder girdle and presence of pain during movement

	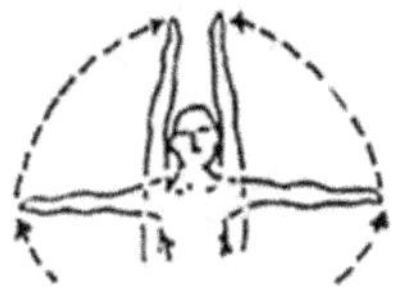	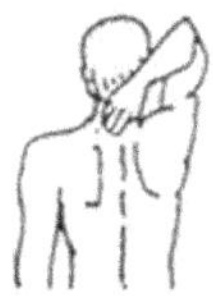	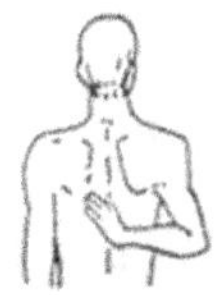
Pain present	Pain present	Pain present	Pain present
☐ R ☐ L	☐ R ☐ L	☐ R ☐ L	☐ R ☐ L

Painful arc (70°–120°)

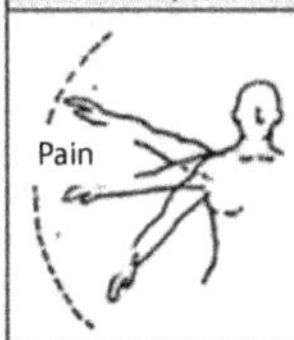

☐ No pain
☐ Pain present right
☐ Pain present left

Test for long head of biceps tendinitis

☐ No pain
☐ Pain in front of shoulder Right
☐ Pain in front of shoulder Left

Instrumental exams performed: ☐ None ☐ Ultrasound (year______) ☐ RX (year_____) ☐ MRI (year_____)

Severity of shoulder condition ☐ Mild ☐ Moderate ☐ Severe

b) Elbow: maneuvers performed ☐ Yes

Elbow disorder right for______ months Elbow disorder left for ______ months

Observations

Localized edema ☐ R ☐ L Where.................. Non-localized edema ☐ R ☐ L

Palpation of lateral epicondyle. Medial epicondyle. Olecranon		**Right**	**Left**
	Pain at lateral epycondyle	☐	☐
	Pain at medial epycondyle	☐	☐
	Pain at olecranon	☐	☐

FIGURE 10.5 (CONTINUED) (b) Medical record: clinical assessment of upper limbs—Part two.

Palpation of medial and lateral epicondyle muscles		**Right**	**Left**
Approx. 2 cm below tendon attachment 	Pain at lateral epicondyle	☐	☐
	Pain at medial epicondyle	☐	☐

Epicondylitis test		**Test for ulnar nerve entrapment at elbow**	
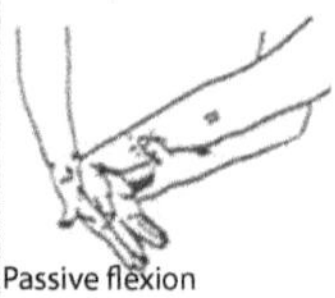 Passive flexion (Wrist flexed and elbow extended)	☐ Lateral elbow pain ☐ R ☐ L		☐ Paresthesia radiating to forearm and/or to 4° and 5° finger: ☐ R ☐ L

Instrumental exams performed: ☐ None ☐ Ultrasound (year_____) ☐ RX (year_____) ☐ MRI (year_____)

Severity of elbow condition ☐ Mild ☐ Moderate ☐ Severe

C) Wrist/hand: maneuvers performed ☐ Yes

Wrist-hand disorder right for________month Wrist/hand disorder left for________month

Observations		
Ganglion cyst	Dorsal ☐ R ☐ L	Palmar ☐ R ☐ L
Edema	Palmar (wrist) ☐ R ☐ L Radial styloid ☐ R ☐ L	Dorsal (wrist) ☐ R ☐ L Ulnar styloid ☐ R ☐ L
Hypotrophy	☐ R Location:	☐ L Location:
Trapezio-metacarpal joint oeformity	☐ R	☐ L
Palmar contracture	☐ R	☐ L

Palpation		**Maneuver for trigger finger**	
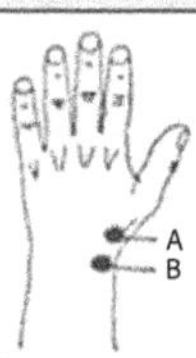	☐ (A) Pain at palpation of trapeziometacarpal joint-right ☐ (A) Pain at palpation of trapeziometacarpal joint-left ☐ (B) Pain at palpation of radial styloid-right ☐ (B) Pain at palpation of radial styloid-left		☐ Trigger during finger extension-right: (☐ 1° ☐ 2° ☐ 3° ☐ 4° ☐ 5°) ☐ Trigger during finger extension-left: (☐ 1° ☐ 2° ☐ 3° ☐ 4° ☐ 5°)

Finkelstein test (ulnar wrist deviation)		**Wrist flexion-extension:** Passive movement against resistance	
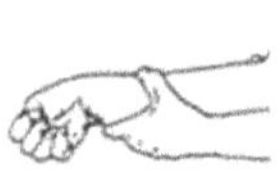	☐ Anatomical snuffbox pain ☐ R ☐ L	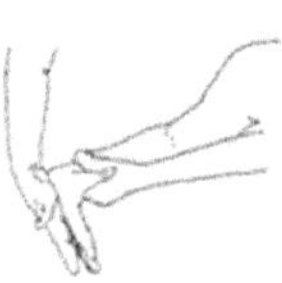	☐ Pain in flexion right ☐ Pain in flexion left ☐ Pain in passive flexion right ☐ Pain in passive flexion left ☐ Pain in extension right ☐ Pain in extension left ☐ Pain in passive extension right ☐ Pain in passive extension left

FIGURE 10.5 (CONTINUED) (c) Medical record: clinical assessment of upper limbs—Part three.

d) Paresthesia: maneuvers performed ☐ Yes

Wrist/hand disorders right for_____months Wrist/hand disorders left for_____months

Proximally

Pressure/palpation: painful	Movement: painful		
Apophyseal spinous process ☐ Superior trapezius M. ☐ Paravertebral M. ☐	Flexion ☐ Extension ☐ Inclination DX ☐ Inclination SX ☐ Rotation DX ☐ Rotation SX ☐	 Fatigability test (30 s) Paresthesia: ☐ R ☐ L	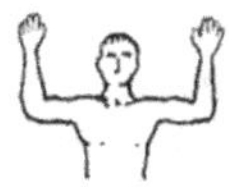 Candlestick test (30 s) Paresthesia: ☐ R ☐ L

Distally

Phalen's test		**Pressure test**	
Maintain position for 60 s 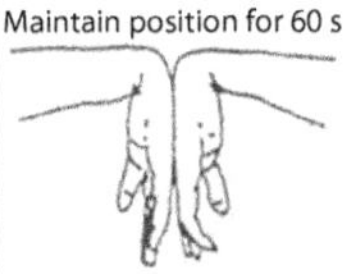	Paresthesia Median nerve ☐R ☐L Ulnar nerve ☐R ☐L Non-localized ☐R ☐L	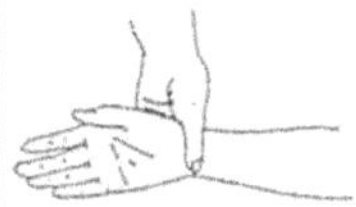 Palmar wrist compression For 30 s	Paresthesia Median nerve ☐R ☐L Ulnar nerve ☐R ☐L Non-localized ☐R ☐L

Instrumental exams performed: ☐ None ☐ Ultrasound (year____) ☐ RX (year____) ☐ MRI (year____)

Severity of wrist/hand condition ☐ Mild ☐ Moderate ☐ Severe

FIGURE 10.5 (CONTINUED) (d) Medical record: clinical assessment of upper limbs—Part four.

10.7 OCCURRENCE OF MUSCULOSKELETAL CHANGES IN WORKING POPULATIONS NOT EXPOSED TO BIOMECHANICAL OVERLOAD OF THE UPPER LIMBS

Upper limb work-related musculoskeletal disorders (UL-WMSDs) are considered to be multifactorial conditions that certain jobs may significantly contribute toward generating, but there may also be other causes. In fact, such conditions may also be found in individuals who are not exposed to work-related risk.

Prevalence and incidence data for nonexposed groups may be very useful for assessing groups of exposed workers, in both risk/injury studies and in programming preventive actions.

The incidence of UL-WMSDs in the general adult working population not exposed to biomechanical overload risk represents the threshold above which the higher percentages of diseases and/or disorders reported in exposed groups can be attributed to specific working conditions. These are the conditions that preventive actions should focus on.

To facilitate comparisons and, generally speaking, better understand the occurrence of the conditions classified according to the anamnestic criteria outlined in Section 10.6.1, and of WMSD in the general adult population, prevalence data relating to a reference group of individuals not professionally exposed to repetitive tasks and/or tasks that strain the upper limbs (De Marco et al., 1996b) was appropriately processed and is briefly presented here.

TABLE 10.6
Conditions Classified According to Severity

	Shoulder	Elbow	Wrist
Mild Conditions	• Slight edema of tendon over apophyseal spinous process • Slight edema of other rotator cuff tendons • Slight edema of the long head of the biceps tendon (LHBT) • Shoulder bursitis • Slight impingement syndrome	• Slightly edematous tendon(s) • Suspected or mild canalicular disease (CD)	• Slightly edematous tendon(s) • Suspected or mild CD • Trigger finger
Moderate Conditions	• Severe edema of tendon over apophyseal spinous process • Severe edema of other rotator cuff tendons • Severe edema of the LHBT • Severe impingement syndrome	• Severe edema of lateral epicondyle attachment • Severe edema of medial epicondyle attachment • Moderate CD • Elbow bursitis • Surgical correction of CD without adverse outcomes	• Severe edematous tendon(s) • Moderate CD • Surgical correction of CD without adverse outcomes
Severe Conditions	• Tendinosis—fibrosis • Calcification or rotator cuff injury • Severe edema of the LHBT—surgical correction with adverse outcomes	• Tendinosis—fibrosis • Surgical correction of CD with adverse outcomes • Severe CD • Surgical correction of disease with adverse outcomes	• Surgical correction of disease with adverse outcomes • Severe CD • Surgical correction of CD with adverse outcomes • Tendinosis—fibrosis

The sample population is composed of kindergarten teachers and local police officers, recruited randomly as part of several health surveillance programs. Medical histories were taken according to the structured protocol described in Section 10.6.1. Based on the disorders reported over the previous 12 months, the study population was given targeted clinical tests and instrumental exams for the purposes of coming to a complete clinical diagnosis.

The results are presented here, broken down by gender and age (above or below 35 years of age).

The composition of the group is shown in Table 10.7. The sample population consisted of 749 subjects: 310 males with an average age of 38.1 years (SD = 8.7) and 439 females with an average age of 36.3 years (SD = 7.6).

Table 10.8 shows the prevalence of subjects with disorders defined as *anamnestic-positive*, broken down by gender, age group, symptom, and joint.

TABLE 10.7
Breakdown of Nonexposed Workers by Gender and Age Group

	Age Groups				
	15–35		>35		
Gender	No.	%	No.	%	Total
Male	*139*	44.8	*171*	55.2	*310*
Female	*176*	40.1	*263*	59.9	*439*

TABLE 10.8
Results of Anamnestic Screening of Upper Limbs in a Nonexposed Population (Above the Threshold): Painful Shoulder, Elbow, Wrist, and Hand and Nocturnal Paresthesia

Age Groups	Female	Male
Anamnestic-Positive: Shoulder (%)		
15–35	0.6	1.4
>35	0	3.5
Anamnestic-Positive: Elbow (%)		
15–35	0	0
>35	0	3.5
Anamnestic-Positive: Wrist (%)		
15–35	1.7	0.7
>35	0	1.8
Anamnestic-Positive: Hand (%)		
15–35	0.6	0.7
>35	0	1.2
Anamnestic-Positive: Nocturnal Paresthesia (%)		
15–35	1.1	0.7
>35	9.1	1.2

The prevalence of pain above the threshold is generally low. The highest prevalence refers to the shoulder and elbow of males aged above 35 years.

Paresthesia above the threshold is primarily nocturnal and reported by women aged above 35 years. Going on to the clinically proven conditions reported by the same sample population, Table 10.9 shows the prevalence of subjects with a diagnosis of UL-WMSD.

It should be noted that nine subjects (all females) presented with more than one condition: two were aged between 15 and 35 and seven were above 35 years.

TABLE 10.9
Prevalence of Subjects in the Reference Group with UL-WMSDs, Broken Down by Condition, Gender, and Age Group

	Male				Female			
Age groups	**15–35**		**>35**		**15–35**		**>35**	
Disease/disorder	**No.**	**%**	**No.**	**%**	**No.**	**%**	**No.**	**%**
TOS	*0*		*1*	0.6	2	0.6	*5*	1.9
Humeral periarthritis	*0*		*1*	0.6	*1*	0.6	*7*	2.7
Lateral epicondylitis	*0*		*1*	0.6	*0*		*1*	0.4
Trapeziometacarpal osteoarthritis	*0*		*0*		*0*		2	0.8
Tendinitis (wrist and hand)	*0*		*3*	1.8	2	1.2	*0*	
CTS	*0*		*0*		*0*		*8*	3.0

TABLE 10.10
Number and Prevalence (Percentage) of Subjects in the Reference Group Affected by One or More UL-WMSDs, Broken Down by Gender and Age Group

	Age Groups					
	15–35		**>35**		**Total**	
Gender	**No.**	**%**	**No.**	**%**	**No.**	**%**
Male	*0*	*0*	*6*	*3.5*	*6*	*1.9*
Female	*4*	*2.3*	*23*	*8.7*	*27*	*6.1*
Total	*4*	*1.3*	*29*	*6.7*	*33*	*4.4*

It was then possible to define the number and prevalence of subjects with at least one UL-WMSD, as can be seen in Table 10.10.

Women have the highest prevalence of disease and the ratio of males to females is 1:3.1.

The data concerning anamnestic-positive disorders and the prevalence of subjects with overt musculoskeletal conditions in a reference group that is not significantly exposed to biomechanical overload can be used to make comparisons with similar results obtained from clinical investigations involving individuals exposed to somewhat high levels of biomechanical overload.

However, comparisons made using routine statistical techniques must factor in the criteria adopted here to classify the diseases and disorders (anamnestic-positive; conditions included). Moreover, due consideration must be given to the unique characteristics, in terms of gender and age, of the reference group. Alternatively, comparisons could be based on specific rates for gender and age groups, or techniques using standardized rates to compare working populations featuring different compositions in terms of gender and age.

10.8 FITNESS FOR WORK ASSESSMENT

According to Italian legislation, after conducting the examinations called for by the health surveillance program, the worker is defined by the company doctor as

- *Fit to work*
- *Partly unfit to work*, temporarily or permanently, with restrictions or limitations
- *Unfit to work*, temporarily (with time restrictions) or permanently

A number of important considerations need to be made with regard to fitness to work.

In difficult cases, the assessment of fit to work seeks to find a reasonable compromise between the worker's individual health status and his or her specific working conditions.

The company's occupational health physician must manage fitness-to-work assessments across the board, from drafting fitness reports to dealing with the various aspects relating to returning difficult cases to the workforce, together with other relevant staff members.

Assuming that all jobs must first and foremost be safe and acceptable for the vast majority of workers, fitness-to-work assessments may present a problem for the occupational health physician especially when called on to judge whether tasks are acceptable for workers who may have been or are currently suffering from medical conditions. It is difficult to decide if a worker should be deemed at risk of developing a UL-WMSD or exacerbating an existing problem.

Fitness-to-work assessments and worker placements must therefore strive to tailor the work (i.e., environment, workstation, and process) to the worker's capabilities and health status, to ensure the worker's ongoing employment, underpinned by criteria of reasonable caution for the worker's health.

In the field of WMSDs due to biomechanical overload there are no hard and fast rules for matching disease and disease severity with levels of biomechanical overload. Therefore, the company doctor needs to make every effort to adopt a cautious and sometimes experimental approach in every case.

The proposals described here have been adopted as guidelines by several Italian regions (including Lombardy and the Veneto region, both in 2009); they are based on the concept that what is broadly acceptable for the majority of a healthy population may still be tolerated by a subject affected by musculoskeletal diseases and disorders (but of course still fit to work) due to, or sensitive to, biomechanical overload.

That being said, health surveillance programs may also give rise to three categories of workers (or cases):

- *Subjects with only anamnestic-positive conditions:* Besides scheduling a "personalized" follow-up for these workers, it is also advisable to look into specific measures with respect to occupational risk exposures.

- *Subjects with overt conditions*, which, as far as can be determined, *are not work related:* In such cases, the worker should be fully or partially removed, either temporarily or permanently, from tasks involving biomechanical overload of the upper limbs.
- *Subjects with overt conditions,* which, as far as can be determined, *are definitely or most likely work related:* In addition to issuing a fitness-to-work assessment stating that exposure must be limited, the case must also be reported to the judicial and supervisory authorities and also, considering the characteristics of the exposure (i.e., duration, intensity), reported as an occupational disease to the insurance system.

When drafting fitness-to-work assessments, if the worker presents with an overt musculoskeletal disease or a disorder of the upper limb (whatever the cause), measures should be adopted to reduce exposure.

For the time being, there is too little expertise in this field to recommend detailed standard guidelines regarding the criteria for such measures in relation to the type and severity of the condition and the exposure level.

On the whole, workers with the following conditions should be permanently removed from tasks entailing repetitive or forceful movements of the upper limbs (bearing in mind that the list is not exhaustive):

- Incapacitating osteoarthritis of the upper limb joints
- Incapacitating outcomes of trauma involving the upper limbs (in relation to the functional demands of the task)
- Radiculopathy due to degeneration and/or malformation of the cervical spine
- Peripheral neuropathy of systemic origin
- Rheumatoid arthritis
- Existing severe mesenchymal disease

The following criteria can be adopted for tendinitis or trapped nerve conditions:

- Temporary removal from tasks entailing biomechanical overload for the duration of acute treatment
- Fit to work assessment subject to reduced exposure to tasks involving biomechanical overload for workers with chronic forms with or without slight functional impairment

In this case, "reduced exposure" needs to be defined on a case-by-case basis. Generally speaking, if the assessment is performed using a synthetic risk index (such as the OCRA index or checklist), reduced exposure may mean a task classified as "green" or acceptable, "yellow," or even "light red" in less serious cases, but always ensuring that postures are compatible with the worker's specific condition.

Ideally, reduced exposure can be defined as

- Frequency less than 20–30 actions/min
- Minimal use of upper limb force, no peak force, generally below 5% of MVC (0.5 on the Borg scale)

- Practically no postures or movements "significantly engaging" the principal joints, especially the affected joint
- Adequate recovery times every hour throughout the shift

It should be emphasized that these workers can be assigned to "yellow" or "light red" tasks (subject to the possibilities offered by the work environment), provided due attention is paid to the affected joint and that the worker is followed up very closely to monitor clinical developments, and that the necessary measures are taken to prevent work activities from further aggravating the condition.

Lastly, care should be taken with how fitness-to-work assessments are issued, and to putting in place procedures for effectively managing cases of limited fitness.

General recommendations such as "not fit to perform repetitive movements" or "not to be assigned to tasks involving biomechanical overload of the upper limbs" should be avoided. Conversely, indications must be as specific as possible about assigning workers to tasks and workplaces, and the case should also be discussed with employers, prevention and protection services, and so on. It is only by adhering to these recommendations that a fitness-to-work assessment will be regarded as *fitness to perform specific tasks.*

Wherever exposure to the risk of biomechanical overload has been clearly assessed through accurate job analysis, a list can be drawn up of jobs that can be assigned to workers with adverse upper limb conditions.

However, besides adopting clearly defined, approved, and technically acceptable management criteria (e.g., meticulous risk analysis for the various tasks and appropriate diagnostic criteria), it is also essential to put together dedicated databases with the necessary IT support.

The entire process of matching jobs with affected workers must necessarily be able to be adjusted quickly to changing needs. In other words, every effort must be made to continue improving both jobs and working conditions insofar as the involvement of the upper limbs is concerned. Accordingly, the matching of tasks with affected workers needs to be periodically analyzed (while never losing sight of the need to safeguard their acquired professional skills).

Importantly, the model for managing the assignment of affected workers to suitable jobs requires close cooperation between the occupational health physician responsible for monitoring the evolution of the condition, and the prevention and protection service or other such bodies responsible for assessing and planning corrective interventions. Such interventions are ongoing and as such depend on many other decision-makers (e.g., employers and HR departments), who cannot and must not feel excluded from managing the placement of workers having "limited" fitness for work.

10.9 MEDICOLEGAL ASPECTS: THE ITALIAN EXPERIENCE

When periodical health surveillance detects and diagnoses overt musculoskeletal pathologies caused by biomechanical overload, as indicated in Section 10.3, three medicolegal reports must be filed.

10.9.1 Reporting Cases to the Judicial Authority

According to article 365 of the Italian Criminal Code and 344 of the Code of Criminal Procedure, health-care operators are legally required to "report" providing services or care in cases that "may" constitute criminal activity (such as serious or very serious personal injury due to work-related causes).

Injury is considered to be serious if the estimated time to recovery is over 40 days or if an organ or sense is left permanently impaired. An estimated time to recovery of over 40 days does not necessarily mean time off work, but the actual clinical duration of the disease based on clinical tests and/or instrumental exams documenting the existence of signs or symptoms indicating the duration of the condition beyond this time frame.

If an overt condition emerges with a time to recovery estimated to exceed 40 days that can be assumed to be due to the specific employment status of the worker, the designated physician is required by law to draft and file a report to this effect to the judicial authority or criminal investigations department of the local health unit.

10.9.2 Reporting Occupational Diseases to the Prevention and Safety Service of the Local Health Unit (National Health Service)

In compliance with Italian law all physicians are required to notify the supervisory authority (Workplace Prevention and Safety Service of the local health unit) and Italian Workers' Compensation Authority (INAIL) of all cases of occupational diseases and disorders included on specific lists. List 1 is a list of conditions very likely to be of occupational origin, List 2 includes conditions less likely to be of occupational origin, and List 3 includes conditions potentially of occupational origin (see Section 3.10).

The purpose of this obligation is to enable the supervisory authority to gather data that can be used to put in place preventive interventions (and take legal action in the case of noncompliance). The system is also designed to help local health units and regional authorities to feed an occupational health surveillance information system called "*MALPROF*."

10.9.3 INAIL Certification

In accordance with Italian law, most diseases and disorders of the upper limb caused by biomechanical overload and repetitive strain are considered in the list of occupational diseases, which, given a certain kind and amount of exposure, are assumed to be of occupational origin and as such are subject to compensation (see Table 10.1).

In such a way, Italy (as most European countries) complies with a specific European recommendation issued on September 19, 2003, which suggested recognizing such diseases as occupational and compensating them accordingly.

In point of fact, the aforementioned lists provide quite a general nonparametric definition of the exposure (i.e., tasks) presumed to have given rise to the condition. Apart from "dramatic" exposure situations, it should be possible to perform satisfactory risk evaluations using the methods (and classification criteria) proposed here to minimize situations liable to generate nonoccasional "significant" exposure.

Based on the aforesaid decree, an occupational physician—or any other physician for that matter—who comes across any of the UL-WMSDs included in the annexes to the decree, and is familiar with the exposure conditions, is obliged to draft a *preliminary occupational disease certificate* and hand it over to the worker, who can then deliver it to the employer (if the worker is still employed) so as to seek to have the condition recognized as an occupational disease and be awarded the relevant compensation. If the worker is no longer employed, the certificate can be sent directly to the local INAIL offices.

Lastly, several important points need to be underlined on the subject of medicolegal obligations:

- The aforementioned procedures must be followed only with respect to cases where the clinical diagnosis is supported by objective findings, to avoid certificates, reports, or claims based exclusively or prevalently on subjective complaints.
- The aforementioned medicolegal measures must be adopted for all cases of tendon or canalicular conditions of the upper limbs in which it is reasonable to assume that the cause or causes are work related.
- The measures can also be adopted for workers who have resigned or left the at-risk occupation, since the condition may continue to progress even after the worker is removed from the exposure.

10.10 MEDICOLEGAL CRITERIA FOR ASSESSING CAUSE-AND-EFFECT: DEFINITIONS AND OTHER BRIEF EXPLANATIONS. DISCUSSION OF CASE STUDIES

10.10.1 Explanatory Notes on the Evaluation of Cause-and-Effect in Forensic Medicine: Definitions

Forensic medicine uses the results of epidemiological studies to investigate cause-and-effect. In order to generate valid and applicable results, epidemiology has developed methods and tools for establishing when it is possible to speak of cause-and-effect. For forensic medicine, the main point is to establish the validity of an epidemiological assumption stating or suggesting a causal relationship.

While the jury is still out on this issue, there is broad consensus concerning the practical utility of the concept of cause and the need for reasonable standard criteria with which to assess causality.

Rothman (1976) has made some useful proposals in this respect that are summarized in Table 10.11.

In forensic medicine, the material or physical relationship between cause-and-effect is based on the principle that every effect must correspond to a cause. While in clinical practice, the cause plays a role in the etiology of the disease, from the legal standpoint a cause-and-effect relationship between the two phenomena is a *sine qua non* for recognizing compensation or damages for certain diseases.

TABLE 10.11
General Model for Determining Cause and Effect in Epidemiology Applied to Forensics

1. **Chronological Criteria**
   The harmful action (exposure) must precede the injury and the length of time must be compatible
2. **Topographical Criteria**
   The anatomical region affected by the harmful action must correspond to the anatomical location of disease onset
3. **Criteria of Phenomenological Continuity**
   In terms of time frame, there must be continuity between symptoms—typical of the disease—and exposure to risk (e.g., tendinitis—tendinosis)
4. **Suitability Criteria**
   Efficiency (congruence with level and duration of exposure). Compatibility (risk effect known for the specific condition)
5. **Exclusion Criteria**
   This means excluding any other cause and circumscribing the etiological factor

Source: Rothman, K.J., *American Journal of Epidemiology*, 104, 1976, 587–592.

Certain factors contribute toward determining an event, others modify the final outcome of a series of related events, and many can enhance or diminish legal relevance. Against this backdrop, the following definitions may be useful:

1. *Cause* (or sufficient cause)*:* That which produces an effect or modifies the state of things; it must take place before the effect, otherwise the event would not take place, and be sufficient to produce the effect. Causation is unique or exclusive when just one cause is sufficient to produce the effect, as opposed to multiple causation or concurrent causation.
2. *Concurrent cause:* This is when several causes determine an effect that otherwise would not take place in the absence of one or more essential but not sufficient causes.
3. *Condition:* A condition is always necessary for the production of an effect (e.g., the force of gravity for falling, pregnancy for miscarriage or abortion, life for murder or infanticide).

   It is sometimes difficult to distinguish a condition from a preexisting concurrent cause, such as diabetes for infected wounds, heart disease for a stress-induced heart attack, an aneurism for a rupture also in the presence of stress. Conditions are static situations, which may, however, evolve spontaneously (e.g., pregnancy), but the concurrent cause is always dynamic and able to modify reality by producing a new event that is legally relevant.
4. *Occasion:* An unnecessary (replaceable) and generally insufficient event that differs from the cause insofar as it is unable to produce the effect that is disproportionate with respect to the action; it is an ordinary, physiological action that usually does not produce any harmful effects, but may bring

to the surface a situation dependent on another cause. Examples include, a pathological hip fracture (osteolytic metastasis) while moving from sitting to standing (occasion). This concept must be kept in mind when the term *at work* is used with reference to occupational accidents, where it has a bearing in terms of the place, time, and purpose of the action causing the accident.

In forensic medicine, causation means ascertaining if a legally relevant medical condition has any relationship with a certain cause, by means of a process requiring not only an understanding of biological science but also an in-depth knowledge of the law governing medicolegal investigations.

Returning to Table 10.11, in forensic medicine, the study of cause and effect is based on the classic criteria of chronology, topography, plausibility, phenomenological continuity, and exclusion.

10.10.1.1 Chronology

This means judging if the period between the harmful action and the onset of the first symptoms of a certain disease is compatible with a cause-and-effect relationship. There may be an immediate, mediated, or delayed relationship, a typical example of the latter being when cancer develops after exposure to carcinogens.

The concept of *latency* may help better understand the concept of chronology, that is, the time from exposure (the cause) to the onset of disease (effect), also known as the *induction period*. It is important to consider the *start of the disease* as the *onset of the first symptom*, and not the diagnosis of the disease by means of clinical and/or instrumental exams.

But how long would a reasonable latency period be for musculoskeletal diseases caused by biomechanical overload due to a certain level of exposure risk (determined using the OCRA method)? For the time being, the literature does not offer any definite answers, merely common sense suggestions.

Table 10.12 shows the *minimum intensity of exposure* and *minimum latency* for the most common tendinopathies of the upper limbs and for CTS, according to the suggestions contained in a document released by the European Commission (2009).

10.10.1.2 Topography

This concerns the relationship between the anatomical location of the condition and the site of onset of the disease. The relationship may be direct, indirect, and/or caused by a *boomerang* effect. The OCRA method may be a useful adjunct for exploring topographical criteria, as it looks at how long the subject spends with upper limbs in awkward postures and provides separate scores for each joint segment. However, it is important to note that the absence of risk for a specific joint does not rule out the existence of a cause-and-effect relationship. For instance, jobs that are at high risk due to frequent and prolonged pinching actions will initially cause the classic appearance of CTS (valid topographical criteria), followed by the onset of epicondylitis, and later shoulder tendinitis. These last conditions should be considered as a worsening of the disease caused by restricting the use of the first joints of the affected wrist.

TABLE 10.12
Minimum Exposure Intensity and Minimum Latency of Main Musculoskeletal Diseases of the Upper Limbs

Tenosynovitis: Inflammation of the flexor and extensor tendon synovial sheaths in the hand. Peritendinitis: Inflammation of the peritendineum

Exposure criteria: Minimum intensity exposure: Individual exposure history with confirmation of prolonged occupational exposure to highly repetitive hand movements. Working with wrist/hand in awkward posture and/or force exerted with hand aggravates exposure. Analysis of repetitive tasks in the workplace (e.g., number of items handled, number of repetitive hand movements), assessment of time spent in awkward wrist/hand postures and assessment of force exerted (e.g., weights handled, force applied) may all add valuable information even if threshold limits for exposure are not established. Highly repetitive procedures (guiding): >10 items handled/min or >20 repetitive movements/min. Considerable force (guiding): objects weighing over 1 kg.

Minimum exposure duration: Days

Maximum latency period: Several days

Induction period: Days

Lateral and Medial Epicondylitis, Biceps Tendinitis, Supraspinatus Tendinitis

Exposure criteria: Minimum intensity exposure: History of exposure with confirmation of prolonged occupational exposure to forceful and repetitive arm motions and/or prolonged periods of work with the arms elevated. Analysis of repetitive tasks in the workplace (e.g., number of items handled, number of repetitive hand movements), assessment of force exerted (e.g., weights handled) and percentage of work time spent with arms elevated may all add valuable information even if threshold limits for exposure are not established. Highly repetitive procedures (guiding): >10 items handled/minor more than 20 repetitive movements/min. Considerable force (guiding): objects weighing over 1 kg. Arm elevation (guiding): arms elevated more than 50–60° for more than 50% of the work time.

Minimum exposure duration: Days

Maximum latency period: Several days

Induction period: Days

Nerve Paralysis Caused by Pressure

Exposure criteria: Minimum intensity exposure: Occupational exposure assessed by history and, if possible, analysis of working conditions showing evidence of prolonged and repeated direct pressure on the affected body part. Repetitive movements with extreme flexion and extension and exposure to vibration (CTS) can worsen the condition (see Annex I, entry no. 506.45 on CTS).

Minimum exposure duration: From several hours to several months.

Maximum latency period: Days

Induction period: Hours

Information Notices on Occupational Diseases: A Guide to Diagnosis (EU COMMISSION, 2009)

Source: EU Commission, *Information Notices on Occupational Diseases: A Guide to Diagnosis*, Office for Official Publications of the European Communities, Luxembourg, 2009.

10.10.1.3 Qualitative and Quantitative Plausibility

This establishes whether a harmful action could plausibly produce a disease. The analysis considers whether cause and effect are proportional, whether the harmful action is compatible with the type of injury, and whether the degree of force applied corresponds to the extent of effects observed. Plausibility may be absolute (i.e., the cause is sufficient) or relative (concurrent causes are necessary).

Table 10.13 shows levels of evidence for indicating a causal relationship (strong, sufficient, or insufficient evidence, etc.) between individual or combined biomechanical overload risk factors and the onset of musculoskeletal diseases or disorders of the upper limb. The data is drawn from a metaanalysis of international epidemiological studies, summarized in a special NIOSH (1997) volume, which the reader should consult for additional details.

These studies show that the highest compatibility between risk factors and diseases or disorders due to biomechanical overload is observed when all factors are combined as a whole, rather than just a single factor. This is the principle underpinning the OCRA method, which generates a final index that summarizes all known risk factors combined. As illustrated in Chapter 4, Section 4.2, the OCRA method:

- Can predict the collective probability of developing a disease or disorder (thanks to epidemiological studies referring to a database of 10,000 clinical entries in subjects with different exposure levels)
- Can determine risk levels (probability of developing a disease or disorder) comparing exposed with nonexposed populations

In light of these features, the OCRA method is rightly regarded as well able to meet the qualitative and quantitative feasibility criteria.

10.10.1.4 Continuity

A bridge syndrome may occur when there is no discontinuity between the harmful action and the appearance of the disease or disorder, or a disease-free interval of variable duration. Essentially, there must be continuity between the symptoms following the harmful action and typical of the disease or disorder and the exposure (e.g., tendinitis—tendinosis) with a consistent clinical course with respect to the harmful exposure.

10.10.1.5 Exclusion

This means excluding any other cause and circumscribing the etiological factor. This requires an accurate differential diagnosis.

In some rulings, exclusion is combined with the presumption of a cause-and-effect relationship, such as between injury and improper professional behavior, in the absence of other plausible preexisting, concurrent, or contingent factors. In other words, the *presence of other causes sufficient per se to cause the disease or disorder* (i.e., trauma and systemic condition).

TABLE 10.13
Level of Evidence Linking Risk Factors for Biomechanical Overload of the Upper Limbs with the Onset of Musculoskeletal Diseases and Disorders

Anatomical Area Analyzed / Risk Factor	Strong Evidence (+++)	Evidence (++)	Insufficient Evidence (+/−)	No Evidence
Neck and Neck/Shoulder				
Frequency		X		
Force		X		
Posture	X			
Vibration			X	
Shoulder				
Frequency		X		
Force			X	
Posture		X		
Vibration			X	
Elbow				
Frequency			X	
Force		X		
Posture			X	
Vibration	X			
Hand-Wrist				
CTS				
Frequency		X		
Force		X		
Posture			X	
Vibration		X		
Combination	X			
Tendinitis				
Frequency		X		
Force		X		
Posture		X		
Combination	X			

Source: NIOSH, *A Critical Review of Epidemiologic Evidence for Work-Related Musculoskeletal Disorders of the Neck, Upper Extremity, and Low Back*, 1997, Publication No. 97B141. (http://www.cdc.gov/niosh.)

10.10.1.6 Hypothetical Statistical and Epidemiological Admissibility

Causes that cannot immediately be ruled out but that are feasible based on the literature or proven statistics must be further investigated (e.g., the link between exposure to chemical or physical agents and occupational cancer). The statistical probability of an event must not be mistaken for the probability of a cause-and-effect relationship. For example, the incidence of cardiac complications in just 1% of effort tests in individuals with heart disease does not exclude the causal link between test and coronary event, which in fact is highly probable if not almost certain.

10.10.1.7 Circumstantial and Anamnestic

Circumstantial data (place, time, etc.) and an accurate history of the disorders and clinical manifestations reported by third parties may confirm or deny a potential causal relationship.

The final judgment regarding the relationship, or link, will refer to

1. *Causal relationship.* A relationship exists between the etiological factor and the disease, insofar as one is the cause and the other is the effect.
2. *Concurrent relationship.* The factor has aggravated an existing disease by intensifying its symptoms, prolonging its course, delaying healing, accelerating its evolution, fostering the onset of complications, or bringing forward a negative outcome.
3. *Occasional relationship.* The harmful action could not feasibly produce the disease or disorder, the effects of which appear disproportionate to the action; therefore, the action predates the condition and is neither necessary nor irreplaceable.
4. *Nonexistent relationship.* There is no relationship between the action and the disease or disorder, merely a concomitance of random events, or the action merely drew attention to a condition caused by something else.

10.10.2 Structured Form for Collecting Data to Define a Cause-and-Effect Relationship: Examples

Figure 10.6 proposes a method or model for collecting data to detect a cause-and-effect relationship between a UL-WMSD reported by the worker and qualitative-quantitative exposure levels.

The form begins with the usual personal details concerning the worker and his or her current duties. Then, there is a section describing the worker's previous employment history and another describing events that occurred in his or her current employment.

It should be stressed that both parts must be compiled *with reference to homogeneous periods of exposure*, along with the level of exposure (determined as accurately as possible) to biomechanical overload of the upper limbs. During each period, the worker may have performed only one task or alternatively could have rotated between different tasks. Once the homogeneous risk exposure periods have been listed in the correct order, the subject's physiological anamnesis (mainly pregnancy

<table>
<tr><td colspan="2">epm international ergonomics school</td><td colspan="4">Annex 10.1: Short form for the definition of causative link for work-related musculoskeletal diseases</td></tr>
<tr><td colspan="3">Date:</td><td colspan="3">By:</td></tr>
<tr><td colspan="3">Name and surname</td><td colspan="3"></td></tr>
<tr><td colspan="3">Date of birth</td><td></td><td>Gender</td><td></td></tr>
<tr><td colspan="3">Place of birth</td><td colspan="3"></td></tr>
<tr><td colspan="3">Address</td><td></td><td></td><td></td></tr>
<tr><td colspan="3">Current employer</td><td></td><td></td><td></td></tr>
<tr><td colspan="3">Department</td><td></td><td>Job description</td><td></td></tr>
<tr><td colspan="6">Current duties</td></tr>
<tr><td colspan="6">Chronological succession of risk exposure and parallel study of physiological and pathological anamnesis in homogeneous periods of exposure</td></tr>
<tr><td>Period</td><td colspan="2">Previous employment</td><td>Physiological anamnesis</td><td colspan="2">Pathological anamnesis
(provide a short description of: onset of first symptom, diagnosis, exacerbations, complaints, for each individual UL-WMSD)</td></tr>
<tr><td></td><td colspan="2"></td><td></td><td colspan="2"></td></tr>
<tr><td></td><td colspan="2"></td><td></td><td colspan="2"></td></tr>
<tr><td></td><td colspan="2"></td><td></td><td colspan="2"></td></tr>
<tr><td>Period</td><td colspan="2">Current employment</td><td>Physiological anamnesis</td><td colspan="2">Pathological anamnesis
(provide a short description of: onset of first symptom, diagnosis, exacerbations, complaints, for each individual UL-WMSD)</td></tr>
<tr><td></td><td colspan="2"></td><td></td><td colspan="2"></td></tr>
<tr><td></td><td colspan="2"></td><td></td><td colspan="2"></td></tr>
<tr><td></td><td colspan="2"></td><td></td><td colspan="2"></td></tr>
<tr><td colspan="6">Notes</td></tr>
<tr><td colspan="6">Conformity criteria for causative link assessment</td></tr>
</table>

FIGURE 10.6 Short form for the definition of causative link for work-related musculoskeletal diseases.

Criteria	Yes	No	Explanation
Chronology Harmful event precedes condition			
Topography Affected area corresponds to harmful action and pathology			
Continuity Continuity between symptoms following the harmful action and symptoms deriving from the disease			
Feasibility Exposure level and duration			
Exclusion Presence of other causes sufficient on their own to cause the pathology			

Conclusions and definition of causative link

Pathology	Absent	Doubtful	Probable

FIGURE 10.6 (CONTINUED) Short form for the definition of causative link for work-related musculoskeletal diseases.

and sport) and medical history can be taken, obviously in summary form. The description of UL-WMSDs must clearly specify when the first symptom of each individual pathology appeared, when the condition became worsened, and the date of the diagnosis.

The form ends by asking if and how the classic five criteria are met (chronology, topography, feasibility, continuity, and exclusion) in order to assess the cause-and-effect relationship for medicolegal purposes.

Tables 10.14 through 10.16 provide examples of these.

TABLE 10.14
Case 1

Period	Present Company—Work History	Physiological Anamnesis	Pathological Anamnesis
Since 1978	Same company Various activities, not clearly defined	Jogging, sport	No problems
Since 2000	Assembly line—motorcycles Prolonged maintenance of posture with arms raised at or above shoulder height		No problems Since 2006, tenderness and functional impairment of the right shoulder (dominant limb) but physical examination consistently negative. Progressive deterioration; physical exam positive for abduction/elevation, painful arc positive on right side; nothing on left side 2012 comparative US—shoulder: right, tendovaginitis of the long head of the biceps, moderate tendinosis of the supraspinatus with 5 mm long calcification Left, minimal tendinosis of the supraspinatus and infraspinatus The company doctor did not specify the level of risk and did not issue any restrictions/limitations as a result of periodic examinations

TABLE 10.15
Case 2

NAME AND SURNAME	L. C.		
DATE OF BIRTH	03.09.1961	**GENDER**	F
PLACE OF BIRTH	XXXXXXX height: 165 cm; weight: 80 kg		
CURRENT EMPLOYER	Engineering		
DEPARTMENT	Assembly of two-wheeled vehicles	**JOB DESCRIPTION**	Assembler

Description of current tasks: Operations at several locations along the assembly line. Workstations often involve the use of vibrating tools (screwdrivers) almost always held with both hands: continuous high-frequency technical actions also requiring the use of force with both hands—frequent use of torque wrenches to tighten bolts (such wrenches can/should be used alternatively with both hands). The worker rotates regularly at five workstations with OCRA checklist indexes ranging from medium to very red.

Period	Present Company—Work History	Physiological Anamnesis	Pathological Anamnesis
From 1979 to 1984	Since 1979, same job—OCRA checklist red (medium/high)	Two pregnancies (absent from work for 2 years	No significant symptoms (even during pregnancy), no significant absences from work
From 1984 to 2011	No change in duties		In 2003 at the age of 43, reported nocturnal paresthesia, both hands, not localized Since 2008, paresthesia also during the day
2011	Changed job (risk level in new job unknown)		Limitations introduced following intervention by local prevention services: assigned to jobs not requiring the use of vibrating tools or wrenches
2012			In 2012, at periodic medical examination, positive Phalen and Tinel tests on both sides. EMG examination: severe bilateral (BIL) CTS

The company doctor did not specify the level of risk and did not issue any restrictions/limitations as a result of periodic examinations

TABLE 10.16
Cases 1 and 2: Cause-and-Effect Relationship

Case 1—Conformity Criteria for Causative Link Assessment

Criteria	Yes	No	Explanation
Chronology: Harmful event precedes condition	x		Exposure since 2000; symptoms first reported in 2006
Topography: Affected area corresponds to harmful action and pathology	x		Shoulder overloaded: Maintained prolonged awkward posture especially of the dominant limb—right
Continuity: Continuity between symptoms following the harmful action and symptoms deriving from the disease	x		Symptom progression since 2006, with the same level of exposure symptoms appeared after about 6 years of exposure
Feasibility: Exposure level and duration	x		Probable high level of shoulder overload and risk exposure at least since 2000, due to prolonged posture of arm at or above shoulder height
Exclusion: Presence of other causes sufficient on their own to cause the pathology		x	

Conclusions and Definition of Causative Link

Pathology	Absent	Doubtful	Probable
Tendovaginalite of the long head of the biceps, tendinosis of the supraspinatus with moderate 5 mm calcification (left), minimal tendinosis of the supraspinatus and infraspinatus			X

Case 2—Conformity Criteria for Causative Link Assessment

Criteria	Yes	No	Explanation
Chronology: Harmful event precedes condition	X		Exposure since 1979; symptoms since 2003
Topography: Affected area corresponds to harmful action and pathology	X		Wrist-hand continuously and bilaterally stressed (dominant limb right), job characterized by high-frequency use of force, with vibrating tools
Continuity: Continuity between symptoms following the harmful action and symptoms deriving from the disease	X		Symptom progression since 2003; considerable and almost continuous symptoms since 2008 including daytime paresthesia). Symptoms appeared after about 24 years of exposure

TABLE 10.16 (CONTINUED)
Cases 1 and 2: Cause-and-Effect Relationship

Feasibility: Exposure level and duration	X		OCRA checklist indexes medium/high with particular emphasis on biomechanical overload risk for the wrist-hand (bilateral). From 1984 to 2011 no change in duties. Following intervention by local prevention services: No use of wrenches: Risk at workstation after 2011 unknown, but the condition predated the instrumental examination performed only in 2012
Exclusion: Presence of other causes sufficient on their own to cause the pathology		X	The two pregnancies definitely predated the onset of the first symptoms; the worker did not report paresthesia during the pregnancies

Conclusions and Definition of Causative Link

Pathology	Absent	Doubtful	Probable
Severe BIL CTS			X

11 Guidelines for Redesigning Jobs and Workstations Featuring Repetitive Tasks, and Criteria for Returning Workers with Musculoskeletal Diseases of the Upper Limbs to the Workplace

11.1 GENERAL REDESIGN CRITERIA

If the risk exposure assessment and analysis of work-related musculoskeletal diseases and disorders reveal significant risk associated with repetitive and/or forceful use of the upper limbs, then it will be necessary to redesign workstations and procedures.

Sometimes, a preliminary or quick assessment will reveal this need, once "critical conditions" have been detected. Besides being urgent, redesigns may occasionally be easy to come up with (but perhaps harder to put in place), since they are often based on common sense rather than complex ergonomic criteria. In this respect, there are various manuals, available both off and online, to help implement simple measures, for example, "Ergonomic checkpoints. Practical and easy-to-implement solutions for improving safety, health and working conditions," and "Ergonomic checkpoints in agriculture," both published by the International Labour Organization (ILO) in cooperation with the International Ergonomics Association (IEA) (ILO, 2010; 2012).

However, in most cases redesigns are anything but simple. And their effectiveness is largely dictated by whether three vital areas are approached simultaneously and in a coordinated manner, such as

- Structural
- Organizational
- Educational

While structural measures are the subject of broad consensus and countless recommendations, organizational changes are hotly debated, and few useful examples can be gleaned from the scientific literature, which often only provides general and extremely obvious advice, such as

- Slow down the pace of high-speed tasks.
- Introduce adequate breaks.
- Find alternative tasks.

This section does not provide actual solutions (a topic that would call for a whole book, let alone a chapter), but will offer guidance as to how to redesign workplaces in order to prevent repetitive diseases and disorders of the upper limbs in the first place, as well as help workers with upper limb work-related musculoskeletal disorders (UL-WMSDs) return to manual jobs.

Suggestions will be offered with respect to the three main areas mentioned in this section, based on a wealth of existing literature on structural measures (largely coinciding with the current harmonized European regulations), and venturing into organizational modifications that have been adopted without substantially impairing productivity, and that have been tested and validated in the field. At the end of this chapter, advice will be provided regarding education and training programs, to support the other two measures (*structural* and *organizational*). (See Tables 11.1 and 11.2.)

Reference should also be made to several European harmonized standards supporting the Machinery Directive (MD), which provide useful criteria and details

TABLE 11.1
Guidelines for Prevention: General Description of Measures

Structural Measures
Optimize the design of workplace, furnishings and lay-out
Choose ergonomically appropriate tools
These measures improve problems relating to:
use of force, awkward postures and movements, and additional factors.
Organizational Measures
Improve the ergonomic design of the work process (pace, breaks, task rotation…)
These measures improve problems relating to:
high-frequency and prolonged movements, prolonged postures and use of force (stereotypy), and inadequate or lack of recovery periods.
Education and Training Measures
Deliver appropriate information regarding specific risks and injuries
Develop specific methods for performing tasks and using techniques
Suggest better scheduling and duration of breaks
These aspects are in addition to the previous two types of measures.

TABLE 11.2
Order of Priorities for Job Redesign Measures

Risk Factor Analysis in order to Prioritize Measures	
1. Identify technical actions with use of force	**Reduce use of force** • Introduce appropriate tools • Improve existing equipment • Improve operating postures
2. Identify actions with awkward postures	**Redesign workplace so that postures and movements are below 50% of the maximum joint range of motion**
3. Detect high-frequency technical actions	**Reduce number of technical actions within each cycle** • Avoid unnecessary actions • Use both limbs to perform actions • Reduce repetition of identical actions at high frequency • Reduce ancillary actions • Increase number of workers
4. Identify additional risk factors	**Reduce harmful effects of additional risk factors** • Use more appropriate tools
5. Detect periods of inadequate functional recovery	**Reschedule breaks and/or rotate repetitive tasks** • Increase duration of recovery time • Distribute recovery periods more effectively

relating to the design of workstations and manual tasks involving new machinery or radically modified existing machinery.

Among them are

- Ergonomic design principles; interactions between the design of machinery and work tasks (EN 614-2)
- Anthropometric measurements applied to the design of workstations at machinery (EN ISO 14738)
- Recommended force limits for machinery operation (EN 1005-3)
- Evaluation of working postures and movements in relation to machinery (EN 1005-4)
- Risk assessment for repetitive handling at high frequency (EN 1005-5)

11.1.1 Structural Measures

Structural measures consist primarily of seeking optimal workstations, furnishings, and layouts, and choosing ergonomically satisfactory work tools. The aim of these measures is generally to improve awkward postures and movements and reduce additional physical factors (including localized compression of anatomical structures of the upper limbs), and the use of high force.

Therefore, structural measures endeavor to lessen the consequences primarily of "posture" and "force" risk factors, but also of many other additional risk factors.

11.1.1.1 Criteria for Reducing "Awkward Postures" as a Risk Factor

Essentially, the aim here is to avoid prolonged movements or postures that involve movements near extreme positions for the joints. In this case, the posture or movement should not force the joint to exceed 50% of its maximum range of motion.

It is often necessary to redesign workstations so that the upper limbs are able to work in the correct position. This should lead to

- Appropriate work surface height for both sitting and standing tasks
- Appropriate chair height for seated tasks
- Work areas tailored to upper limb safety and comfort

The combined criteria laid down in EN ISO 14738 and EN 1005-4 can help to correctly rearrange workstations so as to prevent awkward and/or extreme postures and movements of the upper limbs (see Figure 11.1).

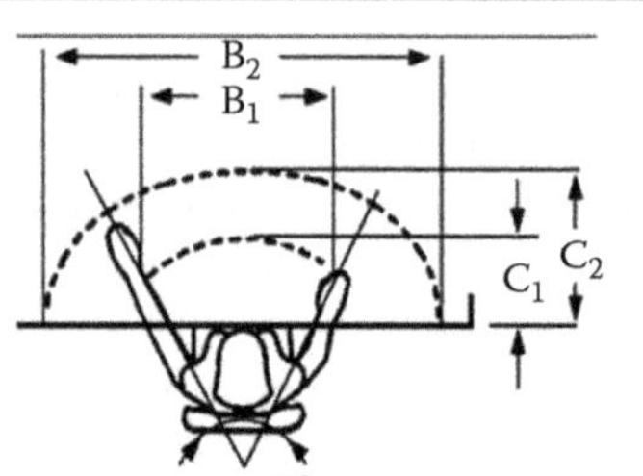

Measure	Value (mm)
B1	480
B2	1300
C1	170
C2	415

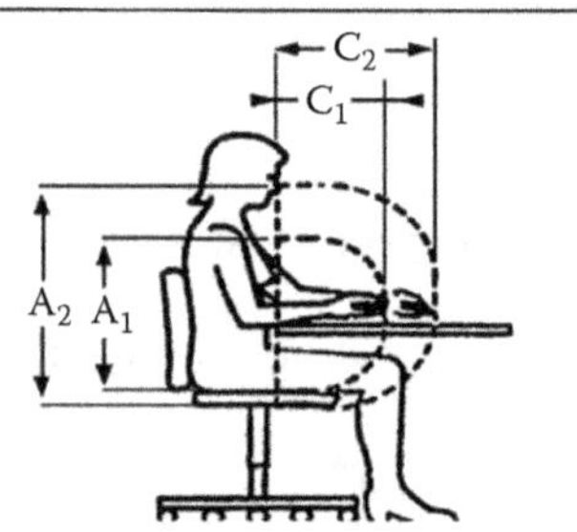

Measure	Value (mm)
A1	505
A2	730
C1	170
C2	415

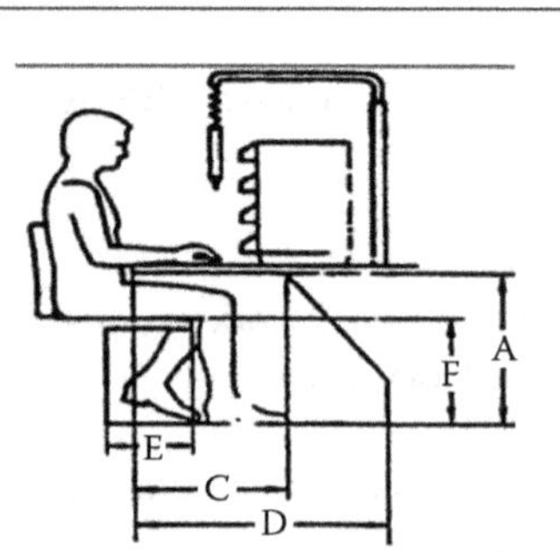

Measure	Value (mm)
A	820–495
A	720
C	520
D	855

FIGURE 11.1 Several examples of measures for redesigning seated workstations (CEN, EN ISO 14738, 2002).

11.1.1.2 Criteria for Reducing "Force" as a Risk Factor

The underlying principle in this case is to avoid excessive muscular exertion or force while performing a task (i.e., in excess of an individual's normal capacity). It is worth remembering that when upper limb joints, especially of the hand and wrist, adopt awkward postures, the ability of the affected segment to apply force is drastically reduced. For example, the amount of force that can be generated with a pinch grip is just 25% of a handed power grip, and similarly grip force gradually diminishes as wrist orientation deviates from neutral.

The following recommendations should be followed to intrinsically reduce excessive use of force:

- Avoid even occasional contractions exceeding 50% of maximum individual capacity.
- No muscle-tendon unit should exceed more than 15% of its maximum capacity during the shift.

The lower the degree of muscular exertion, the longer the permitted duration of the exertion (exponentially). The lower the degree of muscular exertion, the greater the number of technical actions that can be made in performing repetitive tasks, with consequent repercussions also on "productivity levels."

Generally speaking, it is possible to reduce the need for force by using power-driven tools and mechanical grippers and holders that allow positions better suited to the stronger muscle-tendon units, or by automating the entire action.

Tools must meet a series of requirements in order to limit the risks associated with posture and force, thus reducing the risk of workplace accidents.

An ergonomic instrument or tool should allow a worker to

- Avoid having to deviate the wrist by more than 50% of its range.
- Avoid repetitive movements using a single finger.
- Avoid hand pieces requiring grips awkward to the development of force.
- Avoid pulling movements and striking actions.
- Avoid localized compressions.
- Avoid the transmission of mechanical vibrations.

The tool or instrument should also be coated with a slip-proof finish, should not conduct heat, and should not have sharp edges, pointed tips, or a potentially harmful shape.

11.1.2 Organizational Measures

Measures, typically involving changes to the work organization, become necessary when it is ascertained that the job features excessively frequent technical actions and/or inadequate functional recovery periods.

However, improving these two fundamental risk factors, that is, frequency and recovery times, can interfere with productivity, and therefore is less likely to be accepted by employers.

A certain amount of experience has been acquired in this field that may provide useful suggestions for embarking on an organizational restructuring of tasks associated with economic risk.

11.1.2.1 Reduction in the Frequency of Technical Actions

Experience has shown that the most complex issue is undoubtedly the reduction of high-frequency technical actions.

The first and most obvious step (perhaps for ergonomist but not for the employer!) to achieve adequate frequency of action is to slow the pace of the task, consequently increasing cycle time and reducing productivity. However, this solution is the last resort if no other approach has successfully reduced a particularly high level of risk.

Reflecting the common aim of the ergonomist and the production engineer to find ways of reducing the technical actions required to complete a job cycle, it is possible to lower the level of risk without compromising output. In other words, this means *optimizing*, in terms of *quality and quantity*, the technical actions needed to complete the cycle characterizing the task.

Production engineers, especially those involved in designing how tasks are actually performed, are already generally quite experienced in seeking solutions to optimize the performance of actions to complete a task.

Nonetheless, their experience and effort is aimed primarily at reducing the number of actions in order to shorten the time required to perform a task (and hence increase output), rather than to reduce the risk of exposure to biomechanical overload of the upper limbs.

Conversely, merging the principles of ergonomics with the organizational expertise of production engineers in the field has proven to be of invaluable importance not only for enhancing productivity but also for improving working conditions and therefore, significantly, worker health.

Having established this common aim, listed below are the steps involved in an operating procedure designed to optimize outcomes.

11.1.2.1.1 Phase 1: Analysis of "Useless" Technical Actions

During this phase, it is decided whether all the technical actions actually observed are strictly necessary. It is thus possible to single out "useless" actions performed by the operator, and even actions which could be designed out of the task; in other words:

- Detect any useless actions added by the operator, for example, when assembling apart, the operator occasionally strikes more often, or screws more tightly, than required. The operator might only need to strike twice, but actually strikes four, five, or six times. In this case, the operator must be trained to perform no more than the useful actions actually required to complete the task.
- Detect whether any actions added by the operator are entirely arbitrary, or in fact conceal a technical flaw, for example, a faulty pin does not fit snugly, so the operator needs to strike it several times to force it into the correct

position. Once the technical flaw has been identified, it can be eliminated and the number of actions reduced.

- Detect "obsolete" actions, for example, in the course of time, assembly lines may undergo small changes to the machinery, or to the product. Therefore, it is extremely useful to check the way that operators perform their tasks whenever machinery, or products, are modified.

11.1.2.1.2 Phase 2: Analysis of Limb Use When Performing Technical Actions

Once all "useless" actions have been eliminated, the next step is to optimize the distribution of the actions between the two upper limbs.

Analysts often fail to examine both limbs separately, and workers consequently tend to perform all actions with their dominant limb. Simple low-precision actions (e.g., picking up workpieces and placing them on the machining line) may be performed equally well with both limbs, thus reducing the frequency with which the dominant limb is used.

11.1.2.1.3 Phase 3: Analysis of "Identical" Technical Actions

In this phase, it is determined whether workers are repeating identical actions for a significant portion of the job cycle. As mentioned in this section, repetition of identical technical actions may overload specific muscle-tendon units and generate higher risk scores due to the presence of stereotypy.

This can often be avoided by introducing a specific mechanical aid, for example, a worker grasps a handful of screws and positions 10 of them, one at a time, into holes, giving each one a single turn with the fingers.

In this case, a screw gun could be introduced; or, for example, the operator sands a surface manually, but could use a power sander instead.

Because these actions are identical, they are often performed at very high frequency, and therefore may significantly increase the total frequency of action during the job cycle.

Conversely, if evidence is found of identical technical actions, but no suitable tool can be introduced, and at the same time the action frequency has a significant impact on the total frequency, one of the following solutions should be adopted:

- Eliminate the specific manufacturing step altogether by having the part arrive preassembled at another workstation; this is a simple solution, but care should be taken to avoid creating another high-risk job elsewhere.
- Introduce a semiautomatic step to replace the technical actions; this may be a costly solution.
- Reexamine the step technically to find alternative solutions for completely bypassing the specific set of actions; a high tech solution may also enhance the product.

11.1.2.1.4 Phase 4: Analysis of Ancillary Actions

It is necessary to check whether any "ancillary" actions are performed in passing from one cycle to the next. For example: to begin assembling a new part consisting of several components (cycle start), the worker must first pick the part up from a conveyor belt, turn it, place it on the workbench, and then begin the actual assembly. The worker then lifts the completed part, turns it, and replaces it on the conveyor belt. Moving objects from the conveyor belt to the workbench and vice versa may often require physical effort.

It is generally useful to have the conveyor belt and the operating areas intersect each other in such a way as to avoid the worker having to pick up and replace parts; such "ancillary" actions are of no added value to the product. It is also advisable for the part to reach the worker facing the right way to avoid additional handling.

11.1.2.1.5 Phase 5: Situations Requiring Jobs to Be Split

Even after revising actions, sometimes their frequency remains excessively high (more than 80–90 actions per min), calling for the workload at the specific workstation, and/or line to be split.

11.1.2.2 Task Rotation between Different Workstations

The organization of certain work environments may involve one or more workers being assigned multiple repetitive tasks, and alternating these tasks.

It has been claimed that *job rotation* (alternating between several tasks) and *job enlargement* (expanding task content) are measures that may help to minimize repetitive manual tasks and ultimately lower the consequent risk level. However, this may not always suit the environment, or the circumstances. For example, if workers rotate between repetitive tasks over the course of days or weeks, or if the tasks rotated involve the same high-frequency use of force, or posture, then rotation is unlikely to be beneficial, and generally speaking, the Occupational Repetitive Actions (OCRA) index scores for such tasks will be similar.

Conversely, frequent rotation between tasks with different risk levels, requiring a different posture and/or use of force may have a positive impact on risk level.

The effects of daily or multiday task rotation are, in part, considered in the models for calculating the OCRA checklist and index scores. When these calculation models are used to estimate the final index in situations featuring task rotation, they concentrate more on the level and duration of exposure, than on qualitative aspects, such as the action, posture, and level of force used to perform the rotated tasks. With regard to these latter aspects, the suggestion is to appraise situations in which there is an actual variation in the "quality" of upper limb involvement in the rotation of repetitive tasks, in the risk analysis, and at least (for the time being) descriptively in the OCRA index, or checklist. Epidemiological data is currently being collected to validate and "weight" actual improvements generated by rotating tasks with the same risk score, but with an alternation of the upper limbs.

11.1.2.3 Recovery Period Optimization

In addition to introducing task rotation in order to prevent diseases and disorders due to repetitive movements, it is crucial to ensure adequate recovery periods. As has been suggested previously, the ideal ratio of work period to recovery period is 5:1 for each hour of repetitive work.

However, a company may have adequately long recovery periods (actual breaks and/or nonrepetitive tasks), but they may be poorly distributed throughout the repetitive task cycle.

The suggestion is therefore to

- Optimize the distribution of official breaks (provided they are over 10 min long); better still, introduce shorter, but more frequent breaks. For instance, prevent workers out of assembly line from speeding up the pace of their work to have longer breaks, and teach them to increase frequency of breaks instead.
- Schedule breaks if possible at the end of an hour of repetitive work.
- Avoid scheduling breaks too close to meal breaks, or the end of the shift, as these periods are already considered to be "covered" by adequate recovery.
- If breaks take place and are long enough, convert mandatory machine stoppages (for production changeovers, for instance) into actual additional recovery periods.

11.1.3 Education and Training Measures

Training must be tailored to the type of workforce, that is, factory workers, supervisors and production engineers, management, and the company physician, using a different approach for each.

11.1.3.1 Factory Worker Training

Factory workers must be informed of the health risks and disease conditions associated with repetitive tasks, and taught how to avoid them.

The following recommendations should be provided:

- Use both limbs, whenever possible.
- Perform tasks in the required order.
- Eliminate useless actions when performing tasks.
- Grip objects correctly.
- Make proper use of breaks as recovery periods.

It is also essential for workers to learn how to give supervisors suggestions for possibly adopting new operating procedures and technical actions, and for modifying the most challenging steps in job cycles. Workers must also be taught how to inform the company doctor at the onset of symptoms representing early warning signals.

11.1.3.2 Training for Production Engineers and Supervisors

Production engineers and supervisors play a critical role in designing work processes and their training is therefore of paramount importance.

Programs must include learning how to recognize the health risks and disease conditions linked to repetitive tasks performed with the upper limbs, and the medicolegal implications associated with the detection of an occupational disease or disorder.

In order to minimize the risk factors illustrated in other chapters of this book, targeted training programs must cover the following aspects:

- Work cycle analysis and risk factor identification
- New job design
- Modification of existing jobs
- Assessment of the impact of technological innovation on risk factors
- Assessment of the duration, number and distribution of breaks during shifts
- Job rotation and alternative tasks, also with a view to reducing exposure risk

In addition to these general aspects, other more specific topics include

- Training both old and new workers to correctly perform their tasks within the work cycle.
- Periodical inspections to ensure that the work cycle is being correctly performed.
- Checking for issues that arise while tasks are being performed, through periodical meetings with workers.

Production engineers and supervisors must play a major role in the worker training process.

11.1.3.3 Training for Management

Manager training must reflect the responsibilities that management holds toward employees. Training programs must ensure an in-depth understanding of health risks and disease conditions in terms of their medicolegal implications, and include procedures for identifying and controlling risk, adopting effective organizational strategies for the shop floor, and conducting cost-benefit analyses for job design and redesign.

It is only through a thorough understanding of the issue by managers and decision-makers that the necessary measures can be taken to deal effectively with at-risk manufacturing processes and job design in general.

11.2 CRITERIA FOR RETURNING WORKERS TO THE WORKFORCE

Returning workers with musculoskeletal diseases and disorders of the upper limbs to the workforce represents one of the leading problems in work environments featuring numerous tasks entailing potential biomechanical overload of the upper limbs.

Affected workers are generally classified as temporarily or permanently unfit to perform the tasks that in all likelihood caused the condition or worsened it.

In this Section 11.2, a series of procedures and solutions will be suggested for returning partly unfit workers to jobs and workstations with "low exposure" or that have been redesigned. On the one hand, this may help such workers remain to some extent in the workforce, and on the other, safeguard their health.

11.2.1 Identification of Suitable Jobs

First of all, in order to ensure that the measure is, and continues to be, effective over time, it is essential for the entire procedure to be highly cooperative, in so far as it must involve the employer at every level—the company physician, the union representative in charge of safety, and the workers affected by UL-WMSDs.

Staff must be trained to conduct specific risk assessments on the workstations involved, in order to efficiently deal with the issues associated with job redesign, for both healthy workers and those affected by musculoskeletal diseases and disorders.

The following steps will ensure the involvement of technical specialists:

- Training to assess and redesign jobs and repetitive tasks
- Preparation of a list of "potentially suitable" jobs and tasks for workers with UL-WMSDs (risk estimated to be green or yellow)
- Analysis and assessment of identified jobs using the OCRA method
- Proposal of suggestions for possible changes/adaptations of workstations to enable workers with UL-WMSDs to return to the workforce
- Assess the feasibility and scheduling of redesign measures

It should be emphasized that the workstations/tasks potentially suited to workers with severe UL-WMSDs generally feature the following exposure characteristics:

- Frequency less than 20–30 actions/min
- Minimal use of upper limb force, on average below 5% of maximum individual capacity with force peaks of less than 15% of maximum capacity (between *very light* and *light*, on the Borg scale)
- Practically no postures or movements "significantly engaging" the principal joints, especially the affected joint, or present only for limited periods
- Adequate recovery times every hour throughout the shift
- No significant additional risk factors

If posture is not optimal, the suggestion should be to "slow down" the frequency of action, and pay the utmost attention to the clinical condition of the worker being returned to the workforce.

11.2.2 Matching Tasks and Workstations to Individual Conditions

In order to facilitate communications between technical specialists and the company health-care service (company physician), it is advisable to list and classify both the outcomes of the assessment of the selected jobs and workplaces and of the degree/severity of the disease conditions detected. This will make it easier to match affected workers with individual jobs.

The following synthetic classifications are recommended for this purpose:

11.2.2.1 Classification of Task/Work Station

- *Unlimitedly optimal* (no modification required and suitable for all disease conditions)
- *Limitedly optimal* (no modification required but not suitable for certain specific disease conditions)
- *Optimal subject to modification* (suitable for all disease conditions, including severe, but only subject to recommended modifications)
- *Very good subject to modification* (suitable for all moderate disease conditions, subject to recommended modifications)
- *Unsuitable* (difficult to adapt and therefore not recommended for workers with disease conditions)

11.2.2.2 Classification of Disease or Disorder

Table 11.3 shows a classification of UL-WMSDs.

Based on this classification, taking into account the number of suitable jobs, the number of workers partially unfit due to UL-WMSDs, and the workers' skills, it is possible to match modified (but productive) jobs with affected workers.

The main aim of returning workers with "adverse" health conditions back to the workforce with new or redesigned jobs or tasks is to at least reduce the clinical signs and symptoms associated with their specific disease or disorder, while at the same time enabling the worker to remain productive.

Since the recommendations for selecting and redesigning jobs are still in the experimental (implementation) phase, complete cooperation between the company's technical specialists, occupational health service, and above all, affected workers, is essential so that the outcomes of decisions can be almost continuously monitored.

TABLE 11.3
Classification of Conditions for the Purposes of Returning Workers to the Workplace (Protecting Privacy about the Clinical Diagnosis)

Severity	Mild	One joint affected mildly, or more than one joint affected moderately
		Only significant symptoms without clinical or instrumental findings
	Moderate	One joint affected moderately, or more than one joint affected mildly
	Light	Only one joint affected mildly
	Anamnestic questionnaire	Only significant symptoms without clinical or instrumental findings
Joint		Scapulohumeral
		Wrist
		Elbow
		Hand/finger

To this aim, the following activities are just some of those that are highly recommended:

- Classification of the disease or disorder and level of functional abnormality assessed by the company doctor
- Identification of a suitable low/medium exposure task/job, among those available
- Meetings and interviews with the relevant worker before assigning a new job, and then periodically with the company doctor, technical specialists, and union representative for worker safety
- Potentially, supply orthotic devices based on the suggestions of the company doctor, taking into account the worker's condition and the functional demands of the new task
- Adoption of an intense health surveillance program tailored to the individual worker in order to monitor developments

A monitoring program of this kind might include

- A follow-up visit two weeks after starting the new job; this will provide a rough preliminary assessment of adequacy and reinforce the educational message to the worker
- A follow-up visit three months after changing jobs
- Another follow-up visit six months later
- A further follow-up visit 12 months later
- Revision of the new job/task assigned to the worker with UL-WMSD in the event of an unsatisfactory outcome (worsening of the condition, inability to perform the task, and so on)

References

ACGIH (American Conference of Governmental Industrial Hygienists), 2002. *Hand Activity Level (HAL). Threshold Limit Values for Chemical Substances and Physical Agents & Biological Exposure Indices*, 112–114. ACGIH, Cincinnati, OH.

Alexopoulos E.C., Burdorf A., Kalokerinou A., 2003. Risk factors for musculoskeletal disorders among nursing personnel in Greek hospitals. *Int. Arch. Occup. Environ. Health*; 76(4):289–294.

Beltran J., Mosure J.C., 1990. Magnetic resonance imaging of tendons. *Crit. Rev. Diagn. Imaging*; 30(2):111–182.

Bonfiglioli R., Mattioli S., Fiorentini C., Graziosi F., Curti S., Violante F.S., 2007. Relationship between repetitive work and the prevalence of carpal tunnel syndrome in part-time and full-time female supermarket cashiers: A quasi-experimental study. *Int. Arch. Occup. Environ. Health*; 80(3):248–253.

Bonfiglioli R., Venturi S., Graziosi F., Fiorentini C., Mattioli S., 2005. Carpal tunnel syndrome among supermarket cashiers. *G. Ital. Med. Lav. Ergon.*; 27(1):106–111.

Borg G.A.V., 1998. *Borg's Perceived Exertion and Pain Scales*. Human Kinetics, Champaign, IL.

Bystrom S., 1991. *Physiological response and acceptability of isometric intermittent handgrip contractions*, 38. Arbete och alsa—NIOH, Stockholm, Sweden.

Cailliet R., 1978. *Il dolore cervico brachiale*. Ed. Leonardo, Roma, Italy.

CEN (European Committee for Standardization), 1993. EN 1005-3. *Safety of machinery: Human physical performance. Part 3—Recommended force limits for machinery operation*. Unpublished internal draft.

CEN (European Committee for Standardization), 2002. EN ISO 14738. *Safety of Machinery: Anthropometric requirements for the design of workstations at machinery*.

CEN (European Committee for Standardization), 2002. EN 1005-3. *Safety of machinery: Human physical performance. Part 3—Recommended force limits for machinery operation*. CEN Management System. Bruxelles, Belgium.

CEN (European Committee for Standardization), 2003. EN 1005-2. *Safety of machinery: Human physical performance. Part 2—Manual handling of machinery and component parts of machinery*. CEN Management System. Bruxelles, Belgium.

CEN (European Committee for Standardization), 2007. EN 1005-5. *Safety of machinery: Human physical performance. Part 5—Risk assessment for repetitive handling at high frequency*. CEN Management System. Bruxelles, Belgium.

Colombini D., Occhipinti E., 1996. Le applicazioni dell'indice sintetico di esposizione (Ocra) a compiti con movimenti ripetitivi degli arti superiori in diverse realtà produttive: prime esperienze di validazione. *La Medicina del Lavoro*; 87(6):704–718.

Colombini D., Occhipinti E., 2007. Repetitive movements of upper limbs in agriculture: Set up of annual exposure level assessment models starting from OCRA checklist via simple and practical tools. In Ed. Khalid H.M. *Proceedings of Agriculture Ergonomics Development Conference*. IEA Press, Kuala Lumpur, Malaysia.

Colombini D., Occhipinti E., 2008. The OCRA method (OCRA index and checklist). Updates with special focus on multitask analysis. In Eds. W. Karkwoski W., Salvendy G. *Conference Proceedings: AHFE 2008*. Las Vegas, NV.

Colombini D., Occhipinti E., 2011. La valutazione del rischio da sovraccarico biomeccanico del rachide e degli arti superiori con strumenti semplificati: la minichecklist OCRA. Contenuti, campo applicativo e validazione/Risk assessment of biomechanical overload of the spine and upper limbs with simplified tools: The OCRA minichecklist. Content, field of applicability and validation. *La Medicina del Lavoro*; 102(1). http://www.mattioli1885.com/onlinejournals/index.php/lamedicinadellavoro/article/view/768/544 (accessed July, 11, 2016).

Colombini D., Occhipinti E., Fanti M., 2005. *Il metodo OCRA per l'analisi e la prevenzione del rischio da movimenti ripetuti.* Collana Salute e lavoro. Franco Angeli Editore, Milano, Italy.

Colombini D., Occhipinti E., Grieco A., 2002. *Risk Assessment and Management of Repetitive Movements and Exertions of Upper Limbs.* Elsevier Science, Amsterdam, the Netherlands.

De Marco F., Menoni O., Bonaiuti D. et al., 1996a. Studi clinici in popolazioni lavorative: valore e significato dei rilievi anamnestici, dei test clinici e degli esami strumentali nella diagnostica delle patologie muscoloscheletriche degli arti superiori (WMSDs). *La Medicina del Lavoro*; 87(6):561–580.

De Marco F., Menoni O., Colombini D. et al., 1996b. L'occorrenza delle alterazioni muscoloscheletriche degli in popolazioni lavorative non esposte a compiti ripetitivi degli arti superiori. *La Medicina del Lavoro*; 87(6):581–589.

Draicchio F., Silvetti A., Bardellino F., Vinci F., 2007. Biomechanical risk assessment of manual material handling in vegetables and fruit departments of supermarkets. *G. Ital. Med. Lav. Ergon.*; 29(3):573–575.

Durkan J.A., 1991. A new diagnostic test for carpal tunnel syndrome. *J. Bone Joint Surg. Am.*; 73(4):535–538.

EUROFOUND, 2007. *Fourth European working conditions survey.* European Foundation for the Improvement of Living and Working Conditions, Dublin.

EUROFOUND, 2012. *Fifth European working conditions survey.* Publications Office of the European Union, Luxembourg.

European Commission, 2009. *Information notices on occupational diseases: A guide to diagnosis.* Office for Official Publications of the European Communities, Luxembourg.

EUROSTAT, 2010. *Health and safety at work in Europe (1999–2007): A statistical portrait.* Publications Office of the European Union, Luxembourg.

Gagey N., Desmoineaux P., Gagey O. et al., 1991. Apport de l'IRM dans le bilan préchirurgical des lésions de la coiffe des rotateurs/Contribution of MRI to the preoperative evacuation of rotator cuff tears. *Revue de chirurgie orthopédique et réparatrice de l'appareil moteur*; 77(8):521–529.

Grant A.K., Habes D.J., Putz-Anderson V., 1994. Psychophysical and EMG correlates of force exertion in manual work. *Int. J. Ind. Ergonom.*; 13(1):31–39.

Hagberg M., Silverstein B., Richard W. et al., 1995. *Work Related Musculoskeletal Disorders (WMSDs): A Reference Book for Prevention.* Taylor & Francis Group, London, UK.

Hannesschlager G., Riedelberger W., Neumüller H. et al., 1989. Computertomographie der Rotatorenmanschette—Vergleich mit anderen bildgebenden Verfahren/Computed tomography of the rotator cuff: Comparison with other imaging methods. *Rofo Fortschr Röntgenstr*; 150(6):643–649.

ILO, 2010. *Ergonomic checkpoints: Practical and easy-to-implement solutions for improving safety, health and working conditions*, Second edition. International Labour Office, Geneva, Switzerland. http://www.ilo.org/global/publications/ilo-bookstore/order-online/books/WCMS_120133/lang--en/index.htm (accessed July,11,2016).

ILO, 2010. *List of Occupational Diseases (revised 2010).* Occupational Safety and Health Series, No. 74. International Labour Office, Geneva, Switzerland.

ILO, 2012. *Ergonomic checkpoints in agriculture.* International Labour Office, Geneva, Switzerland. http://www.ilo.org/global/publications/ilo-bookstore/order-online/books/WCMS_168042/lang--en/index.htm (accessed July, 11, 2016).

Inaba R., Mirbod S.M., Kurokawa J., et al., 2005. Subjective symptoms among female workers and winter working conditions in a consumer cooperative. *J. Occup. Health*; 47(5):454–465.

ISO, 2000. ISO 11226. *Ergonomics—Evaluation of static working postures*. ISO, Geneva, Switzerland.

ISO, 2003. ISO 11228-1. *Ergonomics: Manual handling—Lifting and carrying*. ISO, Geneva, Switzerland.

ISO, 2007a. ISO 11228-2. *Ergonomics: Manual handling—Pushing and pulling*. ISO, Geneva, Switzerland.

ISO, 2007b. ISO 11228-3. *Ergonomics: Manual handling—Handling of low loads at high frequency*. ISO, Geneva, Switzerland.

ISO, 2014. ISO TR 12295. *Ergonomics—Application document for ISO standards on manual handling (ISO 11228-1, ISO 11228-2 and ISO 11228-3) and evaluation of static working postures (ISO 11226)*. ISO, Geneva, Switzerland.

ISPESL, 2004. *I profili di rischio nei comparti produttivi dell'artigianato, delle piccole e medie industrie e pubblici esercizi. Comparto: imprese di pulizia*. ISPESL Ricerca b28.7.

Kasdan M.L., 1991. *Occupational Hand and Upper Extremity Injuries and Diseases*. Hanley & Belfus, Philadelphia, PA.

Kuschner S.H., Ebramzade H., Johnson D. et al., 1992. Tinel's sign and Phalen's test in carpal tunnel syndrome. *Orthopedics*; 15(11):1297–1302.

Lema Informatica, 2015. *Petra software*. http://www.petrasoftware.it (accessed July, 11, 2016).

Lorusso A., Bruno S., L'Abbate N., 2007. Musculoskeletal complaints among Italian x-ray technologists. *Ind. Health*; 45(5):705–708.

Margolis W., Kraus J.F., 1987. The prevalence of carpal tunnel syndrome symptoms in female supermarket checkers. *J. Occup. Med*; 29(12):953–956.

Marras W.S., 2008. *The Working Back: A Systems View*. Wiley, New York.

Megele R., 1992. Diagnostic tests in carpal tunnel syndrome. *Nervenarzt*; 62(6):354–359.

Menoni O., De Marco F., Colombini D. et al., 1996. Studi clinici in popolazioni lavorative: un modello per l'indagine anamnestica e sue modalità applicative. *La Medicina del Lavoro*; 87(6):549–560.

Meroni M., Battevi N., Vitelli N. et al., 2010. Indagine epidemiologica sulle patologie e sui disturbi muscoloscheletrici degli arti superiori in un gruppo di 2022 videoterminalisti/Epidemiological study of UL-WMSDs in 2022 VDU workers. *La Medicina del Lavoro*; 101(4):276–285.

Missere M., 1998. *L'ecografia dell'arto superiore*. Athena Audio Visuals, Modena, Italy.

Moore J.S., Garg A., 1995. The strain index: A proposed method to analyze jobs for risk of distal upper extremity disorders. *Am. Ind. Hyg. Assoc. J.*; 56(5):443–458.

Myers D., Silverstein B., Nelson N.A., 2002. Predictors of shoulder and back injuries in nursing home workers: A prospective study. *Am. J. Ind. Med.*; 41(6):466–476.

Nicoletti S., Carino M., Di Leone G. et al., 2008a. Prevalenza delle patologie da sovraccarico biomeccanico degli arti superiori negli addetti del comparto del mobile imbottito/Prevalence of upper limb work-related musculoskeletal disorders (UL-WMSDs) in workers of the upholstered furniture industry. *La Medicina del Lavoro*; 99(4):271–280.

Nicoletti S., Consonni D., Carino M. et al., 2008b. Patologie da sovraccarico biomeccanico degli arti superiori: studio retrospettivo di coorte in tre grandi aziende del mobile imbottito/Upper limb work-related musculoskeletal disorders (UL-WMSDs): A retrospective cohort study in three large factories of the upholstered furniture industry. *La Medicina del Lavoro*; 99(4):281–296.

NIOSH, 1997. *A critical review of epidemiologic evidence for work-related musculoskeletal disorders of the neck, upper extremity, and low back*. Publication No. 97B141. U.S. DEPARTMENT OF HEALTH AND HUMAN SERVICES Public Health Service Centers for Disease Control and Prevention National Institute for Occupational Safety and Health. https://www.cdc.gov/niosh/docs/97-141/ (accessed July 11, 2016).

Occhipinti E., 1998. OCRA: A concise index for the assessment of exposure to repetitive movements of the upper limbs. *Ergonomics*; 41(9):1290–1311.

Occhipinti E., Colombini D., 1989. Clinical and functional examination of the spine in working communities: Occurrence of alterations in the male control group. *Clin. Biomech.*; 4(1):25–33.

Occhipinti E., Colombini D., 1996. Proposta di un indice sintetico per la valutazione dell'esposizione a movimenti ripetitivi degli arti superiori (Ocra index). *La Medicina del Lavoro*; 87(6):526–548.

Occhipinti E., Colombini D., 2004a. The occupational repetitive action (OCRA) methods: OCRA index and OCRA checklist. In Eds. Stanton N. et al., *Handbook of Human Factors and Ergonomics Methods*, Chapter 15, 1–14. CRC Press, Boca Raton, FL.

Occhipinti E., Colombini D., 2004b. Metodo Ocra: aggiornamento dei valori di riferimento e dei modelli di previsione dell'occorrenza di patologie muscolo-scheletriche correlate al lavoro degli arti superiori (UL-WMSDs) in popolazioni lavorative esposte a movimenti e sforzi ripetuti degli arti superiori. *La Medicina del Lavoro*; 95(4):305–319.

Occhipinti E., Colombini D., 2007. Updating reference values and predictive models of the OCRA method in the risk assessment of work-related musculoskeletal disorders of the upper limbs. *Ergonomics*; 50(11):1727–1739.

Occhipinti E., Colombini D., 2009. OCRA method: A new procedure for analysing multiple repetitive tasks. *Conference Proceedings 17th IEA Congress of the International Ergonomics Association*. Beijng, China.

Occhipinti E., Colombini D., Occhipinti M., 2009. Metodo OCRA: messa a punto di una nuova procedura per l'analisi di compiti multipli con rotazioni infrequenti/OCRA Method: Development of a new procedures for analysis of multiple tasks subject to infrequent rotation. *La Medicina del Lavoro*; 99(3):234–241.

Osorio A.M., Ames R.G., Jones J. et al., 1994. Carpal tunnel syndrome among grocery store workers. *Am. J. Ind. Med.*; 25(2):229–245.

Panzone I., Melosi A., Carra G. et al, 1996. Repetitive movement of the upper limbs: Results of exposure evaluation and clinical investigation in cash register operators in supermarkets. *La Medicina del Lavoro*; 87(6):634–639.

Punnett L., Fine L.J., Keyserling W. M. et al., 2000. Shoulder disorders and postural stress in automobile assembly work. *Scand. J. Work Environ. Health*; 26(4):283–291.

Putz-Anderson V., 1998. *Cumulative Trauma Disorders—A Manual for Musculoskeletal Disease of the Upper Limbs*. Taylor & Francis Group, London and Philadelphia.

Rothman K.J., 1976. Causes. *Am. J. Epidemiol.*; 104:587–592.

Victorian Occupational HSC (Australia), 1988. *Draft code of practice: Occupational overuse syndrome*.

Wang Y.M., Shih T.T., Jiang C.C. et al., 1994. Magnetic resonance imaging of rotator cuff lesions. *J. Formos. Med. Assoc.*; 93(3):234–239.

Waters T., Lu M.L., Occhipinti E., 2007. Using the NIOSH lifting equation for evaluating sequential manual lifting tasks. *Ergonomics*; 50(11):1761–1770.

Waters T.R., Putz-Anderson V., Garg A., Fine L.J., 1993. Revised NIOSH equation for the design and evaluation of manual lifting tasks. *Ergonomics*; 36(7):749–776.

WHO, 2010. *Healthy workplaces: A model for action—For employers, workers, policymakers and practitioners*. World Health Organization, Geneva, Switzerland.

WHO, 2011. *World health statistics 2011*. WHO, Geneva, Switzerland. http://www.who.int/gho/publications/world_health_statistics/EN_WHS2011_Full.pdf (accessed July 11, 2016).

Williams T.M., MacKinnon S.E., Novak C.B. et al., 1992. Verification of the pressure provocative test in carpal tunnel syndrome. *Ann. Plast. Surg.*; 29(1):8–11.

Yu J.S., 1994. Magnetic resonance imaging of the wrist. *Orthopedics*; 17(11):1041–1048.

Index

D

E

I

J

K

L

N

O

P

U

V

W

For Product Safety Concerns and Information please contact our EU representative GPSR@taylorandfrancis.com
Taylor & Francis Verlag GmbH, Kaufingerstraße 24, 80331 München, Germany

www.ingramcontent.com/pod-product-compliance
Lightning Source LLC
LaVergne TN
LVHW011100300926
849160LV00008B/241

9781498736626